G000168040

ENVIRONMENTAL SCIENCE, ENGINEERING AND TECHNOLOGY

HANDBOOK OF ENVIRONMENTAL POLICY

ENVIRONMENTAL SCIENCE, ENGINEERING AND TECHNOLOGY

Additional books in this series can be found on Nova's website at:

https://www.novapublishers.com/catalog/index.php?cPath=23_29&seriesp=
Environmental+Science,+Engineering+and+Technology

Additional e-books in this series can be found on Nova's website at:

https://www.novapublishers.com/catalog/index.php?cPath=23_29&seriespe=
Environmental+Science,+Engineering+and+Technology

HANDBOOK OF ENVIRONMENTAL POLICY

JOHANNES MEIJER

AND

ARJAN DER BERG

EDITORS

Nova Science Publishers, Inc.

New York

NOTICE TO THE READER

The Publisher has taken reasonable care in the preparation of this book, but makes no expressed or implied warranty of any kind and assumes no responsibility for any errors or omissions. No liability is assumed for incidental or consequential damages in connection with or arising out of information contained in this book. The Publisher shall not be liable for any special, consequential, or exemplary damages resulting, in whole or in part, from the readers' use of, or reliance upon, this material.

Independent verification should be sought for any data, advice or recommendations contained in this book. In addition, no responsibility is assumed by the publisher for any injury and/or damage to persons or property arising from any methods, products, instructions, ideas or otherwise contained in this publication.

This publication is designed to provide accurate and authoritative information with regard to the subject matter covered herein. It is sold with the clear understanding that the Publisher is not engaged in rendering legal or any other professional services. If legal or any other expert assistance is required, the services of a competent person should be sought. FROM A DECLARATION OF PARTICIPANTS JOINTLY ADOPTED BY A COMMITTEE OF THE AMERICAN BAR ASSOCIATION AND A COMMITTEE OF PUBLISHERS.

LIBRARY OF CONGRESS CATALOGING-IN-PUBLICATION DATA

Meijer, Johannes.
 Handbook of environmental policy / Johannes Meijer and Arjan der Berg.
   p. cm.
 Includes index.
 ISBN 978-1-60741-635-7 (hardcover)
 1. Environmental policy--Handbooks, manuals, etc. I. Berg, Arjan der. II. Title.
 GE170.M456 2009
 333.7--dc22
                    2009027486

Published by Nova Science Publishers, Inc. ✦ *New York*

CONTENTS

PREFACE

Today, environmental protection is among the central matters for natural conservation, public health and sustainable business. With advanced technologies and changing lifestyles, the consumption of resources and release of wastes and pollutants are increasing fast. This requires policy makers to design environmental policies that properly guide the development of new products and business operations. The goal of environmental policy is to limit, slow-down, reduce or eliminate environmental damages caused by industrial and human activities. Environmental issues generally addressed by environmental policy include (but are not limited to) air and water pollution, waste management, ecosystem management, biodiversity protection, and the protection of natural resources, wildlife and endangered species. This book gathers the latest research from around the globe in this field.

Chapter 1 - Families who run family business enterprises are the policy recipients the author focuses on in this chapter. They respond to environmental policy either as family members or as business people. In this chapter, the author uses decision-systems theory (DST) to discuss why policy recipients choose to respond in different ways. Understanding policy recipients' decision processes is important for policy developers for two main reasons. First, it can help policy developers devise environmental policies that are accepted as 'legitimate' by policy recipients. This is important because 'legitimate' policies are more likely to be implemented and result in lasting change than policies that are resented. Second, it can help policy developers assess whether or not their proposed policies are going to 'improve society'. Policies that 'improve society' are those that both help future generations satisfy their family's aspirations and also meet global ideals, such as sustainable development ideas.

The author starts the chapter by outlining decision-systems theory (DST) as it provides a theoretical framework for dealing with the question of policy 'legitimacy' and the meaning of 'improving society'. DST indicates that family's aspirations, which the author conceptualizes as a set of five 'motivation-stories', have similar intentions as sustainable development ideas. These intentions are to secure family welfare (human welfare in sustainable development terms) and concomitantly be supported by the environment (conserve the Planet's environment in sustainable development terms). The author uses the '4-Group-Stakeholder model' and DST to argue that the 'ultimate driver' for the current human development system is the desire of families to satisfy their 'motivation-stories'. The 'ultimate *future* driver' is a projection that describes the 'ultimate driver' future generations would have if global ideals (such as sustainable development ideas) were effectively incorporated into their 'motivation-stories'. The author suggests that policy developers ought to create policies that facilitate the 'ultimate *future* driver' for all decisions in the human development system. Developing policies that do this would ensure the legitimacy of the policy and also ensure that the policy developers have done their best to 'improve society'.

The final part of the chapter discusses how policy developers can give priority to the 'ultimate *future* driver' in the policies they create. The 'boxes of influence' concept, which is part of DST, can be used as a policy classification system. It covers policies and programs over the life-cycle of family decision-makers; childhood to retirement. It shows how policy developers can help policy recipients create opportunities that will satisfy their 'motivation-stories' in ways that meet the global ideals of caring for the Planet's environment and caring for people, including their own family and descendents.

Chapter 2 - In recent years, European policy has emphasised the role of green procurement as a policy instrument in efforts to make European markets more environmentally sustainable. By purchasing products and services with low environmental impacts, public bodies and companies may shape the markets of products and services and stimulate their environmental sustainability. Environmental procurement can send signals to producers that these products and services are in demand, thereby helping to reduce the overall environmental impact on society. However, the main question regarding green procurement is what role existing policy plays in driving various actors to integrate environmental criteria into their purchasing decisions.

This chapter provides an overview of the main European and Swedish policies that address public and private procurement and identifies gaps in existing efforts. The differences between the public and private sectors in terms of policy drivers and employed strategies are highlighted. The chapter also discusses the need for further policy efforts to support environmental procurement practices of public bodies and companies.

Chapter 3 - The use of renewable energy sources is a fundamental factor for a possible energy policy in the future. Taking into account the sustainable character of the majority of renewable energy technologies, they are able to preserve resources and to provide security, diversity of energy supply and services, virtually without environmental impact. Sustainability has acquired great importance due to the negative impact of various developments on environment. The rapid growth during the last decade has been accompanied by active construction, which in some instances neglected the impact on the environment and human activities. Policies to promote the rational use of electric energy and to preserve natural non-renewable resources are of paramount importance. Low energy design of urban environment and buildings in densely populated areas requires consideration of wide range of factors, including urban setting, transport planning, energy system design and architectural and engineering details. The focus of the world's attention on environmental issues in recent years has stimulated response in many countries, which have led to a closer examination of energy conservation strategies for conventional fossil fuels. One way of reducing building energy consumption is to design buildings, which are more economical in their use of energy for heating, lighting, cooling, ventilation and hot water supply. Passive measures, particularly natural or hybrid ventilation rather than air-conditioning, can dramatically reduce primary energy consumption. However, exploitation of renewable energy in buildings and agricultural greenhouses can, also, significantly contribute towards reducing dependency on fossil fuels. Therefore, promoting innovative renewable applications and reinforcing the renewable energy market will contribute to preservation of the ecosystem by reducing emissions at local and global levels. This will also contribute to the amelioration of environmental conditions by replacing conventional fuels with renewable energies that produce no air pollution or greenhouse gases. This chapter presents review of energy sources, environment and sustainable development. This includes all the renewable energy technologies, energy savings, energy efficiency systems and measures necessary to reduce climate change.

Chapter 4 - In the realm of environmental policy, it is evident that a widespread shift is taking place from what has traditionally been known as "government" to a new form of "governance". There is a trend throughout literature which refers to government becoming one party equal in power to members of a group comprised of many other parties. That is, government as traditionally defined, is becoming an equal peer amongst nonprofit organizations, nongovernmental organizations, activist groups, citizens, and other entities, as opposed to serving as the dominant ruling body. Globalization has fueled this shift by changing the way in which the world operates – increasing complexity and intricacy of all interactions, especially in the environmental realm. Embodied in the aforementioned trends, *global environmental governance* has emerged as a new paradigm and, for some, a discipline of environmental policy. While no correct rule for global environmental governance has emerged, analysis of the concept is imperative to ensure that the world's environmental problems are addressed in an organized, effective, and mutually-beneficial manner. This chapter identifies challenges associated with the conceptualization of global environmental governance, focusing on the perspectives of authors in relevant fields. It is determined that the challenges of global environmental governance can be accurately described through a handful of overarching themes, including public procurement programs, private certification systems, minority environmental groups (women's groups, among others), and the emergence of post-sovereign environmental governance. This chapter characterizes and compares these four major themes – as well as other minor themes – and demonstrates how various authors have made contributions to the literature supporting or contending the reigning paradigm. Given the abovementioned review of the reigning paradigm, this chapter provides a concise summary of who should be involved in global environmental governance regimes, including a discussion on which parties may hold more power than others and which parties have potential to become more powerful in the future. Lastly, this chapter looks to the future and provides a discussion on potential directions of the field of global environmental governance, including an identification of the primary areas where more work is needed. This chapter concludes that considerable contradiction exists amongst various authors within the global environmental governance debate. It is further concluded that the existing body of literature should serve only as a foundation for what could become a complex discipline. That is, there is room for a tremendous amount of future work to be completed in the realm of governance and the fundamental concepts discussed in this chapter serve as compelling leverage points for future deliberation.

Chapter 5 - Shrimp industry is contested as it is identified with negative social and environmental legacies. Bangladesh, being one of the major shrimp producing countries of the world, has been facing resistance and criticism from local and international environmental NGOs. In response, Bangladesh government along with its donor agencies has come up with a series of environmental agendas and programs to ensure "environmentally sound shrimp aquaculture". This process of institutionalizing environmental domain pertaining shrimp industry have some positive impacts in terms of creating awareness among the people regarding environment, but at the same time it marginalizes others. It benefits a fortunate few, but the fate of people affected by the industry remains almost the same. The study demonstrates a trend and development, which is quite common to all environmental issues today, that is, the trend of moving towards a domain of managerialism, bureaucratization, and governmentality.

Chapter 6 - This chapter compares the public participation regulations and practices in forest policy and forest management in Canada, the USA and two countries in Northern Europe, Finland and Sweden. The countries studied all have extensive forest cover and forestry is important to the economy and local people. They were selected to represent different forest ownership structures. The comparison revealed that the countries with a strong private forest ownership regulate public participation less than the countries where most of the forest is publicly owned. Public participation in relation to private land needs to be based on voluntary processes like certification or it needs to occur at a more political level concerning a relatively large area. Most countries studied had established public participation practices and the participants had a moderate possibility to change the management plan. Only in Sweden is participation in forestry almost nonexistent.

Chapter 7 - Based on the efficiency definition by Koopmans, a case study is presented in this chapter comparing the results of a multi-criteria method and an eco-efficiency analysis for emerging technologies for surface coating. Multi-criteria analysis aims at resolving incomparabilities by incorporating preferential information in the relative measurement of efficiency during the course of an ex-ante decision support process. The outranking approach PROMETHEE is employed in this chapter for the case study of refinish primer application with data from an eco-efficiency analysis. Comprehensive sensitivity and uncertainty analyses (including the first implementation of the PROMETHEE VI sensitivity tool) elucidate the variability in the underlying data and the value judgements of the decision makers. These advanced analyses are considered as the distinct advantage of MCA in comparison to the eco-efficiency analysis, which just comprises various types of normalisation of different criteria.

Chapter 8 - The empirical finding of an inverse U-shaped relationship between per capita income and pollution, the so-called Environmental Kuznets Curve (EKC), suggests that as countries experience economic growth, environmental deterioration decelerates and thus becomes less of an issue. With more or less success, a large number of econometric studies have documented the existence of an EKC for pollutants such as sulfur dioxide, nitrogen oxide and suspended particulate matter. The baseline models estimated in the literature are linear polynomial models that include quadratic (and sometimes also cubic) terms of income as explanatory variables. Recently, these models have been criticized for being too restrictive, and alternative more flexible econometric techniques have been proposed. Focusing on the prime example of carbon emissions, the present chapter provides a critical review of these new econometric techniques. In particular, the authors discuss issues related to functional forms, heterogeneity of income effects across countries (regions), non-stationary ("spurious") regressions and spatial dependence in emissions. As for the functional form issue, some studies have addressed the nonlinearity of the income-emissions relationship by using a spline (piecewise linear) function, Weibull and smooth transition regression models, and more flexible parametric specifications, as alternatives to the polynomial model. The non-parametric models constitute one of the latest econometric tools used. Another important issue in panel data studies is the underlying assumption of homogeneity of income effects across countries. This assumption is too restrictive for large panels of heterogeneous countries. A further econometric criticism of the EKC concerns the issue of "spurious" regressions. As the model includes potentially non-stationary variables such as emissions and GDP, one can only rely on regression results that exhibit the co-integration property. Finally, recent studies allow for spatial dependence in emissions across countries to account for the possibility that countries' emissions are affected by emissions in neighbouring countries. Despite these new approaches, there is still no clear-cut evidence supporting the existence of the EKC for carbon emissions.

Chapter 9 - Concern about the environmental consequences of agricultural development, and studies exploring farmers' awareness of this issue are few. This chapter provides an insight into the environmental consequences of Green Revolution technology diffusion in Bangladesh using selected material evidences, such as, loss of soil fertility and trends in fertilizer and pesticide productivity at the national level, as well as examines farmers' awareness of these adverse environmental impacts and their determinants using a survey data of 406 households from 21 villages in three agro-ecological regions. Results reveal that Bangladesh has lost soil fertility in 11 out of its 30 agro-ecological zones to the tune of 10–70% between 1968 and 1998 due to intensive crop cultivation practices. The intensive HYV rice cultivation pattern (i.e., three rice crops a year: *Boro* rice–Transplanted *Aus* rice–Transplanted *Aman* rice) depletes approximately 333 kg of N, P, K per ha per year. Also, the partial productivity measures clearly demonstrate that productivity from fertilizers and pesticides were declining steadily at a rate of 4.5 % and 7.0 % per year ($p<0.01$) between 1977 and 2002. Farmers are well aware of the adverse environmental consequences of Green Revolution technology, although their awareness remains confined within visible impacts, such as, loss of soil fertility, fish catches, and health effects. Their perception of intangible impacts, such as, toxicity in water and soils is weak. Among the determinants of such awareness, the level and duration of Green Revolution technology adoption directly influence awareness of its adverse effects. Education and extension contacts also play an important role in raising awareness. Awareness is higher among farmers in developed regions, fertile locations and those with access to off-farm income sources. Policy implications include investment in farmers' education, agricultural extension services, rural infrastructure and soil fertility improvements.

Chapter 10 - A contemporary paradigm shift in waste management is to regard wastes as resources. One approach to exploiting the resource value in waste is recycling. Waste recycling in low- and middle- income countries is being driven by the informal sector, often with minimal if any input from institutions of the state. The focus of this chapter is on the often unrecognized and unacknowledged recycling activities of the urban informal sector in Nigeria. Recent experience on Nigerian waste recycling system is reviewed using insights from authors' fieldwork in various cities of Nigeria. The chapter also includes re-analysis of secondary data from published works on informal recycling in Nigeria, and first-hand field experiences. This chapter focuses primarily on the cities of Enugu and Onitsha in southeastern Nigeria, Lagos and Ilorin in the southwest, and Abuja in northern Nigeria. It explores the linkages and contributions to urban governance by the informal recycling sector. The chapter draws attention to some key characteristics of Nigerian recycling systems that have enabled them to cope with the vicissitudes of the recycle trade. Trends and commonalities in research on informal waste recycling in Nigeria are identified. Contributions of ordinary citizens involved in waste recovery and recycling to urban sustainable development, the need to acknowledge and support these contributions through reform of solid waste management, and implications of *integration* of the informal recycling sector into formal solid waste policy and practice in Nigeria are examined. Conclusions of the chapter form the basis of generalization on waste recycling by the informal sector in Nigeria.

Chapter 11 - Although pricing mechanism—especially raising the price of water—has become a high priority in dealing with the problem of water scarcity and the inefficiency of agricultural water management, it is a controversial issue both in developed and developing countries. Based on a linear programming model (LPM), this chapter analyzes the impact of water pricing policy on economic and environmental effects. The examination indicates that farmers will cut their rice planting as a direct response to rising water prices. Due to large capital costs and the labor-intensive characteristics of rice production, the reduction in rice area leads to an increase in farmers' agricultural income and a decrease in agricultural employment and revenue of irrigation districts (IDs). Moreover, together with the reduction of surface water consumption, the effects of changing fertilizers and pesticides resulting from rising water prices on the local environment are negative. In contrast to the findings of other researchers, in China the rising water price will result in an increase of farmers' agricultural income rather than a decrease; water pricing as a single instrument is not a valid means of significantly reducing agricultural water consumption due to the substitution of groundwater for surface water under the current water management institutions.

Chapter 12 - For at least two decades, sustainable development (SD) has been an important concept that promotes development that meets the demands of present and future generations while maintaining essential ecological processes and support systems. Policy makers need to identify environmental activities that limit development of society, configure usable knowledge, and develop sustainable strategies. An important challenge is to manage interactions of people with the environment so as to sustain critical ecological processes. Environmental resources, such as water, forests, and soil, are generally scarce and need to be conserved. Furthermore, ecosystem processes are not completely understood and too uncertain to permit accurate predictions of the environmental impacts of anthropogenic disturbances. Avoiding irreversible losses of species and preserving biodiversity are deemed critical to SD and not readily substitutable by human-produced capital. When one species becomes extinct, it cannot be replaced by another species.

The Precautionary Principle (PP) is an environmental management concept used since the 1980s. It is appropriate in decision-making situations in which there is considerable uncertainty about the environmental impacts of policy actions and such impacts are irreversible. Critical environmental policy decisions cannot be postponed until all scientific information is available. The complexities of natural systems and irreversibility of human impacts justify a commitment of resources to safeguard against serious threats to the integrity of the environment. The PP focuses conservation policy away from the current generation to future generations and changes the burden of proof to the resource developers and users. The PP is criticised as being poorly defined leading to many difficulties in its implementation.

The PP is an integral element in SD. Despite the virtues of this principle, the PP has not been widely used in environmental management in Australia, where its validity has been questionioned. This chapter reviews the PP concept, the different formulations of the principle, and the issues that affect its successful use in managing the Australian environment. Specific objectives of this paper are to: (1) review the major features of the PP; (2) assess the major elements influencing its adoption; and (3) investigate the extent to which the PP is being used in Australia.

Chapter 13 - Involving citizens in the design and provision of government policies and services has never been more in public demand. One remarkable area of progress in environmental policy during the last decade has been a shift from governing by national governments to multilateral participatory processes in which different stakeholders participate together in governance. Such an approach implies that different members of a community or organization can take part in planning and decision-making processes. However, not only who participates, but also in what, why and how they participate are relevant aspects in efforts to increase the legitimacy of environmental politics. In this chapter, the authors introduce a model of Environmental Policy Action as a Social Learning Process. They discuss participation in environmental policy action as a contextual and societal process and study social learning results produced in and by these processes. Such participatory processes are either driven by an actor's inner motivation or organized by society. In addition, the context of participation affects participants meanwhile they—through their action—drive the evolution of the context. They argue that if participants find environmental policy action processes both reasonable for themselves and their communities, and effective in regard to protecting the environment according to their personal perspectives, they experience self and social empowerment.

Chapter 14 - A comprehensive protection and enhancement of ecosystem services cannot be limited to the management of designated protected areas. Much of the earth is privately owned and many governments are now offering incentive payments to these owners (farmers or other land managers) to adapt their management of the land in such a way as to enhance ecosystem service provision. These incentive payments are consistent with the currently dominant model of market-led approaches to environmental management. But while these payments are offered to individual land owners, their decision making is not always the simple product of economic rationality. In the European Union, various agri-environmental schemes (AES) have been implemented and studied since the early 1990s. AES are incentive-based approaches to ecosystem services provision in agricultural landscapes that are rich in semi-natural and cultural features. Farmers' decisions to enter voluntary agri-environmental schemes have been explained in the literature in a number of ways, including the availability of information. Some authors suggest that informal neighbourhood networks impact on the penetration of information through some farming communities, and indicate that certain aspects of community cohesion and social capital are key factors influencing collective attitudes with regards to farm management, especially where management relates to the provision of novelty 'products' such as non-market goods and services. To date little empirical work has been carried out to estimate the extent and relative importance of farmer networks on entry into an AES. This study sets out to detect possible relationships between farm locations and farm entry time for the entrants into a specific agri-environmental scheme; the Environmentally Sensitive Area (ESA) Scheme in Scotland. Using quantitative measures based on Hagerstrand's model of innovation-diffusion as a spatio-temporal model, and GIS as a visualisation tool, clear spatio-temporal uptake patterns are found at different spatial scales and in different types of rural spaces. These findings are critically discussed.

Chapter 15 - Scholars of political economy view that environmental problems are deeply embedded in the reified nature of capitalism. The inherent nature of capitalism is exploitation of labour and nature with a view to maximizing profits. As capitalism expands, different legitimizing agencies for the capitalist class also emerge. State is one of such agencies that, to Marxist sense, serves as *accumulation* and *legitimation*. In the discourse of development, it has been made widely accepted that development activities must go on inspite of its severe environmental costs. "All natural resources now became strategic geo-power asset to be mobilized, not only for growth and wealth production, but also for market domination and power creation. To resist growth is not only to oppose economic prosperity, it is to subvert the political future, national interest, and collective security for the nation state!".

In this context, many believe that ecological consideration can be ignored, or at best, given only meaningless symbolic responses, in the quest to mobilize as many of earth's material resources as possible. Having more material wealth or economic growth in one place, like a particular nation state, means not having it in another places—namely rival foreign nations. It also assumes that material scarcity is a continual constraint; hence, all resources, everywhere and at any time, must be subject to exploitation. Roy Rappaport calls it "subordination of the fundamental to the contingent and instrumental".

Ecological problems are deeply rooted in this ideological underpinning of capitalism. Environment movements emerged out of this environmental problems initiated by capitalist economy. Here the authors are not suggesting that all environmental movements are anti-capitalist and anti-accumulation. There are many environmentalists who would hesitate to define their position clearly as anti-capitalist. The chapter examines the root cause of environmental problems by showing the interconnectedness between capitalist mode of production and environmental problems. Using political economy perspective, the authors will also examine to what extent the environmental movements/NGOs address those issues, and what kind of relationship the environmental movements present today with the capitalist enterprises marked by globalization.

Chapter 16 - With the effective abatement of point source pollution, nonpoint source pollution (NSP) has become a major concern of environmental management in China. Agricultural pollution is predominantly nonpoint due to fertilizer runoff, pesticide runoff, and discharges from intensive animal production enterprises. Environmental quality is a pure public good. While markets are ideal for maintaining incentives to reduce costs and adapt autonomously, the incentives through the market weaken as adaptations require a coordinated response from involved parties. Neither the state nor the market is uniformly successful in enabling individual small farmer to sustain long-term productive use of natural resource systems. It is critical to develop a voluntary incentive program to induce a reduction in nitrogen fertilization levels that also avoids moral hazard and is politically acceptable to the farm community and legally enforceable. Rational economic agents choose a noncooperative strategy to maximize their own well-being. Limitation of group number, establishment of internal rules, group heterogeneity, fairness of norms, expectations of individual efficacy and maintenance of mutuality are possible factors influencing cooperative incentives to conserve environmental public assets.

Chapter 17 - This chapter asks two key questions. What factors should be considered in making environmental policy? And what is the effective process for environmental policy formulation? To answer these questions, this article applies the concept of social demand articulation that has been developed to analyze the drivers and processes of environmental policy-making. Social demand articulation is a systematic approach that stimulates society toward environmental innovation. In particular, knowledge and information flows that raise the technological capability and awareness level of firms and consumers for environmental improvement are analyzed in greater detail than in previous work. Their indicators have been developed and applied to analyze environmental performance improvement cases in the air transportation sector. For effective environmental policy-making, this article emphasizes the steps to establish scientific evidence as well as public awareness regarding an environmental problem. If one of the two steps is missing, mere environmental protection movement could mislead society to achieving the political agenda of a particular interest group. In addition, it is important for environmental policy to contain a clear vision for future society the authors aim to craft. The philosophy that humans should prosper in harmony with nature must underlie environmental policy-making. With this in mind, institutionalizing knowledge and information flows among firms and societal stakeholders can set a path for the environmentally conscious market in which greener products are valued highly and give competitive advantage to environmentally innovative firms.

Chapter 18 - The main aim of this chapter is to show that Kenya has made progress in institutionalising environmental governance, particularly following the Rio Conference on Environment and Development. Prior to the Rio Conference, environmental management was scattered in the line ministries with no clear focus on sustainable development. The paper shows that the country has elaborate legislative framework with instruments that can significantly contribute to sustainable environmental management. However, the implementation of the legislation faces a number of challenges including lack of policy on environmental management and weak capacity. It is evident that serious concerted efforts must be directed at capacity building to make the legislative intentions of a good and healthy environment for all in Kenya a reality.

Chapter 19 - Stakeholder assessment is an important approach in environmental policy analysis. It helps to identify the critical environmental issues in controversy in a given situation, the affected interests, and the appropriate forms of handling the conflicts. Although many studies have highlighted the major prevailing environmental theory of coordination and communication, few have linked this theory with practical environmental policy analysis. The purpose of this chapter is to provide a conceptual framework of stakeholder assessment to reach the goal of environmental dispute resolution and assess the major stakeholders' roles in potential environmental conflicts, particularly in resource-dependent local jurisdictions. A case study was conducted to further examine the theoretical model of stakeholder assessment in environmental conflicts and the policy analysis process. The stakeholder assessment process involves potentially interested stakeholders in order to: assess the causes of the conflict; identify the entities and stakeholders who would be substantively affected by the conflict's outcome; assess those stakeholders' interests and identify a preliminary set of relevant issues; evaluate the feasibility of using a consensus-building or other collaborative process to address these issues; educate interested persons on consensus and collaborative processes; and design the structure of a negotiating committee or other collaborative process to address the conflicts. The results highlighted that stakeholder assessment has been proven valuable as a first step in consensus-building processes and in finding constructive approaches to resolving environmental conflicts.

In: Handbook of Environmental Policy
Editors: Johannes Meijer and Arjan der Berg

ISBN 978-1-60741-635-7
© 2010 Nova Science Publishers, Inc.

Chapter 1

MATCHING ENVIRONMENTAL POLICY TO RECIPIENTS

*Quentin Farmar-Bowers**

Centre for Sustainable Regional Communities, Faculty of Law and Management
La Trobe University, Victoria, Australia

ABSTRACT

Families who run family business enterprises are the policy recipients I focus on in this chapter. They respond to environmental policy either as family members or as business people. In this chapter, I use decision-systems theory (DST) to discuss why policy recipients choose to respond in different ways. Understanding policy recipients' decision processes is important for policy developers for two main reasons. First, it can help policy developers devise environmental policies that are accepted as 'legitimate' by policy recipients. This is important because 'legitimate' policies are more likely to be implemented and result in lasting change than policies that are resented. Second, it can help policy developers assess whether or not their proposed policies are going to 'improve society'. Policies that 'improve society' are those that both help future generations satisfy their family's aspirations and also meet global ideals, such as sustainable development ideas.

I start the chapter by outlining decision-systems theory (DST) as it provides a theoretical framework for dealing with the question of policy 'legitimacy' and the meaning of 'improving society'. DST indicates that family's aspirations, which I conceptualise as a set of five 'motivation-stories', have similar intentions as sustainable development ideas. These intentions are to secure family welfare (human welfare in sustainable development terms) and concomitantly be supported by the environment (conserve the Planet's environment in sustainable development terms). I use the '4-Group-Stakeholder model' and DST to argue that the 'ultimate driver' for the current human development system is the desire of families to satisfy their 'motivation-stories'. The 'ultimate *future* driver' is a projection that describes the 'ultimate driver' future generations would have if global ideals (such as sustainable development ideas) were effectively incorporated into their 'motivation-stories'. I suggest that policy developers

* Tel: +61 3 5444 7464, Fax: +61 3 5444 7998, Email: q.farmar-bowers@latrobe.edu.au, www.latrobe.edu.au/csrc.

ought to create policies that facilitate the 'ultimate *future* driver' for all decisions in the human development system. Developing policies that do this would ensure the legitimacy of the policy and also ensure that the policy developers have done their best to 'improve society'.

The final part of the chapter discusses how policy developers can give priority to the 'ultimate *future* driver' in the policies they create. The 'boxes of influence' concept, which is part of DST, can be used as a policy classification system. It covers policies and programs over the life-cycle of family decision-makers; childhood to retirement. It shows how policy developers can help policy recipients create opportunities that will satisfy their 'motivation-stories' in ways that meet the global ideals of caring for the Planet's environment and caring for people, including their own family and descendents.

INTRODUCTION

Ideally, government policies change existing arrangements to 'improve society' in some way. In this chapter, I will discuss the relevance of people's decisions-making processes to the 'legitimacy' of environmental policy. I will also discuss what 'improve society' means.

New government policies aim to change what some people are doing. New policies discourage, modify or stop an existing activity, or encourage a new activity. The targets of a policy, the policy recipients, are usually an easily identified group of people or organisations. A particular policy might lead to a real improvement in society or it might not. However, the first step in applying a policy is getting it accepted by the recipients as 'legitimate'. Appreciating how policy recipients view policy is important for policy acceptance (Cocklin et al., 2007), and Burton (2004) suggests it is actually critical for the adoption of policy objectives. If the policy recipients do not accept the policy, it may be ignored, compliance costs may rise and advertising may be ineffectual. There might be a political backlash leading to the withdrawal of the policy, and too many unacceptable policies can lead to the downfall of the government. So the legitimacy of a policy is important for technical as well as for political reasons.

Getting a policy accepted as 'legitimate' is likely to be easier when the policy facilitates what the policy recipients want to achieve, or have, in the long-term. In turn, policy developers are likely to achieve 'policy-legitimacy' more easily if they understand the decision processes policy recipients normally use in making-decisions. Policy developers can then select policy instruments that harmonise with these processes and choose policy objectives that facilitate the policy recipients' aspirations. Policy development is then a negotiation between achieving policy-legitimacy and achieving what the government wants and how it wants to go about achieving it.

When the policy recipients are businesses, the profit motive and business ethics (Velasquez 1998) provide the common ground for understanding decision-making and policy development. When policy recipients are families, the common ground for understanding decision-making is more problematical as business ethics are rarely appropriate and other approaches, especially care ethics (Held 2006), but also rights based ethics (Butler 2008) or land ethics (Callicott 1999), are more usual. The differences between family and business decision-making are highlighted when families directly run their own business ventures. Understanding these differences is especially important for environmental policy developers as environmental policy can be directed to change the activities of the business or the family or both.

Policy objectives are politically determined. Every new policy ought to improve society in some way. However, there is a possibility that policy may be parochial, benefiting a particular, already privileged, group in society and not in the long-term interests of all people (Davis 2003). Ross and Dovers (2008), in discussing environmental policy integration in the Australian context, noted that there were few examples of political leadership prepared to override vested interests. Power and ideology are often the most controlling influences on society (Oliga 1996). This parochialism can be reduced by checking new policies against both family aspirations and global ideals such as human rights and ecosystem protection.

Global ideals exist around two topics; how people treat the environment of the Planet and how people treat each other. Many aspects of both topics are covered in ancient religious texts (Szenberg 1997, Friedman 2003). There is now a vast international literature on environmental issues and also a huge literature concerning how people ought to treat each other. International concern about the environment led to the United Nations Conference on the Human Environment held in Stockholm in 1972. This was the first international conference on the environment. Environmental information and a better understanding of the role people have in altering the Planet's environment have grown exponentially in the last four decades. The lead topic at the moment is climate change. Concern about relationships among people is also a major international issue and current failures include wars, genocides, refugees, disappearances and terrorism. The desire to maintain peace led to the creation of the League of Nations after the First World War and the United Nations (UN) after the Second World War. The UN's Universal Declaration of Human Rights (UN 1948) and its conventions* provide a modern starting point and an ideal for societies (Fiss 1999) in terms of how people ought to deal with each other. The idea that both topics (the environment and treatment of people) need to be included in all decisions gained international acknowledgment in the later 1980's with the notion of sustainable development (WCED 1987). The acceptance of this integration as an essential part of sustainable development was confirmed internationally in the Rio Declaration and Agenda 21 (UN 1992)*.

The meaning of 'improve society' could be a scaled-up version of what people want for their families in the very long-term – say for the next two, three or even seven generations (de-Shalit 1995). Alternatively, 'improve society' could be a scaled-down version of global ideals such as sustainable development. Global ideals are not useful unless they are applicable to individuals. Eleanor Roosevelt, the Chair of the UN Commission on Human Rights (1946 – 51), suggested that human rights were only meaningful if they applied to individuals in every situation and that individuals have an obligation to make sure this happens (Meyer 1981). Arias (2000) suggests that 'global thinking' is needed to create the institutions that will deliver human security, economic justice and democratic interactions for people on a daily basis. The same sentiments (that global ideals have to work for the individual to be of value) apply to environmental conditions and also to the idea that development decisions should support both environmental quality and human dignity.

---

* The Declaration and its Conventions and other Instruments contained in General Assembly Resolutions, (1946 onwards), are available in the United Nations Dag Hammarskjöld Library: http://www.un.org/Depts/dhl/resguide/resins.htm
* See also The Earth Charter, available from: http://www.earthcharterinaction.org/content/

I suggest that environmental policy developers require three 'understandings' if they are to create policies that are accepted by policy recipients (families who run businesses) and live up to global ideals on how to treat people and the Planet's environment.

1) Understanding the decision processes of the policy recipients (families who run businesses) whose behaviour the policy developers want to change
2) Understanding what 'improve society' might mean to these policy recipients, and
3) Understanding which global ideals are relevant and how to use these ideals to create policies that will support people in their daily lives and 'improve society' in the long-term.

I will discuss these three 'understandings' in turn. For the first 'understanding', I will outline five of the six concepts in decision-systems theory (DST). DST is an interpretation of the decision-making processes families who run family businesses use (Farmar-Bowers & Lane 2006, 2009). For the second 'understanding', I will discuss how family aspirations relate to 'improve society'. For the third 'understanding' I will consider how global ideals, such as sustainable development, are relevant to policy recipients and policy developers. The '4-Group-Stakeholder model' helps explain the relationship between peoples' aspirations and global ideals. The chapter ends with a discussion on how policy and programs can be classified and harmonised with family aspirations using the 'boxes of influence' concept. This is the sixth concept in DST. I will show how this classification system can help policy developers argue for a portfolio of policies to influence people's behaviours throughout life.

"UNDERSTANDING 1": DECISION-SYSTEMS THEORY; THE DECISION-MAKING PROCESSES FAMILIES USE

Decision-systems theory (DST) is a substantive theory. It was developed from thirty three in-depth interviews with farming families in the State of Victoria, Australia, using the analytical procedure for developing a grounded theory (Strauss & Corbin, 1998, Corbin, & Strauss, 2008). DST is therefore evidence-based. An additional ten in-depth interviews with farming families in New Zealand and thirteen in-depth interviews with farming women in Australia confirm that DST provides a good understanding of family decision making processes.

DST is our interpretation of the processes farming families use when they make strategic decisions. These are the decisions that create changes in their lives such as expanding the farm, selling up, major changes in enterprises, establishing non-farm businesses, getting off-farm employment, conserving native vegetation in perpetuity, creating off-farm investments, volunteering, having children, sending children to fee-paying boarding schools, family holidays, retiring and so on. DST applies to the life span of decisions-makers and provides an interpretation of the processes involved in developing and then satisfying (or trying to satisfy) the family's aspirations (the motivation-stories concept). DST covers the entire spectrum of family activities; it is not just about business decisions. DST provides a way of understanding the relationship between the family and the businesses they run. It provides an interpretation of how families who run businesses make decisions that organise the relationships between business and family. DST provides an understanding of the decision-processes that families use for both their businesses and their family members. Although DST was developed from information on farming families and their farming businesses I think it is relevant to the decision-making processes used by non-farming families who operate family businesses in other industries. Indeed, a number of the families I interviewed ran non-farm businesses in addition to farm enterprises.

Family decision-making over a life-time is very complex and hard to understand. The individual concepts in DST provide an interpretation of different aspects of this complex process that together, I hope, provide a base for further work and sufficient understanding to help environmental policy-developers take a fresh look at the option for policies that advance sustainable development ideas. Understanding these decision processes can help policy developers create policies and programs that influence the major decisions that families make over a life time. DST attempts to provide this 'all of system' understanding to match the issues farming families face, such as community change, climate change, protection of the environment and globalisation all of which are also 'all of system' changes.

DST is not relevant to decision processes when families are not the decision-makers, such as in multi-national corporations, in government departments and organizations or in technical and scientific research establishments. However, DST is relevant if these organisations want to sell to or influence families and family run businesses.

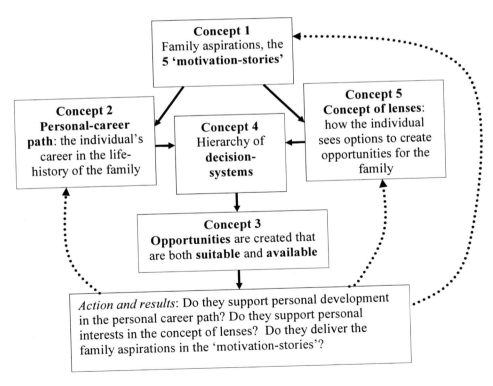

Figure 1. Interaction between the five concepts in Decision-systems theory (DST) including three feedback loops from 'actions and results'.

An Outline of Decision-Systems Theory (DST)

DST comprises six concepts (Farmar-Bowers and Lane 2009). At this point in the chapter I will discuss five of these concepts that relate to different aspects of decision-making. The five concepts are linked and overlap to some extent. They span the 'life-cycle' of the family, that is, they apply to all stages in a life-span. Figure 1 shows the relationships between these five concepts diagrammatically. I will provide an outline of the five concepts in this section then discuss each of the five concepts in detail in the next section. The sixth concept 'boxes of influence' will be discussed towards the end of the chapter.

The five concepts work as follows:

'Motivation-Stories'*

The family has 'family aspirations'. These aspirations are things the family, as a group, wants to achieve or have during their lives. I have conceptualised family aspirations as a set of five 'motivation-stories'. These motivation-stories outline areas or subject matters that energise family members into taking action throughout their lives. The desire to work on these areas or progress them in some way is what gets family members up each morning. I suggest that motivation-story-fulfilment is a major, if not the major, driver in family life. These stories are explained by Farmar-Bowers (2004) and were initially given the title of 'component stories'. The 'motivation-stories' concern 'purposes' not content. For example, one of the motivation-stories concerns 'learning' or 'self-education'. The purpose or focus of this story is 'learning'. It is about making oneself into a competent person. The story is applicable to all adults but the content of this story varies between people. The content of the story – what they mean by competency and how they get competency – is unique to the individual. Some people, for example, may pursue an academic education, others may want to be competent physically, while others may want to develop inter-personal skills and so on. Although the content and goals vary between people, the overall purpose or focus of the story – about becoming competent – is the same. The purposes and especially the contents of 'motivation-stories' are the subject of ongoing negotiations within the family.

Personal-Career Path

Individual family member have their own careers that usually link in with the negotiated family 'motivation-stories'. However, tension can occur between a person's individual career path and the family aspirations as they can be competitive. The personal career path concept allows us to understand the division of labour within a family. It also links the position of the individual to the life-cycle of the family. For example, a person about to retire will have a different view of what is appropriate as a family investment compared to somebody just starting their career.

---

* 'Motivation-stories' is the title of a concept that refers to what farming families told us, through numerous stories, about the purposes of their actions in much of their lives.

Opportunities are Created

Families as a 'group' have to take action in order to satisfy family aspirations and so they have to create opportunities from the range of options they see. The process of selecting an option and creating an opportunity around the option involves applying two criteria – 'suitability' and 'availability'. 'Suitability' is the first criterion in selecting an option to develop into an opportunity. We can understand 'suitability' by the questions the family decision-makers might ask themselves: 'Is the opportunity I could develop from this option likely to satisfy some part of the family's 'motivation-stories'?' 'Does this appear to be an option that would actually deliver one or more of the family's aspirations?' Not every option can be used to develop a suitable opportunity. For example, if the family wants a holiday then options for more work are not going to be suitable. If an individual family member wants to work as an artist then an option to become a plumber, however financially rewarding it might be, will not be suitable. The second criterion is the 'availability' of all the resources necessary to turn the chosen option into a viable opportunity for the family. If an option requires medical degrees and the decision-makers have plumber's tickets then they will not be able to turn the option into an opportunity.

Hierarchy of Decision-Systems

People deal with the large number of decisions that they have to make by grouping them according to the issue or topic they represent. People do this automatically so that they can make a decision about say education, and the next minute make a decision about a holiday, without confusing the two. Different topics fit into different decision-systems and there is a hierarchy of decisions within each decision-system. The decisions in the upper tier of the hierarchy for any decision-system concern 'why-for' issues – 'why are we doing this?' This contrasts with decisions in the lower tiers of the decision-system that concern 'how-to' issues –'how are we going to do this?' The decisions families make in selecting an option and creating opportunities from it, span the whole decision hierarchy in the relevant decision-systems. They first have to decide 'why they want to do it' (why-for issues) and then they have to decide, 'how to do it' (how-to issues). The family may have goals and values throughout decision system hierarchies. 'Why-for' decisions are related directly to satisfying the family's motivation-stories and are justified in terms of caring for family members. 'How-to' decisions are justified in technical and economic terms and concern creating the 'means' needed to progress the 'why-for' issues.

Concept of Lenses

The concept of lenses considers the process of making individual decisions from the decision-maker's perspective. It brings into the option selection process the notion of personal interests and personal knowledge of one's own abilities and resources – what the decision-makers (and family) can contribute to create an opportunity. The concept of lenses provides an understanding of the process the decision-maker uses to decide each particular decision. So decision-makers repeat the process for all major decisions throughout their lives.

Finally, once an opportunity is created and implemented the family starts to get feedback on its success. This is shown as the *Action and results* box in Figure 1. The feedback concerns both the activity itself (are they happy with what is happening, with what they are doing) and on the performance (are the results living up to expectations). The family assesses the action itself, (how they feel about doing this kind of work) and the results of the actions against their 'motivation-stories'. The big question is 'are the family's 'motivation-stories' being satisfied?'

We can think of feedback in two ways; as a static measure and as an exploration. As a static measure the family members judge the outcomes in terms of how closely they deliver the content of their existing 'motivation-stories'. This static measure is a circle: 'family 'motivation-stories' encourage the family to create opportunities to satisfy the stories'. It is static in the sense that they are trying to improve their performance to achieve a defined outcome, much like continuous improvement in total quality management (Deming 1986). But life is not static. There is a large degree of exploration and experimentation. Action is an essential prelude for new ideas (Goa et al., 2003 Harris 1999). Some actions create new ideas about how best the family can satisfy their 'motivation-stories'. Families who run their own businesses are in a good position to experiment by trying out new ventures. They do this by changing the content of the family 'motivation-stories' (e.g. trying a totally new venture) and perhaps also by adding to or changing the stories completely. The process is iterative and the contents of the 'motivation-stories' change as life progresses and people see better ways of achieving things. They may also see new things to achieve – new stories to satisfy. The exploration aspect of feedback is really important for long-term radical environmental change. Trying new things, experimentation, is the only way people will be able to develop new approaches to fulfilling the intentions of sustainable development.

New ideas also change the direction of personal career paths and expand the areas that decision-makers are interested in. Often 'new interests' are rekindled from earlier experiences in life; for example, 'I always wanted to play a musical instrument', 'I always wanted to travel overseas', 'I always wanted to conserve nature'. It seems that some decision-makers wait until near retirement before they feel able to pursue activities that interested them in childhood. For some people, the negotiated family 'motivation-stories' smother their personal aspirations until their family obligations diminish. This suggests that important environmental issues must be supportive of the family 'motivation-stories' if they are going to be in the main stream of family activities and not sidelined until retirement.

The Details of Decision-Systems Theory (DST)

I will discuss the five concepts of DST in turn:
1. 'Motivation-stories'
2. Personal career paths
3. The suitability and availability of opportunities
4. Decision-systems and
5. The concept of lenses.

'Motivation-Stories'

The five 'motivation-stories' obtained from interviewing farming families form a two level hierarchy. Figure 2 shows the relationships between five stories diagrammatically. The stories describe people's 'purposes' in life rather than content; what they actually do. Although different people are likely to have different 'content', their stories have the same 'purposes'. Overall, the purposes in these stories seem to be about caring for oneself and the family. Part of 'caring' is about fitting into society or at least operating successfully in society. We can anticipate from these stories that the decisions families make that they think are really important are those that directly concern caring for themselves and their immediate family.

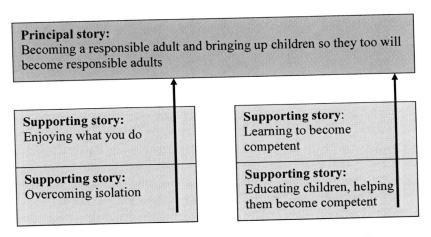

Figure 2. The hierarchy of 'motivation-stories' from interviews with farming families.

Many colleagues who first saw these 'motivation-stories' were surprised because they expected such stories to be about wealth and money. However, the families who were interviewed saw money and wealth as means to an end. So making money and becoming wealthy was not what they were aiming to achieve in their lives. While they often said they were pleased, even very excited, when they made a good profit in the businesses they ran, they suggested that this was because it showed them to be competent business people and because they had things they wanted to do with the money. In addition, they often said they were getting something out of the work itself quite apart from the monetary profit, such as fulfilling a life-long interest. So it was clear that making money and becoming financially wealthy were not the purposes of their lives.

There is some statistical evidence that supports the notion that becoming wealthy is not the main purpose in life. It comes from data that show that 'well-being' does not keep pace with material accumulation. For example, US gross domestic product (GDP) per capita more than doubled between 1950 and 1995 but the genuine progress indicator fell by one quarter (Jenkins 1998). Hamilton (1998) discussed the genuine progress indicator for Australia, and Daly and Cobb's (1989) Index of Sustainable Economic Welfare (ISEW) and Max-Neef's (1995) threshold hypothesis lead to similar inferences.

The five stories that represent family aspirations are about 'purposes'. We can envisage them as empty boxes and each family puts their own content into them. For example, what 'enjoying what you do' means will vary greatly between families – some may like working with livestock while others may enjoy spending time with their family. Or, 'learning to be competent' could mean becoming a competent farmer, business person, nurse, parent and so on.

The 'motivation-stories' are long-term and are satisfied over a life-time through action. This action is usually achieved by setting goals and working towards achieving these goals. People usually have goals that can be achieved in a set time, perhaps within a day or perhaps longer – within ten years. Goals represent milestones in moving towards satisfying their 'motivation-stories'. But of course, people sometimes set inappropriate goals and can have a difficult time as a result. People sometimes talk about 'getting back on track'. We could interpret this as people looking for new things to do that will help them fulfil their 'motivation-stories'. There seems to be an ongoing process of adjustment and personal development throughout life. Like these 'motivation-stories', the goals people set for themselves along the way are also hierarchical, with later goals flowing on from earlier ones. We can visualise starting with a small goal and building on achievements to create new and bigger goals as time proceeds. People often have long-term goals and break them down into a sequence of smaller goals for the next job or for the next day. Goal-setting is a very powerful technique for working efficiently. Although the 'motivation-stories' represent 'good things', the things the family wants to achieve or have during their lives, how they get them can, and usually does, involve hard work and tough decisions that they may not always enjoy.

The 'motivation-stories' and the long-term goals represent what the family wants to achieve and the principal ways in which they want to achieve them. The family's 'motivation-stories' and long-term goals are different from the 'motivation-stories' and long-term goals that an individual family member might have. Agreement on the family's 'motivation-stories' is usually reached through some form of negotiation between family members. These negotiations are very important for the family in the long-term if they are to work together and stay together. The women farmers especially tended to stress the importance of family discussion, negotiation and agreement. Considerations of equity and justice between family members are important in these negotiations for some families.

'Motivation-stories' appear to be created early in life and get more defined as people mature. 'Motivation-stories' do not seem to change much over time but the emphasis among the stories is likely to change as their family's circumstances change. For example, the supporting story of 'educating children' probably exists throughout life but is only in operation for perhaps two decades as their own children are educated at school and college. However, people may continue to use the supporting story in other ways. They may for example, support education improvements politically or they may make donations to educational charities or specific educational organizations, or even take on an intern or apprentice in their business activities. Some farmers noted an improvement in the educational level they and their peers have attained compared to what their parents achieved. Although this may be due to increased opportunities, the overall result is that formal education has been growing in importance for most families.

Personal Career Paths

Personal career paths are a product of the individual's 'motivation-stories' and long-term goals. They are usually a mix of planned action and circumstances. In farming families, the principle family decisions-makers (say a husband and wife) usually collaborate on developing a business and on creating a family. They run the enterprise jointly, although delegation on specific tasks is normal. During this time each may have a series of careers or jobs working in other businesses or in other activities in addition to their farm careers. Some of these interests may continue after they retire from farming. For example, in addition to being a farmer, one member of the family may have a political career in local or state government or may develop off-farm assets or businesses. There is often synergy between parallel careers so that the skills they develop while working off-farm in a company may be helpful in farming and visa versa. Having alternative careers and planning new interests can help maintain personal welfare during the often difficult process of retiring from work (Wythes & Lyons 2006). Retiring and moving from the farm can be especially difficult as the farm provides identity, is a home, a valued place and also a business (Foskey 2005). Off-farm careers / jobs are also very important for maintaining the viability of local communities (Albright 2006).

The concept of personal career paths is important as it highlights the notion that people make different decisions depending on where they are in their own career path and also where the family is in terms of a family 'life-cycle'. For example, people are more likely to take risky decisions at the beginning of their careers because they are at the stage of building a future and have more time to recover should things go wrong (Le´vesque and Minniti, 2006). And again, when the next generation is coming into the family business as an adult, the decision-makers are more likely to respond to the needs of the succeeding generation by retiring or making aggressive decisions to expand the business by developing new opportunities. This reflects the necessity of creating more 'means' to help meet the increased 'needs' as the next generation families come into, and have to be supported by the family businesses. People sometimes talk of 'expansion periods' in their family business followed by 'consolidation periods' that tend to reflect their family's life cycle more than external business cycles as they would in large corporations.

The Suitability and Availability of Opportunities

'Motivation-stories', and the long-goals that are derived from them, are made to happen by creating opportunities. This is an active process; decision-makers build on an option to create a practical opportunity (Brodt et al. 2005). The decisions involved in creating an opportunity depend on two criteria. The first concerns the suitability of the opportunity; it asks the question 'is the opportunity likely to satisfy some of the family's 'motivation-stories'?' The second concerns the availability of the opportunity; it asks the question 'have we got the relevant skills and resources and can we buy in the rest of the skills and resources we need to implement the opportunity successfully?' Suitability and availability questions apply to all opportunities: recreational, business, social, educational, medical, spiritual and environmental and so on.

The availability of an opportunity depends on the presence of three kinds of components. These components are relevant for all kinds of opportunities. *Personal components* are what the decision-maker can bring to create the opportunity; what they can put in from their own resources. These include time, personal energy, enthusiasm, knowledge, skills, resources they already have including the money they have in the bank. *External components* are things the decision-maker has to access from other people, organisations and the environment as they need them to develop the opportunity. They include infrastructure, workers, markets, information, materials, finance, machines, education, clubs, communities, parks, wildlife and friends. The third kind is *random components*. These include things such as fluctuations in prices and markets, weather patterns, and other sorts of happenstances.

For the opportunity to be available, the decision-makers must have an adequate supply of all three kinds of components. If one is missing the opportunity ceases to be available; it is not longer a practical opportunity. Over time, the decision-makers can increase their store of personal components; they can get more training, develop new skills and abilities (Allan 2005), or accumulate more resources. They can also improve their access to external resources; they might improve their credit rating or move closer to infrastructures such as airports or rail links. They might also develop better strategies for dealing with random components to maintain or increase their productivity, such as building more dams, reducing water wastage, improving plant nutrition, using more drought tolerant crops or developing more sophisticated market forecasting techniques. Being aware of relevant research is an important aspect of these strategies. Some farmers establish their own community based research arrangements, such as the Birchip Cropping Group in Victoria, Australia (BCG, 2008).

Decision-Systems

Families organise decision making into a pattern to enable them to gain an understanding of the topics involved. This allows them to make more informed, and more effective decisions as their understanding and experience of the topic grows. Some decisions are comparatively easy to make because they are about technical or economic issues. For example, the purpose of a decision might be to select and build a 'machinery shed' for storing farm equipment. The information needed might be quite complex and difficult to get but essentially there is a solution that provides a good fit between cost and technical performance. Professionals in many areas develop decision-support systems to guide people through the technical and economic issues. Decision-support systems can be used to help decision-makers deal with quite simple to very complex technical issues. For example, Karakosta et al. (2008) devised a system to deal with the country specific transfer of sustainable energy technologies from developed to developing countries under the Kyoto Protocol. Expert systems also provide 'tick-the-box' solutions to enable effective decisions to be made for these kinds of problems. These decisions are made on technical and economic grounds. There is a 'correct decisions' determinable by applying analytical procedures.

There are other decisions where no amount of technical information or analysis can really help. These decisions relate to personal ideas, emotions and feelings. They include issues like 'what career should I pursue', 'should we have children'. These are decisions that really have no correct independent answer although they might be influenced by culture. What make these kinds of decisions correct for the family is how they relate to the family's 'motivation-stories'. It is an appropriate decision if it helps the individual or family obtain some progress on satisfying their 'motivation-stories'. People justify these kinds of personal decisions in a variety of ways but principally in terms of caring for the individual or family. Picking a career that is interesting and rewarding to you on a personal level is a 'caring decision'. So too is helping one's children find suitable careers, even if it means they leave home and decide not to join the farming business. Slote's (2007) arguments, that caring motivation is based on people's capacity for empathy with others and that empathetic caring leads to respecting the autonomy of others, seem relevant to decisions about family members.

This pattern of decision-making by the individual or the family is referred to as 'decision-systems' in DST. In one dimension, decisions are grouped into topics, in another dimension they are grouped according to whether they are justified by business / technical considerations or care ethics. This is shown diagrammatically in Figure 3. Each decision-system involves a cascade of decisions that form a hierarchy of decisions. At least one of these decisions is in the top tier of the hierarchy and is justified in terms of caring for the family. The top tier of the hierarchy functions as a kind of clearing house for the family in which decisions are made that give 'permission' for the rest of the action to take place. Decisions in the 'top tier' of the hierarchy are usually justified in terms of 'caring for the family' (care ethics). Decisions in the rest of the decision-system (i.e. after the decisions in the top tier have been made) are often delegated to an individual to carry through. This is because most of these decisions are justified on technical and economic grounds where expertise, experience, information and analytical skills are required. In farming families, because the businesses are usually small, the decision-makers often make both kinds of decisions. We can imagine farmers saying to themselves; "OK, the family agrees we ought to educate the children in a fee paying school (decision made in the top tier of the decision hierarchy and justified on care) so now I will work out how to generate the necessary income by developing an extended cropping program for this year that should deliver the extra income" (decisions made in the lower tier of the decision hierarchy and justified on economic and technical grounds).

Decision-systems are a concept and are called 'systems' because they have boundaries based on topics. Decision-systems are not totally closed systems in that the topics are linked. Thus one decision or its consequences can influence decisions in other decision-systems. The concept of decision-systems is shown diagrammatically in Figure 3. Families may have a large number of decision-systems; perhaps a dozen or more, one for each major topic in their lives. The decisions within each decision-system form a hierarchy and we can envisage decisions justified in terms of caring for the family at the top and decisions based on technical and economic considerations lower in the hierarchy. The strongest linkages across the decision-systems occur at the top of the hierarchy where the family is working out what priority to give different decision-systems in order to provide the care the family and individual members need. The linkage between different topics at the top of each hierarchy is of paramount importance in the operation of the family and we have sometimes referred to this as the 'family decision-system' to emphasise its importance as the first step towards creating the best mix of opportunities to satisfy the family's 'motivation-stories' (Farmar-Bowers and Lane 2006).

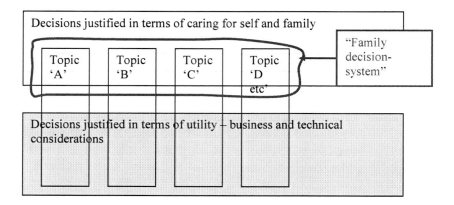

Figure 3. Relations between decision-systems (on topic A, B, C, D, etc) and their justifications.

Once 'permission' has been given within the family decision-system to go ahead with looking at options around a particular topic, the relevant technical and economic investigations can be carried out. As the information is generated in the decision-systems about an opportunity there is some feed-back to the family decision-system to indicate how the reality of creating an opportunity fits in with the family's 'motivation-stories'. It is quite possible that the realty does not suit the family's priorities and the opportunity in the decision-system has to be abandoned or reformed in some other way until it fits. People can have 'second thoughts' about starting a new opportunity. They may feel it is not really going to advance their careers or they are not going to be happy or comfortable doing it (i.e. it is not a suitable opportunity). They may also feel that they don't have enough resources or experience to succeed (i.e. it is not an available opportunity).

The Concept of Lenses

Another way of considering decision-systems is to view them from the perspective of the individual decision-maker creating and implementing opportunities. The 'concept of lenses' is about understanding the processes of decision making for an individual person; the processes they use repeatedly throughout their lives to create opportunities. I use the word 'lenses' because I envisage decision-makers looking through a series of lenses, with each successive lens changing their view of things and eventually helping them decide on a specific action. I discuss, in this section, the five lenses in turn which are shown in Figure 4. Decision-makers would use them iteratively in making important decisions; they might also change the order, or perhaps skip some of them.

The first lens is 'intrinsic interest'. The starting point for dealing with long-term issues, such as 'what to do in life' or 'how to provide for their family', is working out 'what interests me'. If a person is not interested or perhaps even revolted by a topic they are not going to consider it as a career, perhaps not even for a moment. Some people are interested even excited by medicine and will actively look for career options in medicine with great joy. But if a person is not interested in medicine they will not be aware of medical options and will not spend time (waste time) checking them out. So intrinsic interest acts as the first lens though which people view options, most options are excluded as they focus on the topics they find most interesting. So they do not actually choose from what is on offer but rather pre-select options in topics that they find personally interesting. Very often people enjoy doing things that relate to their intrinsic interests – what they find personally interesting. People might change jobs or business venture in order to work on what they are interested in and enjoy doing, even though they will earn less money. There is a close link between this first lens and the supporting motivation-story of 'enjoying what you do'.

The second lens for the decision-maker is 'family considerations'. There are two families involved. The first is the decision-makers' parents and sibling. They have a considerable influence on the decision-maker through the development phases of life and also later on through continuing contact and family talk. The kind of education decision-makers get and the approach they acquire to doing things and being adventurous, or not, are important. The second family is the decision-maker's own family (spouse and children). The need to provide for them can greatly modify and widen the scope of what options decision-makers consider. Having their own family moves the decision-makers from considering only their own 'motivation-stories' and their own career path, to having to negotiate and work on delivering the family's 'motivation-stories'. The negotiated family 'motivation-stories' provide a long-term guide for decision-makers on what purposes they need to pursue. Having a family alters what personal components they can bring to create opportunities; especially what time and mental energy they can devote to business opportunities. A secondary negotiation on each purpose helps decision-makers identify what goals and activities they should pursue to satisfy the family's 'motivation-stories'.

The third lens is 'knowledge of personal components of opportunities'. Previously, we noted that opportunities contained three kinds of components. The personal components of opportunities are the components that the decision-makers provide. These include time, energy, enthusiasm as well as knowledge, skill and the resources they already own. It also involved knowing how to access external components of opportunities and taking account of random components. The external components of opportunities are the components that the decision-makers obtain from other people, organisations and the environment. They include hired help, advice, fertilizers, markets, finance, and waste disposal. I call the third kind of components 'random components'. These include price fluctuation, economic cycles as well as environmental events such as flood, droughts, storms as well as good weather and good rainfall.

An issue for family businesses is time management; spending too much time on one issue and not having enough for other essential areas can become a major problem. Very often not enough time is given for family recreation and care. Decision-makers usually aim to increase the personal components of opportunities by getting more skill and knowledge and also by accumulating more physical resources. For farmers, acting as business people, the choices might be more land, machinery and livestock as well as money. For farmers, acting as family members, the choices might be a fishing boat, more family time, horses and a better family vehicle. However, some personal components, such as personal energy and time, cannot be expanded so people have to become smarter in how they allocate these fixed resources. To some extent they can gear these resources by employing staff and by delegating responsibilities so they can concentrate on the family-decision system (where delegation is not possible).

The fourth lens is 'social considerations'. At one level, social considerations narrow the options available to those that are socially and culturally acceptable. But at another level, they encourage the decision-makers to add actions that are supportive of the community or society rather than just focus on the aspirations of their own family. Overall, social considerations widen the scope of issues the decision-makers considers. Families tend to be associated with a number of different communities. Each community tends to stress different obligations and actions.

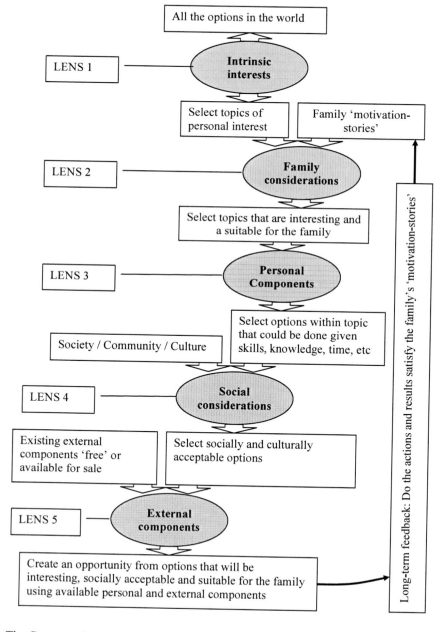

Figure 4. The Concept of Lenses.

The fifth lens is 'external components'. It focuses on the decision-makers access to external components of opportunities. Virtually all opportunities require some external input. This is the final lens and it is about the ability of the decision-makers to get the external resources and arrangements necessary to make a practical opportunity for their family. To a great degree, the ability to get external components depends on their existence within society. Clearly if freight services and markets do not exist then decision-makers cannot access them. However, where they exist, the decision-makers still needs to know about them, how to use them in creating opportunities and must have the ability to pay for them. Accessing external components is often a two-way interaction between the decision-makers and the people who own or manage the external components.

At the end of this process, when all five lenses have been considered successfully, the decision-makers implement the opportunity they have created. Managing the opportunity involves using feedback to improve it. Feedback on the opportunity guides the decision-maker in adjusting what they do next. Improving the technical and economic decisions (in the lower tiers of decision-systems) is relatively easy because performance standards are often available as guides. There are usually many organisations and firms able to help family businesses improve efficiency and profitability. Assessing the feedback on actions in the upper tier of decisions systems (what I am calling the family decision-system), where the decisions are justified in terms of caring for the family, is harder. The important feedback is in terms of the family's 'motivation-stories' and the relevant question is 'do the outcomes of the opportunities help the family satisfy its 'motivation-stories' – fulfil the family's aspirations?' Two things are important in this, the activity involved in the opportunity and the results. For some opportunities, such as recreation, the activity might be the most important consideration while in other opportunities, such as business, the results might be more important.

The concept of lenses is illustrated in Figure 4. Only one feedback loop is illustrated but there are many throughout the concept (Farmar-Bowers and Lane 2009).

"UNDERSTANDING 2" WHAT 'IMPROVE SOCIETY' MEANS TO POLICY RECIPIENTS

It is difficult to get a clear picture of what people want for the very long-term by asking them directly or by considering submissions from groups in society. However, in-depth interviews provide an indirect way of discovering what interviewees think they need for themselves and their immediate family over a life-time (discovering their 'motivation-stories' in terms of SDT). 'Motivation-stories' contain a future orientation and so indicate what families think their grand-children and great-grand-children will need during their lives. 'Motivation-stories' are not a frivolous 'wish list' of items. They have to be taken seriously because the stories are what families are actually working towards fulfilling during their lives.

The principal story for family members is 'Becoming a responsible adult and bringing up children so they too will become responsible adults'. The story sets out two purposes; learning and caring. These two purposes apply to the decision-makers, their immediate family and their descendants. Although individual families differ on the practical meaning of being a 'responsible adult', they know what being 'irresponsible' means, as there are social, cultural as well as legal criteria about irresponsibility, (even though these criteria tend to change over time).

'Learning', in the 'motivation-stories', is a complex matter that contains two supporting stories ('learning to become competent' and 'educating children': see Figure 2). The 'competency' people want is not just technical competency of knowing more things. It is about being more able to understand events and people and being more able to deal with life in ways that support self and others in becoming caring human-beings. Much of this non-technical learning is about moral behaviour especially about being caring and compassionate or empathetic to others. The practical understanding of caring seems to come through the trials and tribulations of life, especially the experience of raising a family and 'doing it tough'. From the 'learning stories' we can surmise that an 'improved society' is a society in which learning is facilitated. Families want to learn not only technical matters but also about the practicalities of caring for themselves and others. Religions, for many people, provide a source of guidance on moral behaviours.

The desire for learning is a powerful sentiment in the general population and is quite influential politically within society. Formal school education is being promoted and people are tending to get more schooling then earlier generations. Very often primary and secondary schooling is provided at public expense to ensure all children can get a basic education.

'Caring', in the 'motivation-stories', is also a complex matter and contains two supporting stories ('enjoying what you do in life' and 'overcoming isolation': see Figure 2). Relatedness and relationships are really important for people's psychological health (Deci and Ryan 2000a). Enjoyment and happiness are important but are difficult to define (Argyle 1987). Societies are tending to become more caring even though the motivations for a particular change may not be entirely altruistic or empathetic. For example, change may be driven by a 'duty of care' notion that suggests legal liability in tort law. Changes towards caring more about people occur politically at all levels from international conventions to local regulations. Changes have been occurring over centuries but particularly since the Second World War and include relationships between states, between the state and the individual and among individuals (Donnelly 1998). These long-term changes include such things as the abolition of slavery, habeas corpus, independent justice systems, anti-discrimination based on gender and race, and improved occupational health and safety regulations.

So in terms of people, the 'motivation-stories' suggest that families want improvements that facilitate their own education and the education of their children. They also want improvements that facilitate the care of people. How exactly these improvements are to be achieved is open to debate. The rate and nature of the improvements are also dependent on resources and on issues of justice. Some groups might want to expedite these improvements for themselves and might restrict progress in other areas. Consequently, overall progress tends to be slow and not always even. Nevertheless, the 'motivation-stories' indicate that the social aspects of an 'improved society' means two things to family decision-makers. One is making changes that facilitate an increase in their own and their families' competency through a range of educational and learning options. The other is making changes that facilitate increasing care for people through relationships and enjoyment. Because each generation of decision-makers in a family want these changes for themselves, for their children and grandchildren, we can take it that 'improving society' is an ongoing desire.

The environmental aspect of 'improving society' is what family decisions-makers want in terms of the environment in their region and collectively the environment of the Planet. While they want to use the resources of the Planet, they also want an environment that supports them physically and psychologically. And they want this for at least the next two generations or more. Since the next generation will also want the same we can assume that people want an environment that supports their descendants in perpetuity. Therefore, the environmental aspect of 'improving society' to the family means developing environmental knowledge and applying this to maintain the full functioning of all ecosystems and the physical maintenance of the planet's environments.

'Improving society', from the point of view of family decision-makers, is a straight forward concept about improving services that facilitates families being able to satisfy their 'motivation-stories'. However, it contains a black box of 'how to achieve these goals'. A large part of the necessary action within this black box is beyond the capacity of individual families to achieve. It requires collective research and collective action. Most of this large scale action is in the realm of governments, organisations and corporations.

The important points for environmental policy developers to understand are (1) the idea that families want an environment that supports them and their descendents in perpetuity, and (2) the idea that linking environmental policy with caring for the future family would increase a policy's legitimacy. Doing this is not easy and requires considerable research. Therefore, long-term environmental research ought to include how to link environmental policy to 'caring for the family'. This includes linking environmental policy to learning and education programs. 'Caring for the family' is not just an add-on to environmental policy based on hard scientific research but is something that has to be undertaken at the same time. This is because the 'caring for the family' aspect is likely to alter what scientific research is relevant for particular issues. Spash (2002) suggested that a wider based research agenda, including science and social issues would help with complex issues like climate change. DST supports this conclusion as opportunities that satisfy 'motivation-stories' require both welfare and technical outputs.

UNDERSTANDING 3: THE GLOBAL IDEALS THAT ARE RELEVANT IN IMPROVING SOCIETY IN THE LONG-TERM

The inclination to improve care for people and the environment contained in 'motivation-stories' is probably a reflection of, but also at the root of, public movements (civil society movements) that change social mores. These movements have often led to national and sometimes global changes, such as the emancipation of women, slavery, the peace movements and environmental or green movements. These movements have required activism from individual members of the public and the creation of organisations to overcome resistance from the powers that be, including government and business interests (McDaniel 2008). These movements have often taken decades to achieve change and most of these movements are ongoing; for example, combating the ongoing slave trade (now called human trafficking, Miller 2008). Their work is not done yet, but they have created a body of information that contributes to global ideals.

The tradition of change through protest will obviously continue to be very important in improving society in terms of caring for people and caring for the Planet. Individual activism provides leadership in specific areas (Couldry 2001). However, the problems of environmental degradation and poverty are now so enormous that speedier approaches to create effective policy are now vital. Protest is essential for problem identification but solving problems requires the support of governments to create policy and fund its implementation. Government support is not always forthcoming even after near universal consensus on issues such as the global environmental issues raised at United Nations' conferences (Barber 2003). Perhaps the 'motivation-stories' of families (who have the power of voting) provide additional evidence to encourage governments to apply the global ideals to improve care of both people and the environment. Family 'motivation-stories' contain good intentions in regard to caring for people and the environment that can lead to change, but I suggest that an outside catalyst is needed to facilitate change. I think this catalyst comes from policy that implements 'global ideals'. If society were improved to enable families to more easily satisfy the good intention in their 'motivation-stories' we would expect to see improvements, over a number of generations, in the care given to people (including the maintenance of cultures) and to the planet's natural environment.

There are two main reasons why the collective actions of families pursuing their 'motivation-stories' may not lead to substantial and universal improvements in the care given to other people and the care given to the Planet's natural environment. The first reason is that families have to meet their needs on a daily basis using the systems that exist on the day. Johns (2003) noted that people base decisions on local, immediate and personal concerns rather than on large-scale social consequences over extended periods of time. This short-term pragmatism relies on information and options immediately available to them. They therefore have to operate within existing systems and generally do not have the time, energy, ability and resources to change an entire system when it fails them. Although people are very adaptive, they can only get what they need within the constraints of the existing systems. Families have to address long-term needs by selecting from what is currently available. They cobble together a series of opportunities from what is available to them and hopefully make these into a satisfying life. These opportunities are packaged so that families have to accept the 'whole deal' to get the principle benefits for their family. Thus a large proportion of their collective impacts to the Planets' systems are a consequence of the external components of opportunities families have to use in creating opportunities. They do not have a lot of discretion; generally they have to use the technologies and services immediately available to them. Roberts (1996) for example, noted that environmental or social concerns expressed by the public do not translate directly into consumer behaviour. Perhaps this is because public concern does not instantly create the appropriate external components people need to create opportunities. Businesses and government create these external components and not the general public.

Substantial improvements in care do not require a change in the 'motivation-stories' of families. However, they do require a different set of external components to help families create opportunities to satisfy their 'motivation-stories' in more sustainable ways. System-wide change would be needed to encourage the development of this set of external components. Changing entire systems is impossible for families to achieve on their own even though they are the ultimate customers.

The second major reason why the collective actions of families pursuing their 'motivation-stories' may not lead to sustainable development outcomes is relevant to businesses as well as families. This is the lack of reliable information, especially cause-effect, systems, and collective-impact information.

Cause-Effect

It is often very difficult to determine the unintended outcomes of actions especially when the outcomes occur in different time periods, in different localities and in different spheres and ecosystems. Unintended outcomes include 'embedded resources and embedded social issues'. For example, rain-forest timber sold in Europe can contain 'loss of rainforest biodiversity', a soft toy made by child labour can contain 'loss of childhood and youth education', and hydro-electricity contains loss of biodiversity in the reservoir area and in the drowned river environment and may also embody the loss of valley towns and communities. Unintended outcomes can also include high running cost from inefficient design leading to high resource consumption. High disposal costs occur if the product contains toxic ingredients such as heavy metals, long lived toxic chemicals and radioactive material. The high disposal costs may not become evident for years and may be expressed in ecosystem damage and in human illness rather than in direct dollar terms. The individual family has to rely on larger organisations (corporations or governments) to give them this information. It is especially difficult when the information is not included in the cost, shown on the products label, taught in schools or is not referred to in the media or in advertising. Globalisation makes it much harder for families to appreciate the cause-effects of the many activities used in the production of goods for sale. This is because the locations, environments and societies that might be damaged are distant, out of sight, foreign or unknown and therefore not so easily appreciated or understood. Consequently the knowledge of what a product or service might contain or embody is not readily available to the family decision-makers.

Systems

It is also difficult to choose the appropriate boundaries to use in decision-making. People make decisions within decision-systems; that is they make decisions topic by topic. A wrong decision can result if the decision is based on information within boundaries that create too small a system (Bates 1997, Midgley 2000). For example, if the decision is to buy an efficient heater, considering the efficiency of the heater itself would lead to one conclusion but considering the efficiency of the heater and the efficiency of the energy source it uses would give a different conclusion. Sometimes it is impossible to select the correct boundaries for a decision because the information is missing. Generally decisions based on larger systems are likely to be more reliable in the long-term than decisions based on smaller systems. Larger systems include the complete set of product cycles, including raw material production, manufacture, running and maintenance, recycling, and degradation. They also contain the social as well as the technical aspect of the product or service.

Collective Impacts

A further problem is the accumulative impact of actions. Often, a single action is acceptable because it only has a few long-term consequences, but when many people carry out the same action, the collective impact on the environment can be substantial.

The problem of 'insufficient information' slows the rate of change in both how families decide to satisfy their 'motivation-stories' and also what external components businesses and governments provide for creating opportunities. It is important to get the right information to families and businesses and avoid relying on the notion that 'might is right'.

It seems important for success that government and civil society organisations use global ideals to help them formulate policy but use the 'motivation-stories' approach to obtain the concurrence of policy recipients. This would ensure that policies care for both the environment and people concomitantly.

THE 4-GROUP-STAKEHOLDER MODEL

The '4-Group-Stakeholder model' provides an overall picture of the responsibility for providing external components including obtaining guidance from global ideals (Farmar-Bowers 2008). The model uses the notion that people usually have several roles in life. For example, a woman can fulfil the role of 'mother' and also be a 'bank manager'. Some of these roles are relatively short term while others, principally those related to personal life, are enduring. The model also uses the notion that we all have a stake in all the roles that other people have. This is because these roles create the society in which we live.

The four groups in the '4-Group-Stakeholder model' are as follows:

- 'Group 1 Stakeholders' are people when they have the role of private individual, a citizen; the personal role of a family members.
- 'Groups 2 and 3 Stakeholders' are people when they have the role of a working person on their own or in organisations such as government or business.
  ○ Group 2 Stakeholders focus mainly on people (e.g. social work, education and health).
  ○ Group 3 Stakeholders focus mainly on the environment (e.g. meteorology, conservation, and environmental policy).
- 'Group 4 Stakeholders' are people in future generations.

Everybody has a Group 1 role. Employed people play Groups 2 and 3 roles. Future generations have Group 4 roles.

The role of Group 1 Stakeholders is to create and implement opportunities that satisfy their own and their family's 'motivation-stories'. They look after themselves and their family. This includes considering their own future and also helping their children, grand-children and perhaps great-grand-children.

The role of Group 2 Stakeholders and also Group 3 Stakeholders is to help Group 1 Stakeholder satisfy their 'motivation-stories'. This includes themselves of course, but as private citizens, as a vital reason for being employed is to get income for yourself and your family to spend. People are employed because they help organisations provide services that customers want and are prepared to pay for. The services they provide are the externals components of opportunities such as education, infrastructure, food, biodiversity, consumer goods, travel, finance, laws, social security, international trade arrangements etc. The (ultimate) customers are Group 1 Stakeholders[*] and they pay for services using the money from working as Group 2 or 3 Stakeholders. The relationships between Group 1, 2 and 3 Stakeholders is shown diagrammatically in Figure 5. The two shaded boxes show the drivers of the systems. We can refer to this as the human development system because it is about how whole societies work together. Group 1 Stakeholders are driven to create opportunities to satisfy their 'motivation-stories'. Group 2 and 3 Stakeholders are driven to create external components that are sold, or paid for in some way, to satisfy their organisational goals which are profit for business organisations, and political power for governments.

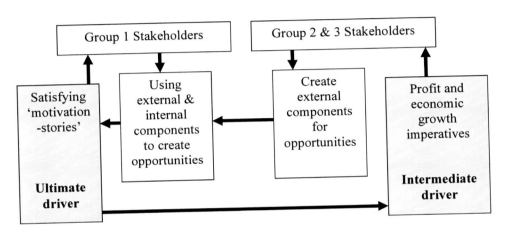

Figure 5. The relationships between the roles of Stakeholders Groups 1, 2 & 3.

Group 2 and 3 Stakeholders have the most political, economic, resource management and ownership power as their roles involve managing organisations, including governments and businesses, and maintaining the institutions that govern society. Indeed many corporations have larger economies than whole countries (UN Human Development Report 1999). They can behave, and usually do, as if they are in a closed loop aiming to make profits, grow in financial terms and retain power in society. The information flow in society among Group 2 and 3 Stakeholders, from élites to élites as it were, is significant in retaining political and economic power (Davis 2003). Mercer and Jotkowitz (2000) noted that Australians do not have access to high quality information about local environments, nor do they have 'right-to-know' legislation. They tend to be treated as 'customers' by local governments rather than as citizens. Ross and Dovers's (2008) review of the government strategies, structure and process of environmental policy integration in Australia tends to confirm that Group 2 and 3 Stakeholders talk among themselves. The talk seems to be mainly about Group 3 Stakeholders trying to have Group 2 Stakeholder take account of environmental issues in economic policy development. Mestrum (2003) noted that this talk has taken us full circle in 30 years, from calling for 'limits to growth'[*] in order to protect the environment, to environmental protection in order to preserve the growth process. So although Figure 5 shows the goals of Groups 2 and 3 as the 'intermediate driver' in the whole system, it is a very large

---

[*] The organisations in Groups 2 and 3 trade amongst themselves of course (when one business sells to the next in a supply chain) but the ultimate customers are Group 1 Stakeholders.

driver and the amount of information, time and effort given to it tends to swamp the notion that there is an 'ultimate driver'.

There is an implicit understanding in society that the longer-lived services, such as laws, education, infrastructure and biodiversity conservation, should also be of value to Group 4 Stakeholders (future generations). This was put more formally, but negatively, in the Brundtland report on sustainable development as, "development that meets the needs of the present without compromising the ability of future generations to meet their own needs" (WCED, 1987, p 87). The implication is that if we want Group 4 Stakeholders to have a bright future something has to change.

The two drivers in the system (shown in Figure 5) are unlikely to change. Group 1 Stakeholders will always want to satisfy their 'motivation-stories'. Business will continue to require profits. What can change is how they satisfy their 'motivation-stories' and how they make profits. To enable these changes to occur there has to be a change in the available 'external components of opportunities'. External components are the items and arrangements that decision-makers have to access from other people, organisations and from the environment. They require external components to help them create opportunities (outlined in the DST concept of suitability and availability of opportunities).

An important aspect of both DST and the 4-Group-Stakeholder model is that they are both dynamic. External components of opportunities can lead to the creation of personal components when the external components improve personal capacity. For example, an external component might be the establishment of a training program. Participation in the training program might improve the students' knowledge and allow them to bring more personal knowledge into the creation of opportunities in future.

To get a brighter (sustainable) future for future generations, Stakeholder Groups 2 and 3 would have to produce external components that are more supportive of *critical arrangements* and more able to maintain *critical resources* than the external components they currently produce (Farmar-Bowers and Ward 1995).

Critical arrangements are those arrangements in society that are essential for fairness and justice in the delivery of people's physical and psychological needs. Perhaps distributive justice principles, such as those suggested by Rawls (1971, 1993), provide a guide for arrangements to reduce unfair inequalities. An understanding of the Universal Declaration on Human Rights would also be important tool for policy developers (Wronka, 1998). Ruttan (1999) suggested that failure to become 'sustainable' will be more to do with the failure to develop innovative institutions (*critical arrangements*) than because of resource and environmental constraints (*critical resources*).

Critical resources are the resources needed to maintain the life forms and life systems on the Planet (all biodiversity; ecosystems, species and genes, see: Co. A. 1996). The benefits people derive from *critical resources* are sometimes referred to as ecosystem services (Millennium Ecosystem Assessment, 2005). It is important to note that resources are needed not just for the physical needs of people but also for their psychological and spiritual needs. Even some economists agree ".... nature is not something to be subjugated, but instead something we depend upon absolutely to meet both physical and spiritual needs" (Farley and Costanza 2002 p.247). One of the most important ethical questions of our times is how much

---

* Meadows et al., 1972.

of the Planet's resources should be for humans and how much for other forms of life (Meyer 1998).

It is the role of Stakeholder Groups 2 and 4 to identify what these *critical resources* and *critical arrangements* are. They can obtain guidance from global ideals on how to treat people and deal with the environment. Guidance on people matters, such as justice and rights, and on planet matters, such as biodiversity conservation and pollution abatement, is substantial. Information on people and planet matters is constantly being developed, expanded and improved.

Critical resources, the materials and systems that human beings need, should not be confused with resources that are 'critical' for a technology. For example, plutonium is critical for atomic bomb technology but is not a critical resource for human needs. Petroleum and its derivatives are critical for a range of technologies, notably the internal combustion engine, but petroleum is not critical for human needs. In fact, burning petroleum damages the critical resources of air quality and climate stability.

An important issue, that has the greatest impact on Group 4 Stakeholders, is how to bring people and planet matters together in each decision. Developments are required that improve *critical arrangement* and also maintain *critical resources*. These kinds of developments are 'real' sustainable developments because they improve the future for people and maintain the Planet's environment and functioning of its systems. There is not much value for future generations (Group 4 Stakeholders) in developments that improve one at the expense of the other.

An important job for policy officers (including politicians, legislators and CEO's of corporations) in Stakeholder 2 and 3 roles is to assist the ongoing development of global ideals. This can be done in international forums such as the United Nations, in regional and also in bilateral agreements. But perhaps a more urgent task for policy officers is translating the existing global ideals on people and planet matters into usable policies at a national, regional, and local level for governments and for businesses. The aim is to help future generations satisfy their 'motivation-stories'. To do this, the *critical arrangements* have to support the rights of people. They also have to ensure the maintenance of *critical resources*. The way policy can do this is by developing 'external components of opportunities' that will help people create suitable opportunities. These opportunities need to be in all the areas of influence. We will consider 'areas of influence' later on in this chapter but essentially external components need to be available to help families create opportunities that are suitable to satisfy their 'motivation-stories'.

Unfortunately the global ideals in sustainable development are not the only global influences. Other sets of ideas such as commercialism and militarism are influential guides for policy development. While these policies may be contributing to economic growth and its distribution now, they may not benefit Group 4 Stakeholders as much as those policies that maintain *critical arrangement* and *critical resources*.

In summary, the '4-Group-Stakeholder model' suggest that the role of Stakeholder Groups 2 and 3 is to help Stakeholder Group 1 satisfy their 'motivation-stories'. They do this by providing the external components needed to create and implement opportunities. Government policies play an important part of the process in providing external components. Group 2 and 3 Stakeholders provide all kinds external components. However, only those that can be used to create opportunities that protect *critical resources* and maintain *critical arrangements* will be useful in the future, because these are the opportunities that will help

Group 4 Stakeholders satisfy their 'motivation-stories' when their time comes. We can refer to these components as 'new external components' (see Figure 6).

It is the job of employed people (Stakeholders in Groups 2 and 3) to ensure that the rate at which 'new external components' are created matches the needs of the world's population. People differ greatly on what level of protection they think is acceptable or desirable for people and the environment. However, it seems that the current level of protection for *critical resources* and effectiveness of *critical arrangements* is not meeting the needs of a substantial proportion of the human population. The current world population growth together with the growth of resource use indicates that human-kind is likely to degrade much of the environment of the planet before environmental stability is reached. Schumacher (1977 p. 160) asked, "Can we rely on it that a 'turning around' will be accomplished by enough people quickly enough to save the modern world?" His answer was to get down to work. It seems quite possible that without considerable effort from Group 2 and 3 Stakeholders the projections from the 'Limits to Growth' (Meadows et al., 1972) and 'Beyond the Limits of Growth' (Meadows et al., 1992) will prove substantially correct. It is difficult not to be sceptical, as 30 years on from Schumacher we still need to get down to work.

Since the 1970's, when the Club of Rome's projections were formulated, global warming has been added as an environmental issue. Although there is considerable effort from Group 2 and 3 Stakeholders on a global scale to cap greenhouse gas emissions it seems likely that global warming climate will have a considerable impact on all living creatures including humans. Moran et al. (2008) suggest that the Human Development Index (HDI) and ecological footprint paint a clear picture of near universal trends in development away from sustainable development. This is despite the growing global adoption of sustainable development as a policy goal.

The global ideals of how to deal with the planet's environment and how people ought to behave to one another are available as guidelines for policy people. Global ideals can assist policy developers' moral imagination (Werhane 2002). This will help them develop the policies and programs needed for families to satisfy their 'motivation-stories' in ways that protect the planet's environment and ensure fairness, justice among people and nations.

Environmental policy developers are Group 3 Stakeholders. They might be able to make more effective environmental policy if they understand the global ideals about the treatment of people and know how to integrate these ideals in the environmental policies they develop. This would move society towards creating sustainable developments. In terms of delivering sustainable development ideas, the notion of specialising in environmental policy is an old fashioned idea that was obsolete by the late 1980's. Robinson (2004, p. 378) noted "[it is]…increasingly obvious that that solutions that address only environmental, only social or only economic concerns are radically insufficient". Environmental policies can only really deliver sustainable development ideas if they maintain the viability of *critical resources* and are teamed with social policies that maintain *critical arrangements*. Ross and Dovers (2008) noted that what they refer to as 'policy integration' (elsewhere called whole-of-government or joined-up government) is a challenge for specialised, hierarchical public administrations. Perhaps this implies that environmental policy is unlikely to be 'sustainable development policy' without organisational change in government even if individual policy development officers become familiar with *critical arrangements*. At the moment, it seems that most policies revolve around Group 2 and 3 Stakeholders as if their objectives are the 'ends' rather than the 'means' to sustainable development for people. Perhaps this is why administrations have been confused over the interpretation of sustainable development ideas (Orr 2002, Daly 1996). The role of global ideals in human development is shown diagrammatically in Figure 6.

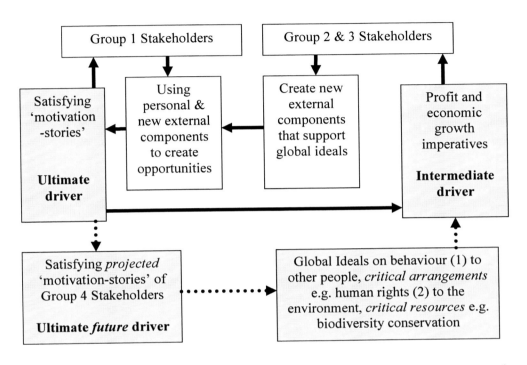

Figure 6. The Stakeholder Groups and the relevance of global ideals for the creation of 'new external components' of opportunities.

Figure 6 suggests that at any one moment global ideals are only relevant for the future (so are irrelevant for the present). This situation could continue indefinitely were it not for the fact that we are moving into the future. Current consumption of *critical resources* will have a negative impact on people's livelihoods in the future. Also, without improvements in *critical arrangements* conflict between people and exploitation will continue. Conflicts lead to critical resource loss quite apart from the direct human suffering involved. This suggests that it will be harder to satisfy the 'motivation-stories' of Group 4 Stakeholders (future generations) than it is to satisfy Group 1 Stakeholders (existing people). In addition, population is growing so there will be more people in Group 4 wanting satisfaction. Because of these changes, it would seem that satisfying future generations (Group 4 Stakeholders) is a more stringent requirement than satisfying the Group 1 Stakeholders.

If we are interested in trying to give future generations the same ability for them to satisfy their 'motivation-stories' as we have now, then we should be using "satisfying the *projected* 'motivation-stories' of Group 4 Stakeholders" as the ultimate *future* driver of human development. To do this, Group 2 and 3 Stakeholders should be trying to develop 'new external components' that facilitate the application of global ideals to every development decision.

We can view sustainable development ideas as a summary of the global ideals we have been referring to. Quite often sustainable development is interpreted as the 'triple bottom line'

(Elkington 1999: CoA 1992). That is the economic, social and environmental outcomes of developments. The triple bottom line concept is at odds with the 4-Group-Stakeholder model and we can see this from Figure 6. Figure 6 shows that one of the drivers is profit. This fits with the economic part of the triple bottom line, but in Figure 6 it is an 'intermediate' rather than an 'ultimate driver'. Nevertheless, profit is a result (albeit an intermediate result) that companies require in order to remain in business. Profit is a goal for business but it is only an intermediate driver in the whole system. And similarly, earning money in the 'motivation-stories' is a 'means' to an 'end' and not the ultimate purpose in life.

The 'social' part of the triple bottom line could be viewed as *critical arrangements* and the 'environmental' part as *critical resources*. *Critical resources* and *critical arrangements* are requirements not drivers in the system. 'True profits' can only be made after these two requirements are met. Counting something as profit before these requirements are met is tantamount to including a loss as a contribution to profit. This is because a dollar profit is being made at the expense of the Planet's environment or at the expense of people's rights or freedoms etc. Neither *critical resources* nor *critical arrangements* are bottom lines in Figure 6. The 'bottom line' in the 4-Group-Stakeholder model is the desire of Group 1 Stakeholders and eventually Group 4 Stakeholders to satisfy their 'motivation-stories'. These are the 'ultimate driver' and 'ultimate *future* driver' of human development.

Success in satisfying the 'motivation-stories' for all families (for each generation) can only be claimed if there *critical arrangements* and *critical resources* are not diminished. This is because part of the principal motivation story for families is about becoming a responsible person and helping children become the same. Failing to make improvements in *critical arrangements* and failing to make progress in maintaining *critical resources* could be viewed as failing to match (live up to) the 'responsible person' criterion. If this criterion is accepted, then it is the responsibility of Group 2 and 3 Stakeholders to make suitable external components available to Group 1 Stakeholders and it is the responsibility of Group 1 Stakeholders to use these external components to create suitable opportunities.

USING DST FOR POLICY DEVELOPMENT

DST provides an interpretation of the processes farming families use to create opportunities across business and non-business areas in order to satisfy family aspirations. We can infer that families who run other kinds of family businesses use similar processes to make decisions. DST is useful for policy developers because it highlights areas that influence the decisions that families make. Changing things in these 'areas of influence' can result in different decisions being made. Some of these new decisions can result in life-long changes.

Policy developers have always aimed to concentrated their policies and programs in these 'influential' areas. In regard to agriculture and rural issues, most policy concerns have focused on technical and economic matters such as agronomy, land degradation, water quality, markets and efficient resource use. These issues are very important for farming practice and are dealt with by farming family decision-makers in the lower tier of hierarchy in decision-systems where decisions are justified on technical and business terms. This gives the policy developers and policy recipients a common language and easily understood objectives. Long-term sustainable development issues, including conservation of native biodiversity and climate change mitigation, are growing policy areas that are more likely to be dealt with by farming families in the top tier of the decision-systems hierarchy where decisions are justified in terms of caring rather than in terms of business and technology. Differing language and objectives make communication between family decision-makers and policy developers problematical for these issues. And of course, the policy developers are likely to be different from those concentrating on the technical and business aspects of farming. DST provides a logical explanation of the policy recipients' decision processes which gives the policy developer the insight to develop new kinds of policies to address these issues and some confidence that they will work. However, policy developers still need to understand the details of the people they want to influence and how they are currently being influenced by existing programs and policies.

A useful step in understanding the current influence of existing policies and programs is to classify them in terms of DST. This can be done using the 'boxes of influence concept' which is the sixth concept in DST. The 'boxes of influence concept' was developed by amalgamating the 'personal career path concept' and the 'concept of lenses'. The boxes of influence concept identifies seven areas of influence during the life of a decision-maker (the seven boxes in this concept). Policies and programs can be classified according to which box they fit into. This classification provides an overall picture of what family decision-makers face in terms of influence and how the policy or programs work to create change. Policy developers can use the completed classification in two ways. They can focus on improving policy in particular boxes of influence, or they can review all boxes of influence and improve policy in response to the current pattern of existing policies. What they choose will depend on what changes they want policy recipients to make, especially between short-term specific actions and longer-term general system change.

Policy developers can also use this classification system to assess the relevance and effectiveness of policies and programs in each of the seven boxes of influence for advancing sustainable development ideas. Policies and programs create change by altering the available external components of opportunities (from the concept of suitability and availability of opportunities). Those policy changes that create external components that encourage decision-makers to develop opportunities in line with the 'ultimate *future* driver' (see figure 6) meet the sustainable development criteria. That is, the opportunities created as a consequence of the policy support *critical arrangement* and also maintain *critical resources*. Policy legitimacy can also be assessed as this is an important aspect of policy effectiveness. Legitimate policies are those that policy recipients (family decision-makers) think will help them create an opportunity that will satisfy some aspect of their family's 'motivation-stories'. If a policy is not accepted by policy recipients as legitimate, than it will not be influential even if it is relevant in terms of potentially creating external components of opportunities that could help families create sustainable development opportunities.

The seven boxes of influence are as follows:
1. Start of career
2. Knowledge – secondary and tertiary education
3. Obligations to parents' family
4. Obligations to decision-makers own family
5. Social obligations
6. Business and commerce

7.   End of career.

BOX 1. Start of Career

People's intrinsic interests start early in life. Very frequently interviewees could remember incidents when they were only 4 or 5 years old that sparked an interest in some particular topic. Early education (formal and informal) can nurture or alternatively bury this interest. This means that preschool and primary school policies and programs can be very influential in establishing and developing life-long interests. This is in addition to what happens at home with parents and close relatives.

Early education cannot be used to encourage specific activities in later life. However, it is possible to develop a future population in which more people have an interest and are knowledgeable in specific topics relevant to sustainable development. These topics could be conservation, ecology, wildlife, environmental justice and so on.

To undertake a 'boxes of influence' classification, the policy analyst would investigate all current early education programs and policies (public and private) that relate to the topics they are interested in promoting. This would be a two stage process, first listing all relevant early education policies then enquire from education experts about what the policies actually do and how effective they are at doing it. This analysis might lead quite logically to investigating what other countries are doing and eventually developing new programs and policies to nurture young people's interests in the range of topics that are important for them, their families and society in the future.

Early education policies would have a long lead-time before pay-off. Perhaps this might be two or more decades. This long lead-time suggests the need for careful planning. For example, information and opinion on what kind of expertise might be in high demand in future could be collected on a systematic basis from a wide range of people.

Environmental policy developers could use early education policies to influence future populations' attitudes to major topics such as the joint importance of *critical resources* and *critical arrangements*.

BOX 2. Knowledge – Secondary and Tertiary Education

Moving from a general childhood interest in a topic to a professional approach requires further education and again this can only happen if appropriate policies and programs are in place. While expanding the number of professionals in a range of topics is important, the skill to imagine and then develop new and perhaps revolutionary ideas and put them into practice is critical. This is especially important as the world and world events are changing so rapidly. While education cannot create these people, it can at least facilitate their development and encourage the acceptance of change in society.

The analytical approach of investigating programs and policies in this area is the same as in BOX 1. First putting all the relevant programs and policies into the box, then working out their relative effectiveness, then looking at what other people (including in other countries) might be doing and finally creating new programs and policies as necessary.

A major element in policies in Box 2 would be research. The creation of new knowledge is a fundamental requirement for progress. Knowledge in itself is a commodity but its application is critical. Policies in this box need to facilitate the creation of external components to allow (1) the development of educational opportunities (2) the creation of knowledge and (3) application and utilization of this knowledge.

Environmental policy developers could use policies in secondary and tertiary education to follow through on the changes they wanted to occur in BOX 1. For example if they wanted more professional 'green architects' they could establish curricula that contained the appropriate subjects and increase funding for more university places.

BOX 3. Obligations to Parents' Family

The decision-maker's parents' family includes siblings as well as their parents and perhaps other relations who have an influence within the family sphere. What happens within families, although seemingly private, has a huge influence on what people do and become in later life. The obligations that people have to siblings and their parents often influence behaviour for a very long time, probably for their entire lives. The caring provided by the parent's family is usually especially important for older people (retirees) and younger people (children) and for people with special needs.

The influence of the private sector on families is huge as families represent a major commercial market. The public sector has direct influence on families and also indirect influence via the private sector through a range of policies (expressed as legislation and regulations). For example, governments usually regulate contracts and private providers through fair trade and consumer protection policies. Government policies also influence where people live and how they live, through planning, land zoning, infrastructure development, energy standards and various support programs.

The analytical approach is the same as for other boxes but the influence paths may be very complex as both public and private sectors programs and policies are involved. The solution might be to use in-depth interviews with families in order to identify areas of special significance in terms of influence and analyse the programs and policy relevant to these areas. This might keep the analysis workable.

Environmental policy developers could use policies and programs that influence people's obligations to their parent's family to encourage environmental awareness and its application to resource use and consumption within the home environment. Environmental policy could include how family members, especially older people and children, help each other understand and deal with environmental issues by facilitating good care through training and funding arrangements. Changes in the family's home environment could have very substantial impacts of many important environmental issues because so much of the family income is spent on consumption. The main policy goal might be the combination of using fewer *critical resources* yet increase the care given to family members.

BOX 4. Obligations to Decision-Makers Own Family

Obligations to one's own family – spouse and children, greatly influence decisions. Like BOX 3, how these obligations are carried through into actions is influenced by policies and programs in both the private and public sectors. These policies and programs do not alter the basis for the obligation which is 'care', rather they can change what people do – what actions they take in the name of caring.

Caring for family concerns the consumption of material goods as well as maintaining the environmental and social conditions that family members need. 'Care' can include ensuring that these things exist in perpetuity or are developed if they are absent. The two big areas are fully functioning global ecosystems and a supportive society. On a more detailed level, many of the material things that people need such as food, transport, medical and energy can be delivered in ways that damage the environment and / or other people's rights. Better alternatives are sometimes available but the normal accounting system does not involve social and environmental accounting which makes it difficult for people to appreciate the damage. Functionally identical materials can embody very different amounts of global ecosystem destruction, social dysfunction and human rights abuse. Information is one of the keys for making choices to reduce the negative impacts of consumption on the environment and people. One can conclude that the ability of individuals to meet these obligations to their family depends greatly on their society providing the appropriate information. This is a function of policy on information programs. Point of sale information is often effective. Family obligations to the succeeding generations might include reducing consumption of products that pollute for decades. Quite often governments allow, or use themselves, pricing arrangement that stimulate consumption and reduce equity. These include arrangements that allow price splitting; the separation of infrastructure costs and consumption fees*. Policies that ensure all charges relate proportionately to consumption are needed. Progressive pricing policies might be better at limiting the consumption of products that cause pollution or health problems (Goodman and Anise 2006).

The analytical approach involves identifying the policies and programs that influence how the decision-makers interpret their obligation to their own families. Policy research can help indicate how governments can help decision-makers deliver their obligations effectively but in a more sustainable way. Again the mechanism is for policy to provide 'new external components' (which might be *critical arrangements*) that will allow families to create opportunities that deliver family obligations (their 'motivation-stories') using fewer *critical resources*.

There is no clear demarcation line around the notion of 'family'. It is possible the notion could be extended outwards into society and forward to future generations. This suggests there is a continuum running from family obligations to social obligations more generally. Policies and programs can alter how people define 'family' and what practical obligations people feel are due to more distant parts of the family, society and humanity. Generally, these wider obligations to society are administered through wealth redistribution and social policies of governments on behalf of the public.

---

* For example, utilities such as gas, electricity and water are sometimes billed in two parts; a connection fee and a consumption charge.

Environmental policies related to the decision-makers' family can be used by policy developers to implement a very large range of changes, from nutrition of families to housing, location, resource consumption, education levels and subjects studies, technologies used, conservation issues, and even family size. Some changes can be made quite specific while other can be more general and longer lasting. An important aspect of boxes 1 to 4 is the opportunity they present of joining care for people with care for the Planet's environment.

BOX 5. Social Obligations

Families do not exist in isolation but rather exist within communities and society as a whole. They do not just relate to one community, a geographical one, but rather are involved in a number of communities of practice. Some of these are social / recreational, others relate to various family or personal activities such as schools, religion, politics and volunteering. Policies and programs can greatly influence participation and community action.

Environmental policy developers can use this area for quite specific changes to improve community activities or cohesion and to meet regional objectives. The interviews with women farmers suggested that they had a wider involvement with local communities than male farmers. Environmental policy developers might use women's networks to help develop and also help implement changes that require community action and long-term changes in the opportunities that families create. Women's networks might be very effective in securing regional conservation programs.

BOX 6. Business and Commerce

Most of the influences in this box come from the private sector as they concern the operation of business. However, public sector policy and programs influence the operation of business via legislation, regulation, taxation and permits as well as through training and the provision of advice. In the longer term, policies on research and extension are very important as they can help businesses prosper.

The more knowledge that individual decision-makers have of the arrangements within this box the more likely they will be successful in business. Over time, business decision-makers usually become more familiar with the arrangements in all areas associate with their business interests such as markets and finance. They become more able to refine their business activities as a result. It would seem important for the public sector to understand the impact of policies and programs on the operation of family businesses within the context of all arrangement. Most of these arrangements are organised in the private sector. One of the difficulties may be the impact of non-business programs on business activities. These can range from occupational health to environmental policies and programs. An interesting approach for environmental programs has been to couch them in business terms as if they are business ventures (Stoneham et al. 2003).

Environmental policy developers can use business policies and programs to influence two areas. The first area to influence is treatment of people and the consumption of resources in the work environment. In this area policies can be put in place that influence how people deal with their colleagues in the work place and in commerce more generally. That is, they can influence how people are treated in the work environment[*]. This area deals with the activities among Stakeholders in Groups 2 and 3 and with the organisation in these two Groups. These policies can support and improve the *critical arrangements* that are relevant to the work environment. Policies can influence how business decision-makers go about organising the production, consumption and degradation systems within supply chains. These policies can target the protection of *critical resources*. The second area to influence is the production of good and services for the ultimate customers (families). It concerns what products and services are produced for the public (ultimate consumers) to buy. This is about how Stakeholders in Groups 2 and 3 deal with Stakeholders in Group 1 (the general public), and by inference, future generations.

For both areas, a systems-thinking approach would greatly help policy developers deal with the scope of issues involved, from resource procurement cycle, through production, to the degradation cycle.

BOX 7. End of Career

There are many policies and programs that influence the end of career such as pensions, health care programs and taxation as well as private sector arrangements such as retirement accommodation, insurance, pension plans and succession planning. Information is an important part of being able to use the range of programs and arrangements that relate to retirement. Quite often these programs have to be organised years before retirement occurs.

Environmental policy developers can use this area for influencing how people care for each other and also for influencing the consumption patterns of older people. Pension and inheritance policies can influence the financial planning in a large part of people's lives. Policies aimed at making changes towards the end of life are less likely to make large changes to the future than policies aimed at very young people. However, considerable resources can be involved as some retirees have accumulated significant wealth. On the other hand, supporting retirees can be expensive to the public purse. The significance of older people's actions is likely to increase as the percentage of older people in the population increases.

Overview of Boxes of Influence Concept

Figure 7 shows these seven boxes of influence diagrammatically in relation to a lifespan of a decision-maker and in terms of the kind of policies that are relevant in each box.

Classifying policies into the seven boxes allows the policy developer the excuse of looking across the entire range of government and private sector policies and programs. Very often government policy development occurs in a single department and it is very difficult for departmental policy officers to really see how their policy or program interacts with other programs from other government departments and from the private sector. The 'boxes of influence' concept provides a systematic way of reviewing all relevant policies.

---

[*] The treatment of people at work is dealt with in the ILO's Conventions and recommendations which are available on their web site, http://www.ilo.org/global/lang--en/index.htm and for a summary see 'Rules of the game' ILO 2009.

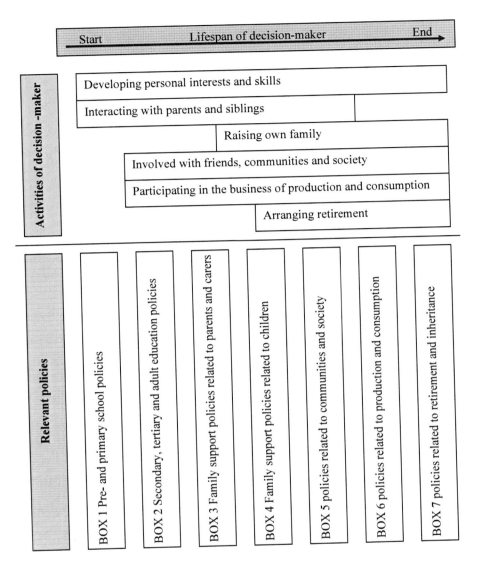

Figure 7. The seven boxes of influence and relevant policy areas.

The approach for gathering information in the 'boxes of influence' is to deal directly with the policy recipients. The policy recipients are usually a particular group or easily identified communities. This means talking with people and finding out what existing programs and policies they are concerned about and what things they are doing automatically because of some long-standing policies or programs. This will provide a list of the most important policies from the policy recipients' perspective which can be compared with the list of policies from agencies. This classification would show where policies and programs are thickest on the ground and where they are sparse. It would indicate which policies are accepted as legitimate by policy recipients and which are not.

Figure 8 gives a fictional example of what this listing might look like in terms of numbers of policies and programs that influence environmental issues in some way. It seems obvious, but policies and programs from the private sector would dominate in BOX 6 where money is to being made through production and consumption of goods and services.

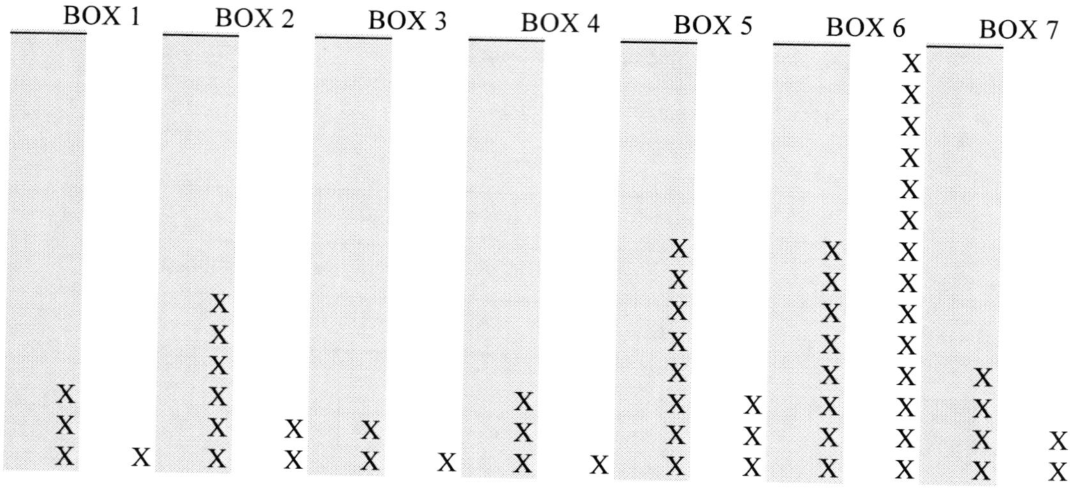

BOX 1	BOX 2	BOX 3	BOX 4	BOX 5	BOX 6	BOX 7
					X	
					X	
					X	
					X	
					X	
					X	
				X	X	
				X	X	
	X			X	X	
	X			X	X	
	X			X	X	X
X	X		X	X X	X X	X
X	X X X	X	X X	X X	X X	X X

Figure 8. The relative number of policies and programs in the seven 'boxes of influence' from the public (shaded column) and private sectors (clear column) that concern or influence environmental issues (positively or negatively). This is a fictional example.

This overall view of policy and programs may highlight areas where new policies and programs would be lost in a sea of existing policies and programs. There may be scope in these areas for policy reform and reduction. It may also indicate other areas where there are very few policies and programs.

The kind of influence obtained by policies and programs is not the same in each Box. In broad terms, policies that enhance activities in Boxes 1, 2, 3 and 4 would tend to have a long lasting but general impact on the behaviour of policy recipients. For example, improving agricultural education in universities and colleges and increasing the number of places, scholarships and prizes would stimulate people, who have an interest in food, agriculture and rural affairs, to make their careers in these areas. This would have a positive impact on numerous rural issues for decades to come. However, the change would be across the whole field of interest and not specific to any one issue. Policies and programs in Boxes 5, 6 and 7 could lead to more specific actions but actions that were dependent on the continuance of the policy or program. For example, if the government wanted specific work undertaken, such as trees planted in aquifer recharge areas to reduce dryland salinity, they would achieve this most effectively by funding projects in Box 5, 6 and 7. Most private-sector policies and programs occur in Box 6 specifically because the returns are specific and quick.

We could view policies and programs in terms of the kind of motivations various programs would engender in the policy recipients. Deci, & Ryan, (2000 a, b) suggested that there is a motivation continuum from intrinsic through extrinsic to amotivation. The intrinsic to extrinsic part of this continuum appears to match the lower to the higher numbered boxes. For instance, programs in the boxes 1, 2, 3 and 4 would tend to stimulate people's intrinsic interests in a topic and engender an intrinsic motivation to understand or work on the topic. People work on a topic because they find it stimulating and exciting. Programs in boxes 5, 6 and 7 would tend to create an extrinsic motivation such as seeking a monetary reward or incentive for working on a topic. They would work on a topic because they get paid. Incentive programs are essential when the government needs to buy very specific items or services such as the conservation of a particular species or the management of a specific area. The service will continue so long as the government continues to provide the incentive. Unfortunately providing a payment for an activity that previously has been undertaken voluntarily for free, because of an intrinsic interest, creates an extrinsic motivation that can 'crowd out' the intrinsic motivation (Reeson and Tisdell 2006).

CONCLUSION

Decision-systems theory's (DST) first five concepts provide an interpretation of the decision making processes farming families use in making personal and business decisions. DST is a substantive theory but can be generalised to apply to other family-run businesses. It gives an insight into the relationship between the family and the businesses they operate. Understanding the decision processes used by family-business operators is important for organisations or people wanting to influence the strategic decisions made by families and family businesses. It may also help families improve the effectiveness of their decision making.

The overall aim of families is to create opportunities that satisfy their family's aspirations (conceptualised as five 'motivation-stories'). The DST provides the logical framework for understanding how environmental policy developers can harmonise new policies to the kinds of decisions and the kinds of justifications that families use in making decisions. Policies that help families create opportunities that satisfy some aspect of their 'motivation-stories' are considered legitimate by policy recipients.

Family 'motivation-stories' have the same kinds of intentions as sustainable development ideas (caring for people and for the environment). However, the opportunities business families actually create do not generally lead to these two intentions being realised. This is because the decision-makers have to create opportunities from the components that are currently available to them. They have difficulty including sustainable development ideas in creating new opportunities and progressing existing developments using the available 'external components'. Families and family businesses need 'new external components' if they are to create opportunities that deliver their family's aspirations and the intentions of sustainable development ideas. To do this, 'new external components' have to facilitate the creation of opportunities that maintain *critical resources* and support *critical arrangements*. 'Critical' in this context is critical for the delivery of human needs, not critical for the operation of particular technologies. Families on their own are not able to make all the necessary changes. Developing 'new external components' is a task for governments and industries.

The 4-Group-Stakeholder model outlines the responsibilities different stakeholders have in making and using these 'new external components'. It also indicates what the two drivers of the human development system are. The intermediate driver of the business part of the system is profit, while the ultimate driver is the desire of families to satisfy their 'motivation-stories'. Given the growth in population and resource use, future generations will find it increasingly harder to satisfy their 'motivation-stories'. Policy and programs developed now can help future generations if they allow families in the future to create opportunities that care for both people and the environment. Environmental policy developers can do this by providing 'new external components' for opportunities that would satisfy the 'ultimate *future* driver' of the human development system. The 'ultimate *future* driver' differs from the current 'ultimate driver' by including global ideals.

Global ideals on how to treat other people and deal with the environment provide well established guidance for environmental policy developers. Both kinds of ideals have to be advanced in every decision to move society towards a more sustainable future. This implies that environmental policy will be more successful if it is compatible with policies that advance human welfare and vice versa. The concept of 'boxes of influence', which is the sixth concept in DST, provides a way of classifying policies and programs according to how they influence policy recipients. This is an important step in working out how policies can influence policy recipients throughout their lives and what kinds of policies are required for different kinds of tasks. Environmental policy developers can use DST and the 4-Group-Stakeholder model to help governments and the private sectors develop the external components that will allow family decision-makers to create opportunities that satisfy their 'motivation-stories' and deliver sustainable development ideas. Such policies would 'improve society'.

REFERENCES

Albright, C. (2006). Who's running the farm? Changes and characteristics of Arkansas women in agriculture. *American Journal of Agricultural Economics, 88* (5), 1315–1322.

Allan, J. (2005). Farmers as learners: evolving identity, disposition and mastery through diverse social practices. *Rural Society, 15* (1), 4 (18).

Argyle, M. (1987). *The Psychology of Happiness*. London and New York: Methuen and Co. Ltd.

Arias, O. (2000). Confronting Debt, Poverty and Militarism: A Humane Program of Support for the Developing World. *Journal of Third World Studies, 17* (1), 13 – 20.

Barber, J. (2003). Production, Consumption and the World Summit on Sustainable Development, *Environment, Development and Sustainability, 5*, 63 – 93.

Bates, F. L. (1997). *Sociopolitical Ecology, Human Systems and Ecological Fields*. New York and London: Plenum Press,

BCR (2008). Birchip Cropping Group, Agricultural Community Innovators. web site: http://www.bcg.org.au/ (Accesses 24 October 2008)

Brodt, S., Klonsky, K., & Tourte L. (2006). Farmer Goals and Management Styles: Implications for Advancing Biologically Based Agriculture. *Agricultural Systems, 89* (1), 90-105.

Butler, C. (2008). *Human Rights Ethics: A Rational Approach*. West Lafayette, Indiana: Purdue University Press.

Burton, R.J.F. (2004). Seeing through the 'good farmer's eyes': towards developing an understanding of the social symbolic value of 'productivist' behaviour. *Sociologia Ruralis 44* (2), 195–215.

Callicott, J. Baird (1999). *Beyond the Land Ethic, More Essays in Environmental Philosophy*. Albany, New York: State University of New York Press.

Co. A. (Commonwealth of Australia) (1992). *The National Strategy for Ecologically Sustainable Development,* Canberra, Australian Capital Territory: AGPS. http://www.environment.gov.au/esd/national/nsesd/strategy/index.html

Co. A. (Commonwealth of Australia) (1996). *National Strategy for the Conservation of Australia's Biological Diversity.* Canberra, Australian Capital Territory: AGPS, Department of the Environment, Sport and Territories. http://www.environment.gov.au/biodiversity/publications/strategy/index.html

Cocklin, C., Mautner, N., & Dibden, J. (2007). Public policy, private landholders: perspectives on policy mechanisms for sustainable land management. *Journal of Environmental Management 85,* 986–998.

Corbin, J., & Strauss, A. (2008). *Basics of Qualitative Research, Techniques and Procedures for Developing Grounded Theory.* Los Angeles, London, New Delhi, Singapore: Sage Publications.

Couldry, N., (2001), The Umbrella Man: crossing a landscape of speech and silence, *European Journal of Cultural Studies, 4* (2), 131 (22).

Daly, H. E., & Cobb, J. B. (1989). *For the Common Good, Redirecting the Economy Towards Community, the Environment and a Sustainable Future.* Boston: Beacon Press.

Daly, H. E. (1996). *Beyond Growth: The Economics of Sustainable Development.* Boston, Massachusetts: Beacon Press.

Davis, A. (2003). Whither mass media and power? Evidence for a critical elite alternative. *Media, Culture and Society, 25,* 669 – 690.

Deci, E.L., & Ryan, R.M. (2000a). The ''what'' and ''why'' of goal pursuits: human needs and the self-determination of behaviour. *Psychological Inquiry 11* (4), 227–268.

Deci, E.L., & Ryan, R.M. (2000b). The darker and brighter side of human existence: basic psychological needs as a unifying concept. *Psychological Inquiry 11* (4), 319 –338.

Deming, Edwards, W. (1986). *Out of Crisis, Quality, Productivity and Competitive Position.* Cambridge, Melbourne, Sydney: Cambridge University Press.

de-Shalit, A. (1995). *Why posterity matters: Environmental policies and future generations.* London and New York: Routledge.

Donnelly, J. (1998). *International Human Rights, Second Edition.* Boulder, Colorado & Cumnor Hill, Oxford: Westview Press.

Elkington, J. (1999). *Cannibals with forks: the triple bottom line of 21st Century business.* Oxford: Capstone.

Farley, J., & Costanza, R. (2002). Envisioning shared goals for humanity: a detailed, shared vision of a sustainable and desirable USA in 2100. *Ecological Economics 43,* 245 – 259.

Farmar-Bowers, Q. (2004). Personal Drivers – Interviews: Background Report 5. In J. Crosthwaite, J. Callaghan, Q. Farmar-Bowers, C. Hollier, & A. Straker (Eds.), *Land Use Changes: Their Drivers and Impact on Native Biodiversity – Driver Research Phase 1: Overview Report,* (CD ROM). Melbourne: Department of Sustainability and Environment. Available from: www.dse.vic.gov.au. http://www.dse.vic.gov.au/CA256F310024B628/0/B097A38A262BC176CA25738E00 1C782F/$File/DLUC+5++Interviews.pdf

Farmar-Bowers, Q. (2008), *Making Sustainable Development Ideas Operational: A General Technique for Policy Development.* Saarbrücken, Germany: VDM Verlag Dr. Müller.

Farmar-Bowers, Q., & Lane, R. (2006). *Understanding Farmer Decision-systems that Relate to Land Use: Report to the Department of Sustainability and Environment.* Melbourne: RMIT University. Available from: http://eprints.infodiv.unimelb.edu.au/archive/00001842/.

Farmar-Bowers, Q., & Lane, R. (2009). Understanding farmers' strategic decision-making processes and the implications for biodiversity conservation policy. *Journal of Environmental Management, 90,* 1135–1144.

Farmar-Bowers, Q., & Ward, B. (1995). *Platypuses have Power: Ecologically Sustainable Development and Road Works: Special Report No 51.* Vermont South, Victoria: ARRB Transport Research.

Fiss, O. M. (1999). Human Rights as Social Ideals. In C. Hesse & R. Post (Eds.), *Human Rights in Political Transitions: Gettysburg to Bosnia* (pp. 263 – 276). New York: Zone Books.

Foskey, R. (2005), *Older Farmers and Retirement*: A Report for the Rural Industries Research and Development Corporation, January 2005. *RIRDC Publication No 05/006,* Project No. UNE 68A,. Canberra: RIRDC.

Friedman, H. H. (2003). Creating a Company Code of Ethics: Using the Bible as a Guide. *Electronic Journal of Business Ethics and Organisation Studies 8* (1) (publishing date: 2003 -04-06). Available from:
http://academic.brooklyn.cuny.edu/economic/friedman/CodeOfEthics.htm

Gao, F., Li, M., & Nakamori, Y. (2003). Critical Systems Thinking as a Way to Manage Knowledge, *Systems Research and Behavioural Science, 20,* 3 – 19.

Goodman, C., & Anise, A. (2006). *What is known about the effectiveness of economic instruments to reduce consumption of foods high in saturated fats and other energy-dense foods for preventing and treating obesity?* Copenhagen, WHO Regional Office for Europe, Health Evidence Network report:
http://www.euro.who.int/document/e88909.pdf, accessed November 19, 2008.

Hamilton, C. (1998). Measuring Changes in Economic Welfare, The Genuine Progress Indicator for Australia, in R. Eckersley (Ed.), *Measuring Progress, Is Life Getting Better?* (pp. 69 – 92). Collingwood, Victoria: CSIRO Publishing.

Harris, M. (1999). *Theories of Culture in Postmodern Times.* Walnut Creek, London, New Delhi: AltaMira Press, A division of Sage Publishing Inc.

Held, V. (2006). *The Ethics of Care, Personal, Political, and Global.* New York: Oxford University Press.

ILO, International Labour Organisation (2009). *Rules of the game: a brief introduction to International Labour Standards, Revised Edition.* Geneva, Switzerland: International Labour Organisation.

Jenkins, T. N. (1998). Economics and the environment: a case of ethical neglect, *Ecological Economics 26,* 151 – 163.

Johns, D. M. (2003). Growth, Conservation, and the Necessity of New Alliances. *Conservation Biology,* 12 (5), 1229 – 1237.

Karakosta, C., Doukas, H., & Psarras, J. (2008). A Decision Support Approach for the Sustainable Transfer of Energy Technologies under the Kyoto Protocol. *American Journal of Applied Sciences, 5* (12), 1720 – 1729.

Le´vesque, M., & Minniti, M. (2006). The Effect of Aging on Entrepreneurial Behaviour. *Journal of Business Venturing, 21,* 177–194.

Max-Neef, M. A. (1995). Economic growth and quality of life: a threshold hypothesis. *Ecological Economics, 15,* 115 – 118.

McDaniel, W. C. (2008). Repealing unions: American abolitionists, Irish repeal, and the origins of Garrisonian disunionism. *Journal of the Early Republic,* 28 (2), 243 (27).

Meadows, D. H., Meadows, D. L., Randers, J., & Behrens, W.W. (1972). *The Limits of Growth.* New York: Universe Books.

Meadows, D. H., Meadows, D., L., & Randers, J. (1992). *Beyond the Limits: Global Collapse or Sustainable Future.* London: Earthscan Publications Ltd.

Mercer, D., & Jotkwitz, B. (2000). Local Agenda 21 and Barriers to Sustainability at the Local Government Level in Victoria, Australia. *Australian Geographer, 31* (2), 163 – 181.

Mestrum, F. (2003). Poverty Reduction and Sustainable Development. *Environment, Development and Sustainability 5*, 41 – 61.

Midgley, G. (2000). *Systems Intervention, Philosophy, Methodology and Practice.* New York, Boston, Dordrecht, London, Moscow: Kluwer Academic / Plenum Publishers.

Millennium Ecosystem Assessment (2005). *Ecosystems and Human Well-being: Synthesis.* Washington, DC: Island Press: http://www.maweb.org/documents/document.356.aspx.pdf

Meyer, P. (1981). The International Bill: A Brief History. In P. Williams (ED.), *The International Bill of Human Rights* (pp. XXIII – XLVII). Glen Ellen, Ca: Entwhistle Books.

Meyer, J. L. (1998). The State of the Global Environment. In D. G. Dallmeyer, & A. F. Ike (Eds.), *Ethics and the Global Marketplace* (pp. 7 – 15). Athens and London: The University of Georgia Press.

Moran, D. D., Wackernagel, M., Kitzes, J. A., Goldfinger, S.H., & Boutaud, A. (2008). Measuring sustainable development – Nation by nation. *Ecological Economics 64* (3), 470 – 474.

Oliga, J.C. (1996). *Power, ideology, and control.* New York: Plenum Press

Orr, D.W. (2002). Four Challenges of Sustainability, *Conservation Biology, 16,* (6), 1457 – 1460.

Rawls, J. (1971). *A Theory of Justice.* Cambridge, MA: Harvard University Press.

Rawls, J. (1993). *Political Liberalis.* New York: Columbia University Press.

Reeson, A., & Tisdell, J. (2006). When good incentives go bad: an experimental study of institutions, motivations and crowding out. In *Australian Agricultural and Resource Economics Society (AARES) 50th Annual Conference*, Sydney: http://www.ecosystemservicesproject.org/html/publications/markets.html

Roberts, J. A. (1996). Will the real socially responsible consumer please step forward? *Business Horizons, 39* (1), 79 (5).

Robinson, J. (2004). Squaring the circle? Some thoughts on the idea of sustainable development. *Ecological Economics 48*, 369 - 384

Ruttan, V. W. (1999). The transition to agricultural sustainability. *Proceedings of the National Academy of Science USA, 96*, 5960 – 5967.

Schumacher, E. F. (1977). *A Guide for the Perplexed*, Great Britain: Jonathan Cape Ltd.

Slote, M. A. (2007). *The Ethics of Care and Empathy.* New York: Routledge.

Spash, C. L. (2002). *Greenhouse Economics, Value and Ethics.* London and New York: Routledge.

Stoneham, G., Chaudhri, V., & Strappazzon, L. (2003). Selecting policy mechanisms for biodiversity conservation on private land. In J. Crosthwaite, Q. Farmar-Bowers, & C. Hollier, (Eds.), *Land Use - Yes!- But Will Biodiversity Be OK? Proceedings of a Conference at Attwood, Victoria, August 2002*, CD ROM. Melbourne, Victoria: Department of Sustainability and Environment. [www.dse.vic.gov.au]. Available from: http://www.dse.vic.gov.au/CA256F310024B628/0/B1C11258B3805974CA256FFD002 85F4F/$File/Stoneham+Chaudhri+and+Strappazzon.pdf

Strauss, A., & Corbin, J. (1998). *Basics of Qualitative Research, Techniques and Procedures for Developing Grounded Theory.* London, New Delhi, Thousand Oaks: Sage Publications.

Szenberg, M. (1997). Managing God's estate: current environmental policies and the Biblical tradition. *International Journal of Social Economics, 24* (6), 628 (15).

UNDP, (United Nations Development Program) (1999). *Human Development Report 1999: Published for the United Nations Development Program,* (UNDP). New York, Oxford: Oxford University Press.

UN, (United Nations) (1948). *Universal Declaration of Human Right:, Adopted and proclaimed by General Assembly resolution 217 A (III) of 10 December 1948.* New York: United Nations. http://www.un.org/Overview/rights.html

UN, (United Nations) (1992). *Agenda 21: The UN programme of Action from Rio.* New York: United Nations.

Velasquez, M. G. (1998). *Business Ethics, Concepts and Cases, Fourth Edition.* London: Prentice-Hall International (UK) Limited.

Werhane, P. H. (2002). Moral Imagination and Systems Thinking, *Journal of Business Ethics, 38,* 33 – 42.

WCED, (World Commission on Environment and Development) (1987). *Our Common Future,* Oxford: Oxford University Press.

Wronka, J. (1998). *Human Rights and Social Policy in the 21st Century* (Revised Edition). Lanham, New York, Oxford: University Press of America Inc.

Wythes, A. J., & Lyons, M. (2006). Leaving the Land: An exploratory study of retirement for a small group of Australian Men. *The International Electronic Journal of Rural and Remote Health Research, Education, Practice and Policy, 6:531.* (online) Available from: http://rrh.deakin.edu.au Access from:
http://www.rrh.org.au/publishedarticles/article_print_531.pdf

In: Handbook of Environmental Policy
Editors: Johannes Meijer and Arjan der Berg

ISBN 978-1-60741-635-7
© 2010 Nova Science Publishers, Inc.

Chapter 2

GREEN PROCUREMENT POLICIES AND PRACTICES: SWEDISH PERSPECTIVES FROM THE PUBLIC AND PRIVATE SECTORS

Charlotte Leire, Oksana Mont and Carl Dalhammar*

International Institute for Industrial Environmental Economics at Lund University, Lund, Sweden

ABSTRACT

In recent years, European policy has emphasised the role of green procurement as a policy instrument in efforts to make European markets more environmentally sustainable. By purchasing products and services with low environmental impacts, public bodies and companies may shape the markets of products and services and stimulate their environmental sustainability. Environmental procurement can send signals to producers that these products and services are in demand, thereby helping to reduce the overall environmental impact on society. However, the main question regarding green procurement is what role existing policy plays in driving various actors to integrate environmental criteria into their purchasing decisions.

This chapter provides an overview of the main European and Swedish policies that address public and private procurement and identifies gaps in existing efforts. The differences between the public and private sectors in terms of policy drivers and employed strategies are highlighted. The chapter also discusses the need for further policy efforts to support environmental procurement practices of public bodies and companies.

* P.O Box 196, Tegnérsplatsen 4, SE-221 00 Lund, Sweden Tel: +46 46 222 0281; Fax: +46 46 222 0240; E-mail: charlotte.leire@iiiee.lu.se.

INTRODUCTION

The potential of green procurement as a policy instrument has been increasingly recognised in recent years and the political support for it has been growing at international, EU and national European levels. For example, OECD has published recommendation on green public procurement in 2002. In 2006, a Marrakech Task Force on sustainable procurement was created with the aim of disseminating sustainable public procurement around the world.[*] Green and sustainable procurement policies have been developed in many OECD countries, including the United States, Japan, Canada, Australia, and South Korea, as well as in newly-developed countries, e.g., China, Thailand, and the Philippines (Commission Proposal COM(2008) 400 final).

These developments have reinforced the recognition of the prominent role of the public sector as a driver in environmental work (Swedish Energy Agency, 2008). Therefore, in recent years, expectations for the potential of green procurement to become a tool to stimulate changes in the European market and provide producers with incentives to develop more environmentally sound products and services have increased.

The governments are powerful actors on the market through their own purchasing decisions and have a high influence over the markets that supply products to the governmental sectors (e.g., office equipment, transport vehicles, electricity, etc.). The power of governmental actors is significant also because unlike other market players, public organisations at various levels have a high potential to make coordinated efforts in their purchasing strategies in order to achieve policy aims. While green procurement has indeed a high potential to become an instrument of change, there is increasing recognition that green public procurement practices tend to vary among regions and sectors, making the dissemination of best practices difficult (Kogg, 2009).

Despite the potential of the public sector in making changes in green procurement, the main part of procurement practices (50%) are, however, performed by companies (Erdmenger, 2003; EuroFutures, 2004; Falk, 2001). Green corporate procurement could undoubtedly provide significant environmental improvements and changes in the market that in turn could lead to reduced environmental impacts from production and use of products and services. Compared to public procurement, however, there is rather little research done that analyses the current status of procurement practices in businesses and their outcomes. There seems to be quite a difference between prerequisites for public and private sectors to develop green procurement practices. Private organisations aim to create improved performance among the many actors in the supply chain and increase competitiveness in the market by taking into consideration interests of primary and secondary stakeholders (Clarkson, 1995). Public organisations, on the other hand, have to contribute to better achievement of social and environment policy objectives through, among other things, purchasing (European Commission, 2000). As stated in an EC directive, *"contracting authorities and contracting entities may be called upon to implement various aspects of social policy when awarding their contracts, as public procurement is a tool that can be used to influence significantly the behaviour of economic operators. As examples of the pursuit of social policy objectives, one*

---

[*] The goal is for 14 countries, members of the Task Force, from all regions of the world to have Sustainable Public Procurement in place by 2010 FOEN (2008). Marrakech Task Force on Sustainable Public Procurement, Federal Office for the Environment (FOEN)..

can mention legal obligations relating to employment protection and working conditions..."
(European Commission, 2000). Policy demands on non-economic considerations taken in business activities seem to be stronger in public sector procurement than in private sector procurement (Neill and Batchelor, 1999). However, despite being under heavier policy pressure, public procurement tends to be more restrictive with utilising possibilities to pose non-economic demands on suppliers.

Taking into consideration the aforementioned problematic areas within green procurement, the aim of this contribution is to analyse the state of affairs in green public and private procurement in Europe with a special emphasis on Sweden. Sweden has been chosen as one of the seven EU countries at the forefront of European green procurement, together with Austria, Denmark, Finland, Germany, the Netherlands, and the UK (Bouwer, de Jong et al., 2006). These countries have a strong political commitment to green procurement; they have developed national guidelines and programmes supporting dissemination of green procurement; they use life cycle thinking and eco-labelling as support tools in green procurement; and they have open-access information sources and Web sites that support both public and private organisations in their purchasing activities. The goal of this chapter is to learn from European and in particular Swedish experiences in green procurement and to provide suggestions for future policy directions.

The following section provides background to green procurement and outlines definition, drivers and barriers, as well as the differences between public and private green procurement practices. Section 3 gives an overview of relevant EU policies, mandatory and information policy instruments for green procurement and law cases that influence the interpretation and praxis of public procurement practices. It also discusses policy relevance for public and private sectors. Section 4 describes Swedish regulatory framework for green procurement following the same structure as in Section 3. Section 5 summarises the status of the current green procurement practices in Sweden. Finally, section 6 sums up the challenges and opportunities of green procurement and discusses their implications for policy and practictioners in Sweden.

BACKGROUND TO GREEN PROCUREMENT

Green procurement has emerged as a way to promote markets for environmentally benign products by stimulating the purchasing and consumption of these products and services by focusing on the buyer (whether private or professional) as the prime actor and by using market forces as the driver for change.

Definition of Green Procurement

The names and definitions that are used to denote green procurement vary. Examples include green purchasing, environmentally responsible purchasing, environmentally responsible procurement, sustainable procurement and sustainable purchasing. For public procurement, terms such as green public procurement and green public purchasing (EFTA, 2007) are used.

Green procurement implies the incorporation of environmental considerations in all phases of the purchasing process, starting from identifying more environmentally sound products, services, materials and technologies to defining specifications for contracts and calculating life cycle costs. Even decisions about avoiding unnecessary purchases are included: *"Green Public Procurement is the approach by which Public Authorities integrate environmental criteria into all stages of their procurement process, thus encouraging the spread of environmental technologies and the development of environmentally sound products, by seeking and choosing outcomes and solutions that have the least possible impact on the environment throughout their whole life-cycle"* (Bouwer, de Jong et al., 2006). In the recent communication, green public procurement is defined as *"...a process whereby public authorities seek to procure goods, services and works with a reduced environmental impact throughout their life cycle when compared to goods, services and works with the same primary function that would otherwise be procured"(Commission Proposal COM(2008) 400 final).* As for the private sector, no specific definition for green private procurement has been found in any official documents. In management literature, however, green purchasing is defined as an environmentally-conscious purchasing practice that reduces sources of waste and promotes recycling and reclamation of purchased materials without adversely affecting performance requirements of such materials (Min and Galle, 2001). A number of different focus areas relevant for private companies have been identified (Preuss, 2005), including looking at environmental criteria/standards related to the products or supply chain, supplier assessment, internal environmental protection initiatives, downstream initiatives, or inbound and outbound logistics. These alternatives can been considered as different avenues for action in green procurement in the corporate sector (Kogg, 2009).

The span of activities in green procurement is quite diverse. In the early days of green procurement practices, the scope of products under consideration in green procurement was rather narrow; one such consideration that received much attention was recycled paper. Nowadays, the range of products is extended to include alsoservices..Examples of products and services that are often addressed in green procurement are office equipment, computers and other IT products, papers, construction, transport vehicles and services, furniture and many other products (Bouwer, de Jong et al., 2006; Leire, 2009).

The definition of green procurement is rather broad, and the distinction between social, ethical and environmental concerns is sometimes difficult to make. However, there are some important differences between aspects and methods used in green procurement and socially responsible purchasing. For example, socially responsible purchasing tends to focus on upstream life cycle stages and mainly on production methods and conditions, in which production takes place, such as workers' rights, health and safety issues, wages, workforce issues related to disabled workers, racial equality, minorities, ethnicity, gender equality and human rights (Lobel, 2006). Green procurement, on the other hand, focuses not only on environmental aspects of production, but also on environmental features of products and the use phase. Thus, socially responsible purchasing appears to be more focused on supplier performance and compliance compared to environmentally responsible procurement that also gives substantial attention to product performance. In both cases, however, organisations develop certain criteria that suppliers need to fulfil if they want to sell their products to the organisation. The processes for developing and setting social and environmental criteria and the steps for following-up supplier performance have many similarities (McCrudden, 2004), but the origins of the criteria can differ significantly. For example, many companies use ILO standards and conventions as a starting point for developing social criteria, while environmental criteria for green public procurement stem from, e.g., external product rankings (prominent in the construction industry), life cycle assessments (LCA)', eco-labelling criteria or criteria developed in the Directive on Ecodesign for energy-using products, as well as black and grey product lists developed within organisations.

Drivers for Green Procurement

There is a great belief in the use of green procurement as a driver for creation of markets for more sustainable goods and services. Due to the large quantities of products purchased, organisations have a great potential to influence the production and consumption patterns in the society. In macroeconomic terms, purchasing volumes in the public sector in OECD countries are equivalent to about 16% of the total GDP in Europe (Commission Proposal COM(2008) 400 final), but this figure varies among countries (Russel, 1998). In Sweden, public purchasing makes up about 25% of the GDP (Falk, Frenander et al., 2004a). Corporate purchasing, in comparison, equals approximately 50% of the GDP (Erdmenger, 2003; EuroFutures, 2004; Falk, 2001) and up to 25% of the total GDP in Sweden (Falk, Frenander et al., 2004b).

The main internal and external drivers of green purchasing practices in both the public and private sectors are regulation, possibility to gain competitive advantage or respond to societal and stakeholder pressures (Walker, Di Sisto et al., 2008).

Green procurement in both public and private sector can bring about significant benefits to the society (MTF-SPP, 2006):

- By sending a clear message to various stakeholders to buy sustainably and by stimulating their behaviour change by setting the example
- By directly effecting the environment, regional development and social conditions by advancing economic performance through capitalising on efficiency opportunities and improving profile of public spending.
- By stimulating the market for sustainable products and services, making these more economic to produce, and hence increasing the general demand for them.

Specifically, by choosing more environmentally sound products and services, green procurement can stimulate markets and production in a more environmentally sound manner, which could lead to reduced climate change impacts, conservation and preservation of limited natural resources, creation of markets for recycled and reused products and for reduction of volume of waste for landfills, leading to the reduction of CO_2 emissions from landfills.

The main argument with the environmental potential of green public procurement is that the aggregated public purchase expenditures are substantial[*] and could as such serve as a considerable driver for the greening of products on the market, both to stimulate the penetration of labels on the product market, and also to contribute to the a higher adoption of product and process improvement (Commission Proposal COM(2008) 400 final). Moreover,

---

[*] In Sweden, purchasing in the public sector is between SEK 300-400 billion which equals 25% of the Swedish GDP. In Europe, the value of public purchasing is on averge 14% of a country's GDP.

green procurement is coupled with other advantages for various stakeholders in the society, such as effective partnerships and knowledge development and transfer; it can help encourage innovation, which is claimed to be another critical competitiveness factor for organisations facing environmental challenges (UNEP, 2003).

Some of the potential environmental savings from green procurement in the public sector have been investigated in a study funded by the European Commission. The study claims that if all public authorities across the EU demanded green electricity, 18% of the EU's GHG reduction commitment under the Kyoto Protocol (60 million tons of CO_2) could be saved. Nearly the same saving could be achieved if public authorities opted for buildings of high environmental quality. Moreover, if all European public authorities demanded more energy-efficient computers, this could lead to the significant changes on the European computer market and would result in saving of 830 000 tons of CO2. Finally, if all European public authorities chose efficient toilets and taps in their buildings, this would reduce water consumption by 200 million tons (equivalent to 0.6% of total household consumption in the EU) (Commission Proposal COM(2008) 400 final).

Besides the environmental benefits, governments increasingly justify green public procurement as a way to internalise the external costs of their purchases, compensating for the lack of other policy instruments that would account for these externalities. A recent study on costs and benefits of green public procurement in Europe revealed that joint procurement initiatives of public authorities typically have a positive impact on the purchase price and to some extent also on life cycle related costs of products and services, including maintenance and energy consumption (Rüdenauer, Quack et al., 2007). Often, higher purchasing prices are compensated for by lower operating costs in the use phase. For example, 95% of the total costs of pumps in heating installations are determined by operating costs. Thus, purchasing decisions solely based on the price can cause "mis-investment", while consideration of the life cycle costs can lead to the purchase of more efficient heat pumps, thereby stimulating the market of more environmentally sound products. Buying green products and services that cost less in the long run may help save taxpayer's money.

Green requirements posed by public procurement on industry and businesses have shown to be able to result in green products design and to promote the development of green technologies (Nader, Lewis et al., 1992). One early example of this is how recycled paper became the standard office supply in many European countries. It was the cumulative demand of public authorities that stimulated the wide use of recycled paper (Otto-Zimmerman, 2001). This development caused a spin-off effect in the electronics industry, which started to design and market office equipment that was compatible with the use of recycled paper. At the same time, acceptance of recycled paper by domestic consumers also increased. In other words, the choice of recycled paper over conventional paper in the public sector affected domestic as well as industrial consumers, and has also given the supply of recycled paper a significant competitive edge. Since then, recycled paper has not only become a standard supply, but also cheaper than the previously conventional (chlorine bleached) paper. The demand from public organisations allowed producers to lower costs through scale economies and learning-by-doing (Marron, 1997).

Green purchasing may also open up opportunities for the purchasing organisation itself. These are mainly business improvements, such as reduced operation and environmental costs, higher level of knowledge and competence, the prevention of disturbances in operations, the provision of assistance with environmental problem solving in the customer firm, higher trustworthiness and better image. Other opportunities can be competitive advantages and possibilities for business development (Leire, 2006).

Barriers to Green Procurement

So far, the potential of green procurement has been just tapped in, but not fully explored and utilised for the societal good. According to a recent study, only 14 Member States had adopted National Action Plans (NAPs) for greening their public procurement – (Council Directive 2004/18/EC), with 12 Member States still working on developing them.[*]

Several barriers have been identified for more prominent success of green procurement – (Council Directive 2004/18/EC). Some of the barriers are relevant for both public and private sectors, while others are more specific for certain types of organisations.

A recent study has identified six main barriers for public organisations to engage with public procurement (Council Directive 2004/18/EC). The first barrier pertains to the availability of information for developing criteria for green procurement, which is relevant for both public organisations and private companies. It has been demonstrated that there is still lack of clear and comprehensive information sources, such as databases, which can be used by various purchasers for setting up the right environmental criteria in tender documents – (Council Directive 2004/18/EC).

Another barrier relates to the availability of information concerns lack of information about and consequently insufficient awareness of the benefits of environmentally sound products and services. In green procurement buyers require significant amount of information in order to make informed purchasing choices. They need to know what environment impacts to focus on, how to translate them into purchasing criteria, what product alternatives exist on the market and what is their environmental profile, as well as how to compare these product or service alternatives (Russel, 1998). In addition, they need to know suppliers and their environmental practices, they need to be aware about general environmental issues of concern in the society, and have access to operational green procurement procedures and tools.

The third barrier is the low general awareness of both buyers and sellers about the benefits of environmentally benign products and services, which stems from lacking understanding about life cycle costs of products and services. This is a significant barrier, especially considering the existing perception in society that green products are more expensive than traditional products.

Legal uncertainty about possibilities to include environmental criteria in tender documents has been identified as yet another important barrier. There has been a number of Directives and other policy documents at the EU and national levels aiming to clarify the legal boundaries, for example (European Commission, 2004; IEFE, 2005). Still, a lot of uncertainty regarding the legal boundaries remains at the operational level (Falk, 2001).

The lack of political commitment and support needed for facilitating progress in green procurement has also been identified as an important barrier. It has been highlighted that training of purchasers in public and private organisations as well as internal managerial support and practical tools need to become more widespread, (Bouwer, de Jong et al., 2006).

---

[*] More information on the state of National GPP Action Plans can be found at: http://ec.europa.eu/ environment/gpp/national_gpp_strategies_en.htm .

There is furthermore a lacking coordination and dissemination of best practices in various organisations and levels: national governmental procurement, local municipal procurement practices, as well as success cases from businesses from different sectors.

For many public and private organisations, one of the main challenges is the lack of knowledge and expertise for evaluating different alternatives in terms of their environmental aspects and impacts. This may lead to that purchasers feel reluctant to priorities green procurement because they need concrete knowledge of which environmental requirements are relevant for a particular product group, as found in Christensen & Staalgaard (2004). Challenges related to the evaluation include the uncertainty about how to define a green product and how to weight the relative importance of different life-cycle performance indicators (Handfield, Walton et al., 2002). Also, there is a perceived lack of concrete product selection guidance, resulting in problems in identifying greener product alternatives (Russel, 1998). In addition, there is a perception of lack of know-how or resources for possible verification and follow up of the life cycle oriented information.

Insufficient individual capacity is another information-related challenge. The capacity aspect can be related to knowledge, insights about environmental issues, environmental education, and can have a bearing on the usefulness of a particular type of information. The feeling of inability or inadequacy can also stem from a lack of enthusiasm or intellectual understanding (New, Green et al., 2002). Purchasing managers can have a variety of attitudes toward environmental issues, and sometimes also have an ambivalent perception regarding the potential and immediate costs and gains of green purchasing initiatives (Bowen, Cousins et al., 2001b).

Compared to single criteria considerations, the life cycle perspective adds to the complexity of green procurement in that the number and scope of purchasing criteria increase and need to cover various stages of a product life cycle. The scope is extended to include not only the characteristics of the product *per se*, but also how it has been produced and distributed, as well as its environmental impact during use and disposal stages.

In addition to lacking awareness, cost issues and lacking clarity in regulation, business companies mention poor supplier commitment and industry specific barriers (Walker, Di Sisto et al., 2008). Moreover Among som businesses report that green procurement process may lead to decreased lead-times and decreased flexibility (Handfield, Walton et al., 2002). Finally, perhaps the largest barrier for private companies to implement green procurement practices is to execute a comparison of alternatives and to follow up on supplier performance.

Green Procurement in Public and Private Sectors

The identification of drivers and barriers for green procurement in the preceding sections imply differences in purchasing processes between private and public organisations. In the EU, it has been declared that public organisations have to contribute to a better achievement of social and environment policy objectives through, among other things, purchasing: *"contracting authorities and contracting entities may be called upon to implement various aspects of social policy when awarding their contracts, as public procurement is a tool that can be used to influence significantly the behaviour of economic operators. As examples of the pursuit of social policy objectives, one can mention legal obligations relating to employment protection and working conditions..."* (European Commission, 2000). Private organisations, on the other hand, do not have such legal requirements. They may use green procurement to reach a competitive advantage, secure market shares, extend the customer base, create an improved performance among many actors in the supply chain and increase competitiveness on the market (Clarkson, 1995). Despite the heavier policy pressure on non-economic considerations in public sector procurement, the possibilities to pose non-economic demands on suppliers in this sector are more restricted (Leire and Thidell, 2009).

Table 1. A comparison of private sector and local government purchasing (adopted from (Murray, 1999).

	Private sector purchasing	Public sector purchasing
Primary stakeholders	Shareholders Board Employees Customers Suppliers Local community	Central, European and global government Elected members Officers Customers Suppliers Ratepayers Local electorate
Corporate objectives	Profit satisficing survival Market share Image	Democratic and customer focused delivery of public services Political advocacy Sustainability (Local economic development and environmental and social sustainability)
Purchasing objectives	Cost reduction Quality improvement Innovation transfer Supply chain management	Value for Money/ "Best Value" Local economic development Social improvements Profile promotion Cost reduction Quality improvement Innovation transfer
Purchasing legislative framework	Code of Ethics Internal Purchasing Manuals	EU Public Procurement legislation Domestic Procurement legislation Standing Orders, Financial Regulations, Scheme of Delegation Code of Ethics Internal Purchasing manuals
Reporting structure	Board of Directors	Committee of Elected Members
Control/enforcement mechanisms	Audit	Local Government Auditor/Audit Commission Internal Audit

The public sector operates under the constraint of having to be seen to behave in a rational and fair manner, which means that there is generally a greater degree of traceability and structured procedure in public sector purchasing, not only for the fairness, but also for the transparency of it (New, Green et al., 2002). This also means that in practice, the procurement procedure can be lengthy and complicated for the purchasing officers.

Other points of differences are types of product groups, with the public sector typically buying a great deal of finished goods and services, and the private sector purchasing also materials and semi-finished products. Finally, there is a large difference between sectors in terms of their ability to actually influence suppliers, not only through posing demands, but also through monitoring the performance and fulfilment of the demands among the suppliers. The private sector tends to have more resources that can be allocated for supplier relations that involve monitoring, supplier audits and even elaboration of improvement plans. Public purchasers typically do not have necessary resources for these activities and can at best combine resources with other public organisations to check supplier performance.

EUROPEAN POLICIES FOR GREEN PROCUREMENT

Traditionally, purchasing organisations (and units within organisations) did not take environmental features of procured products or services into account. In recent years, a number of policy and legislative documents have been developed that provide basis for incorporating environmental criteria in procurement process of different organisations, most often public organisations.

The concept of green public procurement was devised already in the late 1980s in some European countries and in Japan. A number of European countries recognised green public procurement as a policy instrument in the early 1990s. Since then, some European countries have put green procurement into practice. Denmark established its first action plan for green public procurement already in 1994, followed by implementation of a green public procurement plan in Japan in 1995.

The OECD started working on green public procurement in 1996 and later, in 2002, OECD member countries, and among them many European countries, agreed on a Council recommendation "to improve the environmental performance of public procurement" (OECD, 2002).

A number of EU treaties and strategies support green procurement. For example, Article 6 of the Amsterdam Treaty (1997) specified that *"environmental protection requirements should be integrated into the definition and implementation of the Community policies and activities [...] in particular with a view to promoting sustainable development"*. The European Sustainable Development Strategy (SDS) from 2001 added the environmental dimension to the "Lisbon Process", which originally was developed with the goal of making Europe the world's most competitive, knowledge-based society by 2010.

Internationally, there has also been some development of public procurement. The "Plan of implementation" of the World Summit on Sustainable Development, held in Johannesburg, encouraged *"relevant authorities at all levels to take sustainable development considerations into account in decision-making"* and to *"promote public procurement policies that encourage development and diffusion of environmentally sound goods and services"* (UN, 2002).

Policies Promoting Green Procurement in the EU

In 2001, the European Commission developed an interpretative communication, in which it outlined the possibilities provided by Community law for integrating environmental criteria into public procurement (Commission Proposal COM(2002) 274 final).

Later, a Communication on Integrated Product Policy (Commission Communication COM(2003) 302 final), developed within the Sixth Environmental Action Programme (2001), for the first time specifically highlighted the potential of green public procurement in reaching environmental goals. It outlined the need to determine the extent of green public procurement in the Member States and to assess the potential impact of green procurement on the environment and on facilitation of markets of environmentally sound products and services. The Communication recommended the EU member states to develop National Action Plans on green public procurement by the end of 2006 that should be revised every three years. It was envisaged that the plans would not be legally binding, but that they would nevertheless increase awareness and trigger implementation of green public procurement in the countries across Europe. It was envisioned that the National Action Plans would also facilitate an exchange of best practices across countries. However, by the fall in 2008, only 14 countries had developed National Action Plans on green public procurement. In addition to the National Action Plans, the Communication elaborated on information measures for public authorities to green their procurement activities including a handbook for public procurement, an online product database, and a website that would offer much of the necessary information.

After the Communication on Integrated Product Policy, the Commission developed two new Directives - (Council Directive 2004/17/EC) and (Council Directive 2004/18/EC) - that further specified legal framework for public procurement. These Directives explained how public procurers can include environmental criteria in their procurement process, in particular in technical specifications, in selection and award criteria and in contract performance clauses.

In 2004, the Commission launched an action plan to stimulate the development and use of environmental technologies (Environmental Technology Action Plan - ETAP) that aims to reduce economic, financial and institutional barriers to the development and promotion of environmentally sound technologies. Public procurement is mentioned in this Communication as an important measure for improving market conditions and facilitating the uptake of environmental technologies. Since then, two reports on ETAP implementation have been published, which detailed results during 2004-2006 period and outlined priority actions for the future (Commission Communication [SEC(2007) 413]). The report also underlined that action plans should *"establish objectives and benchmarks for enhancing green public procurement as well as guidance and practical tools for public procurers"*.

The same year, the High Level Group on the revision of Lisbon strategy for growth and employment produced a report that also recommended, similarly to the Communication on Integrated Product Policy, that national and local authorities draft action plans by the end of 2006 to make their public procurement more environmentally sound with a specific focus on new vehicle fuels and renewable energy technology (Kok, 2004).

In 2006, the EU Sustainable Development Strategy set the goal that by 2010 the average level of EU green public procurement should be brought up to the standard that was achieved in 2006 by the best performing member states. This Communication became a part of the Sustainable Consumption and Production and Sustainable Industrial Policy (SCP/SIP) Action Plan that set out a framework for the implementation of instrument packages aimed at improving the energy efficiency and environmental performances of products (European Commission, 2008a).

In 2008, in the Commission Proposal on Public procurement for a better environment the Commission presented a suggestion to set targets for green public procurement as part of the action plan for "sustainable consumption and production" (Commission Proposal COM(2008) 400 final). It was also decided to conduct an impact assessment of the costs and benefits of various policy measures to increase the level of green public procurement in Member States (European Commission, 2008b). Moreover, a common political target of 50% of the total green public procurement to be "green" by 2010[*] was proposed (Commission Proposal COM(2008) 400 final).

Table 2. Laws and policies directly affecting green procurement.

Name	Year
Commission Interpretative Communication on the Community law applicable to public procurement and the possibilities for integrating environmental considerations into public procurement	**(Commission Proposal COM(2002) 274 final)**
Communication on Integrated Product Policy	**(Commission Communication COM(2003) 302 final)**
Council Directive 2004/17/EC of 31 March 2004 coordinating the procurement procedures of entities operating in the water, energy, transport and postal services sectors	**(Council Directive 2004/17/EC)**
Council Directive 2004/18/EC of 31 March 2004 on the coordination of procedures for the award of public works contracts, public supply contracts and public service contracts	**(Council Directive 2004/18/EC)**
Report of the Environmental Technologies Action Plan (2005–2006)	**(Commission Communication [SEC(2007) 413])**
EU Sustainable Development Strategy	**2006**
Sustainable Consumption and Production and Sustainable Industrial Policy (SCP/SIP) Action Plan	**2008**
Public procurement for a better environment	**(Commission Proposal COM(2008) 400 final)**

Information Tools for Green Procurement in the EU

In order to put green procurement ambitions into practice, procurement officers from public authorities and contracting parties, as well as from private companies, need information regarding how the legislative framework should be followed, the operational extent of their mandate to set green criteria, and support with the development of

---

[*] The political target is not legally binding, but the Commission has indicated that if EU member states do not improve their GPP practices the targets may become mandatory in the near future.

specification criteria. The assumption is that product-related environmental information, in combination with preconditions such as environmental awareness, knowledge and attitudes, will lead procurers to make informed choices when purchasing products and services. (Leire and Thidell, 2005). Therefore, information instruments are important as a complement to regulatory instruments (Ernst & Young, 1998; Stern, 1999). This is the reason why a number of initiatives have been undertaken in the last decades for developing information and tools aiming to practically assist purchasing officers with green public procurement.

Many actors, including governments, NGOs and industry associations, have contributed to collection and dissemination of environmental information regarding a large number of products. One of the most advanced information policy instruments that can greatly assist purchasers, are environmental labels and declarations. These labels are informative instruments intended to change the behaviour of manufacturers and consumers through the supply of various kinds of environmental product information. They can either be voluntary or mandatory, and in recent years a number of them have been developed in the EU. Voluntary labels include those that are developed and administrated by a third party organisation, and that signify that the product meets certain environmental criteria. Mandatory labels are most often energy labels that provide information to end-users about the energy consumption and performance of office equipment and other domestic white goods and appliances (e.g., fridges and freezers, washing machines, electric tumble dryers, combined washer-dryers, dishwashers, lamps, air conditioners and electric ovens). A recent initiative by the European eco-labelling organisation is to develop and continuously update the Green Store, the online catalogue of eco-labelled products that enables users to search and find eco-labelled products according to their producer and their availability in a specific country (European Eco-labelling Organisation, 2008). It also provides regularly updated information on the number of companies awarded the Eco-label for every product group in every European member state.

European and national eco-labels can also be very useful for public organisations and contracting authorities because they provide background information about environmental criteria for products and services, which can be used in technical specifications in tender documents. For example, environmental criteria are developed within the framework of the European eco-label and the Energy Star Regulation. In addition to these sources, requirements on producers to provide more green information on their products will probably be adopted under the Energy-Using Products (EuP) Directive (2005/32/EC) in the near future. Some recent proposals for the revision of a number of EU Directives aim at developing criteria that would be more useful for green public procurement. For example, it is suggested that minimum requirements and advanced performance benchmarks should be set within the Ecodesign Directive.

Another information tool to support green public procurement is a handbook on green public procurement (GPP) helping public administrations to use environmental criteria when buying products or services (European Commission, 2004). It explains in clear, non-technical terms how public purchasers, such as schools, hospitals and national and local administrations, can take into account the environment when buying goods, services and works. It also describes how to introduce green criteria into the different stages of the public procurement procedure (Council Directive 2004/17/EC), (Council Directive 2004/18/EC). It also clarifies legal possibilities of integrating environmental criteria into procurement process according the recent procurement Directives. The handbook is a significant step forward in facilitating green procurement among public organisations, as it takes into consideration the recent decisions of the European Court of Justice, which helps further specify boundaries for green public procurement for practitioners. The handbook also serves to disseminate best practices as it includes good examples of environmental tenders from local authorities in Europe.

Eco-Procurement Initiative from the European network ICLEI established a Procura+ Exchange, an information and exchange mailing list for procurement officers from local authorities in Europe that aims to spread good green procurement practices, sharing expertise and purchasing know-how (ICLEI, 2008). In addition, in many countries, work on various levels in society has been undertaken to provide guidance and answers to purchasing officer in need of information or in doubt about the legal boundaries of green purchasing.

EPEAT (Electronics Products Environmental Assessment Tool) is an online programme of the Green Electronics Council. It represents another example of an information tool that assists institutional purchasers to identify and evaluate environmentally preferable electronic products (Green Electronics Council, 2008). It establishes a clear set of performance criteria for desktop computers, laptops, and monitors, and recognises higher levels of environmental performance. The tool was developed by a multi-stakeholder group composed of equipment manufacturers, governmental and private purchasers, non-governmental organisations, and environmental professionals. It is designed to be easy to use, with an interactive Web site to speed product registration. The tool encompasses a range of environmental attributes, including toxic materials, material selection, life cycle extension, energy use, design for end-of-life and end-of-life management, and packaging (Katz, Rifer et al., 2005).

Table 3. Information policy instrument that indirectly facilitate the implementation of green procurement.

Voluntary labels	ISO Type I ecolabels, the Green Store—the European Eco-label catalogue
Mandatory labels	Energy labels
EPD	Environmental product declarations
GPP Handbook	Buying green! A handbook on environmental public procurement
Procura+ Exchange	Eco-Procurement Initiative from the European network ICLEI
EPEAT	Electronics Products Environmental Assessment Tool

Table 4 summarises the aforementioned information policy instruments that can stimulate green procurement practices.

CASE LAW AND ITS INFLUENCE ON PROCUREMENT LAWS IN THE EU

In addition to information policy instruments, also case law has served as a trigger in the emergence of green procurement practices. Before the adoption of the current procurement Directives (i.e., (Council Directive 2004/17/EC, ; Council Directive 2004/18/EC), public procurement was regulated by a number of Directives adopted in the early 1990s.* These Directives provided little guidance on the scope for environmental criteria in public tenders, and the rules were interpreted rather strictly in the Commission guidelines, which restricted the scope for green public procurement. The common understanding was that authorities should award public contracts to those parties, which offered the *lowest price or the most economically advantageous tender*. The criterion of the lowest price was not difficult to apply, but it was more difficult to establish criteria for establishing the most economically advantageous tender. Criteria used to evaluate which tender that was most advantageous could include price, delivery date, running costs, cost-effectiveness, quality, aesthetic and functional characteristics, technical merit, and after-sales service and technical assistance. There was a lot of scepticism regarding whether—and if so, how—environmental criteria could be applied, unless they had obvious benefits, e.g., potential health implications or similar. A further problem was that the Commission's view was that it should be the "most economically advantageous tender" for *the purchasing unit*, i.e., there was little room for socio-economic benefits, such as the estimated benefits of a cleaner environment stemming from the procurement of greener goods and services.

Thus, it appeared as if there was limited scope for environmental criteria in green public procurement. However, the ultimate interpretation of EU law lies with the European Court of Justice. While the Commission had favoured a rather strict interpretation on the relevant Directives the EC court expanded the potential for green public procurement through case law. One influential legal decision was taken in Case C-513/99 (Concordia Bus Finland Oy Ab v. Helsingin kaupunki and HKL-Bussiliikenne [2002] ECR I-7213), as it widened the discretion to use environmental criteria in public procurement. In this case, the Court stated that the award criteria used by contracting authorities to identify the most economically advantageous tender need *not necessarily be of a purely economic nature* (European Court of Justice, 2002). Thus, the ruling opened up the door for a wider use of environmental criteria in green public procurement and meant that "the most advantageous offer" could include also socio-economic factors, and that an estimation of benefits need not be restricted to the purchasing unit.

Other cases have further opened the door for the more progressive use of environmental criteria in purchasing. For example, through case law it was established that in some cases, for example in connection to the purchase of renewable electricity, processing and production methods could be considered in procurement even if the processing methods did not affect the characteristics of the purchased product. Case law has also established that social criteria relating to important societal objectives are allowed in public purchasing, for instance demands for using unemployed labour in building contracts.

Consequently, when Directives 2004/17/EC and 2004/18/EC were adopted, the case law was taken into account, and therefore the Directives contained a number of references to environmental and social criteria, and also more detailed rules concerning the relevance of environmental management systems and eco-labels in procurement.

---

* Current and previous (obsolete) Directives can be found at: http://ec.europa.eu/internal_market/ publicprocurement/legislation_en.htm#current.

Policy Relevance for the Corporate Sector

EU procurement policies that regard green procurement mainly focus on public purchasing. The focus on the corporate sector is more limited. However, some policies highlight the vast potential for environmental improvement through corporate purchasing, as exemplified by the EU policies on Corporate Social Responsibility. Also, the latest green procurement communication from the Commission (Commission Proposal COM(2008) 400 final) addresses the linkage between the green procurement that takes place in the public sector with the one taking place in the private sector: *"The definition and criteria used for identifying and promoting "greener" goods are based on a life cycle approach and cover elements which affect the whole supply chain, ranging from the use of raw materials and production methods to the types of packaging used and the respect of certain take-back conditions. These criteria can equally inform private procurement practices. Member States and Community Institutions are encouraged to strengthen this link between Green public and private procurement."*

The Communication on IPP has also highlighted the role for corporate purchasing (Commission Communication COM(2003) 302 final): *"The private sector can demand greener products and greener production processes from their suppliers. They have considerable potential to influence the market for greener products, should they choose to do so, through, for example, demanding a certified environmental management system, such as EMAS."* Moreover, the Communication states that *"The tools being developed for greening public procurement and listed above should also facilitate greener corporate purchasing. In addition, the different types of labeling mentioned below will also be of use. The Commission has also begun working to stimulate the large corporate purchasing market by pushing for corporate purchasing practices to be more transparent through reporting."*

Still, while several EU policy documents encourage companies to adopt green procurement practices, they mainly highlight the role for national policies and public organisations for achieving this aim. Some states have started to adopt more coordinated efforts, including the London green procurement code for businesses (London Development Agency, 2008). The next section elaborates on such efforts, putting the focus on the case of green procurement in Sweden.

SWEDISH POLICIES FOR GREEN PROCUREMENT

In its most recent Communication on green public procurement (Commission Proposal COM(2008) 400 final) the Commission proposes that, by 2010, 50% of all tendering procedures in the EU member states should be "green". However, because green procurement practices among different EU member states and in various European regions differ, there is a vast number of different and uncoordinated requirements set within green procurement practices throughout Europe, which may result in market disturbances. This is one of the reasons for the EU to undertake efforts to coordinate public procurement in the Member States and at the EU level. This is one reason why the European Commission is developing common green procurement criteria for several product groups. The ambition is to harmonise requirements throughout Europe and to encourage the EU "laggard" states to take efforts to catch up with the frontrunners.

Sweden is considered to be one of the frontrunners in green public procurement practices (Bouwer, de Jong et al., 2006). Progressive green procurement practices can be found among - inter alia - central government agencies, regional authorities, municipalities, and universities. Sweden has furthermore been a pioneer in technology procurement, which is public procurement that intends to stimulate innovation through invitation to tenders that demand innovative solutions not yet available on the market. Technology procurement can aid both the market introduction and commercialisation of new products.

The Swedish industry is usually considered to be in the forefront regarding sustainability efforts in general,[*] and Swedish consumers are known to have high environmental awareness. Environmental demands on products procured emerged in Sweden in the 1980s and started with discussions on chlorine bleached paper. At the time, demands stated by the authorities were lagging behind the demands posed by individual consumers, which in turn triggered a pull effect on the market (Enander, 2000). In the late 1980s, several policy signals indicated that public procurement was to be "greened". One of the first explicit calls was the financial government bill 1989/1990, followed by an array of other official documents, including Agenda 21 and several action plans and strategies, most notably the Swedish Environmental Protection Agency's action plan "A more environmental public procurement – proposed action plan" (Swedish EPA, 2005), which proposed targets and measures for a three year period.

Swedish legal developments of green procurement have been heavily influenced by the EU rules and relevant case law, as discussed in the previous section. In 1994, the Swedish Public Procurement Act came into force. It introduced several restrictions on the possibility to pose green demands in public procurement and brought on a debate on the ways and extent that this could be done. The legal boundaries for green procurement in the public sector were as a result considered to be a significant hurdle for practices to progress. The Committee for Public Procurement gave recommendations that environmental purchasing criteria that have a bearing on upstream life cycles of products should not be employed. Thus, environmental impacts from the production phase could not be targeted in green purchasing. This created problems also for the use of eco-labels Type I in green purchasing (Leire and Thidell, 2009). Later, however, the Committee for Public Procurement stated that production methods could also be considered in green procurement, obviously influenced by the developments in the EU procurement law. Sweden enacted two new public procurement laws in 2007, which corresponded to the new EU procurement Directives.

Policies Promoting Green Public Procurement in Sweden

In Sweden, green public procurement as a policy has gone from a strategy applied by a limited number of authorities to a more general strategy applied by most government agencies and regional and local authorities. Currently, green public procurement can be considered as a

---

[*] For instance, Sweden has the highest number of EMS-certified companies per capita, see Statistics from the Swedish Agency for Economic and Regional Growth [NUTEK], available at: http://www.nutek. se/sb/d/215/a/762.

"strong recommendation by government" (European Commission, 2008c), and this has stimulated some genuine change. For instance, in 2006, 35% of the cars purchased by state agencies were compliant with environmental criteria.

Despite this, recent events suggest that the Swedish government is about to pursue a more "coercive" approach. In a recent statement, the Swedish government proposed an ordinance, which will require that from February 2009, government agencies will only be allowed to purchase environmental vehicles (Näringsdepartementet, 2009). Further, authorities should only use environmental cars and green taxi services, and there are qualifying rules for purchases of trucks and other vehicles. The proposed ordinance will further require an increased use of alcohol ignition locks for purchased and rented vehicles.

The Swedish Energy Agency has further proposed that central authorities should be obliged to purchase only energy-using products, which fulfil criteria for energy efficiency developed by the Swedish Environmental Management Council (Swedish Energy Agency, 2008).

Table 4. Swedish policies that promote green public procurement.

Policy	Implications
Framework purchasing	In 1998 the Swedish government established a co-ordination function for government procurement. The objective was to increase efficiency in public spending by co-ordinating the central government's purchasing and procurement. The main idea is that government authorities should be able to buy supplies or services at better terms than each authority could reach by acting separately. Often, environmental criteria has been included in the criteria applied for tenders for framework purchasing.
Purchasing action plans, recommendations and guidelines	In Sweden, a National Action Plan including targets was endorsed by government on 8 March 2007. Moreover, environmental purchasing criteria guidelines for some 75 product groups have been developed in collaboration with various stakeholders (state-local-industry), and by various national/regional/local networks. These documents strongly influence purchasing among state agencies and other authorities.
The Environmental Quality Objectives	Sweden has created a system with 16 general environmental quality objectives and 72 sub-targets. Regional and local municipalities set their own regional and local targets, and design action plans to reach the objectives. GPP is highlighted as an important strategy for goal achievement at central/regional/ municipal levels.
The Environmental Code	The Code contains a number of rules and principles that can be relevant in GPP, such as the use of best available technique and the Substitution Principle, which mandates the replacement of harmful chemicals for less harmful ones.

The Swedish Government has endorsed a three-year National Action Plan (NAP) for Green Public Procurement 2007-2010 (Official letter 2006/07:54). Targets include that *"the proportion of public bodies that regularly stipulate environmental requirements shall increase"* and that *"the proportion of government framework agreements contained properly formulated environmental requirements having an impact on product and services shall increase"* (MSR, 2009). Measures include setting national targets for public consumption, tightening the control of government agencies, involving local politicians and leaders, ensuring the requisite skill among purchasers and offering an efficient and simple purchasing tool (MSR, 2009). Operational responsibility for implementing the action plan rests with the Swedish Environment Management Council, the key actor for promoting the green procurement developments that also represents Sweden in international and EU collaboration on environmental requirements for public procurement. The Swedish EPA is now looking into how to make the target quantitative, and how to stipulate that public authorities and organisations should measure and report on green purchasing targets.

Policies Promoting Green Corporate Procurement in Sweden

While most policies aim at green procurement in the public sector, increasing attention has being paid to corporate procurement and its potential. In Table 5, a number of main policy drivers for sustainable procurement among companies are outlined. Some innovative initiatives include Future Trade[*] and the Swan Club[+]. Future Trade targets large retailers that represent a significant share of the Swedish retail sales, and these retailers can exercise a large influence over manufacturers, transport companies and other relevant economic actors. The Swan Club provides an opportunity for all types of corporations who purchase a significant amount of eco-labelled products and services to market their efforts. It is operated by the eco-labelling organisation the Nordic Swan.

Information Policy Instruments and Practitioner Tools for Green Procurement in Sweden

A number of information tools, both volunteer and mandatory, have gained a prominent status on the Swedish market today. Among the volunteer types of labels are the NGO-operated "Good Environmental Choice", the official "Nordic Swan", the organic food "KRAV", the trade union operated office equipment label "TCO", the "Forest Stewardship Council" (FSC), the official European eco-label "EU-Flower", the NGO-operated "WWF", the German "Blue Angel", the official "Energy Star". The mandatory label is primarily exemplified by the EU energy label. Among the lifecycle oriented labels, the "Nordic Swan" has so far produced the largest number of criteria documents[*] (Norden, 2008). The types of product groups that the ecolabelling covered initially consisted of domestic products, rather than products for professional use, however with time this changed.

In 1998, the Swedish government appointed a Committee that was to help further the use of public procurement as a tool for an ecologically sustainable societal development. One task of the committee was to develop a common, internet-based guide for the entire public sector.[+] The Committee developed a tool (the EKU-tool) that was based on cooperation between the state, the council of federation of county councils and the municipalities. In 2003, the

---

[*] See http://www.framtidahandel.se/.
[+] http://www.svanen.nu/Default.aspx?tabName=Svanenklubben&menuItemID=7098.
[*] In October 2007, 66 product groups were represented by the Nordic Swan.
[+] EKU-delegationen (2001). Ett levande verktyg för ekologiskt hållbar offentlig upphandling.

administration of the EKU-tool was handed over to the Swedish Environmental Management Council, who administrated the EKU-tool for four years before it was renamed the MSR purchasing criteria[§]. The environmental criteria used within the MSR purchasing criteria were inherited from the previous version, the EKU-tool (SFTI, 2001).

Table 5. Swedish policies that promote green procurement in the private sector.

Policy	Implications
Green Public Procurement (GPP)	GPP has "ripple" effects; it stimulates corporations who wish to get public contracts to green their own purchasing.
Future Trade (Framtida Handel)	A huge number of large Swedish retailers and producers have agreed to green their procurement and to provide proper documentation of this. They have also signed up to stringent targets for reduced environmental impacts from transport and reduced energy use in the food chains. In order to reach the targets, further procurement measures will be necessary.
The Environmental Code	The Environmental Code has several principles, which have implications for corporate procurement. For instance, the Substitution Principle mandates all economic actors to replace harmful chemicals with less harmful ones, and it has stimulated a number of green efforts.
Licensing/supervision	When working with the "traditional" legal instruments permitting and supervision, Swedish authorities have put pressure on manufacturers to green their transport activities and reduce environmental impacts from chemicals use. This provides incentives for manufacturers to purchase greener transport, less hazardous chemicals etc. (Dalhammar, C. (2008).
The Swan Club (Svanenklubben)	The Swan Club provides an opportunity for corporations that purchase a significant amount of eco-labelled products and services to market their efforts.
Other support tools	Other support tools include databases that enable choice of environmentally superior products. One example concerns the BASTA database,[§] which helps procurers to identify sustainable construction material. The Swedish Chemicals Agency previously provided lists of chemicals that should preferably be substituted, and this was often used by the different authorities and enterprises, also in procurement. Those lists are now replaced with a new tool, PRIO, and are also available in English.[±]
Certification and diploma systems	Certification and diploma systems are often a driver for green procurement among companies. Such systems include both the well-known ISO 14001 and EMAS standards, and the wide number of diploma schemes run by municipalities and business organisations in Sweden.

Status of Green Procurement Activities in Sweden

The implementation of green procurement activities in public and private sectors in Sweden is generally a rather recent phenomenon. As such, the activities are not always well

---

§ "Miljöstyrningsrådets Upphandlingskriterier". http://www.msr.se/sv/Upphandling/Kriterier/.
§ See www.bastaonline.se [incl. English version].
± See . See: http://www.kemi.se/templates/PRIOEngframes.aspx?id=970.

systematised and are rarely documented internally in the organisations. Moreover, there seems to be an unsystematic inclusion of the environmental, social and ethical considerations, to which companies often refer under the umbrella term of Corporate Social Responsibility.

However, various reports (surveys or studies) have been conducted with the aim to measure the progress or status of green procurement activities in the organisations.

Environmental considerations applied in procurement can refer to both the environmental performance of products and suppliers. Comparing the two types of approaches, an English study found that the most popular green procurement approach among corporations is the product based one (Bowen, Cousins et al., 2001a). However, among Swedish companies it is nowadays very common to also qualify suppliers based on environmental criteria in the initial qualification procedure (Leire, 2009).

Supplier related criteria include how the purchasing organisation complies with environmental laws and regulations, the existence or status of an environmental management system, the presence of an internal environmental policy and/or action plan (GRIP, 1997). As such, although environmental criteria targeting suppliers tend to be similar in different purchasing situations, the product related criteria are likely to vary according to product properties. As implied from research on socially responsible procurement, actors in the public sector seem to have less means than companies in the private sector to focus on the supplier performance, mainly due to lack of resources to follow up on the posed demands that go beyond asking for proof of EMS or internal environmental policy (Mont and Leire, 2009).

In the following sections, the main focus will be on documented practices that regard environmental considerations of products, rather than suppliers, since environmental policies tend to be more relevant for this type of practices.

Reports on Activities in the Swedish Public Sector

As stated in previous sections, the public sector is often seen as being responsible for societal development and well-being, and as such green procurement in public purchasing is considered to be an important instrument for reaching various political, social and economic goals (Mielisch, 2000), including sustainable development. Therefore, the public sector is often expected to be at the forefront of green procurement contributing to more sustainable consumption and production patterns. Quite a few European and Swedish studies have been executed in recent years investigating the progress in bringing green procurement policy into practice among public purchasing agents.

The current status of green procurement practices in Sweden can be compared with the European average, which seems to be significantly lower. A European study found that countries comprising the "Green-7 group" – Austria, Denmark, Finland, Germany, the Netherlands, Sweden and the UK – have more purchasing tenders that include green criteria than do other EU countries (Bouwer, de Jong et al., 2006). The study focused on local authorities and demonstrated that 54% of them had medium green procurement activity, while 30% had an intensive experience and one tenth only had low experience. Among the local authorities, 57% always or usually specified environmental requirements when procuring with tender, and 25% always or usually specified environmental requirements in direct procurement. According to the findings, it is 25–35% more common to include green criteria in Sweden compared to the EU average (Bouwer, de Jong et al., 2006). Similarly, another

European study showed that on average, only 19% of all public administrations in the EU15 member states used environmental criteria in more than half of their purchases (Ochoa and Erdmenger, 2003).

The findings from a Swedish study show similar results; a national authorities and their engagement in green procurement found that 37% of Swedish public procurers chose the most environmentally sound products when they placed call-off orders under a framework agreement (Söderström, 2008). As many as 46% of the procurement officers used eco-labelling criteria of some kind to achieve sustainable procurement and half the organisations used the MSR tool for sustainable procurement. However, only 11% claimed to always or usually follow up completed procurement procedures from an environmental viewpoint (Söderström, 2008).

Similarly, a Nordic study showed that in Sweden, public purchasers seek to consider environmental aspects relatively often: 58% of the studied purchases included some kind of environmental criteria (Kippo-Edlund, Hauta-Heikkilä et al., 2005).

Another study from the Swedish EPA investigated 270 procurement procedures carried out by municipalities, county councils and government agencies in 2006, and covering purchasing in 27 product groups (Sjöholm and Sunnermalm, 2008). The results demonstrated that environmental criteria were specified in 78% of the surveyed procedures. The greatest proportion of environmental criteria was found in county council procurement procedures. Here environmental criteria were specified in 92% of cases, while in the municipal procurement environmental criteria were included in 76% of procurement cases. Government agency procurement displayed the lowest frequency of environmental criteria. With regards to the size of the organisation, when looking at the municipalities, the study found only marginal differences between large, medium-sized and small municipalities. However, when it came to large municipalities, all of them applied environmental criteria in their procurement. The study showed that there was no difference in the frequency of environmental criteria in procurement of goods compared to services. The type of environmental criteria most frequently applied, in almost half of the 270 procurement procedures, was compulsory requirements. The most frequently used environmental requirements related to substances, environmental management systems, environmental performance and producer responsibility for packaging (Sjöholm and Sunnermalm, 2008). Compared to the study that preceded the aforementioned one (Jonsson, 2004), the study from 2008 showed that not much progress had taken place in the 4 years. One aspect that differed, however, was that in recent times, there tends to be a more even distribution of the type of environmental criteria and requirements used by various organisations.

In addition to the frequency, the progress in green procurement can also be judged by how much of an impact the environmental criteria have in the tenders. For this reason, the specificity of the environmental criteria has been investigated. One study investigated the greener public procurement situation in the Nordic countries and focused on the larger public purchases (Kippo-Edlund, Hauta-Heikkilä et al., 2005). The findings include that in only 19% of all studied purchases the environmental criteria were well specified (Kippo-Edlund, Hauta-Heikkilä et al., 2005).

Similar findings are available from a study by the Swedish EPA (Swedish EPA, 2007): 10% of the purchasing procedures that made use of environmental criteria had "ill-defined" criteria. Moreover, in 16% of the cases there was a mixture of well and poorly defined criteria (Sjöholm and Sunnermalm, 2008).

A comparative case study of eight Swedish municipalities investigated the process of including environmental demands in their procurement practices (Frank and Karlsson, 2007). The study concluded that the municipalities were well aware of that they should take environmental concern into consideration in the procurement procedure; however, the actual implementation was inhibited, since they needed to ensure that they were not discriminating their suppliers. All the municipalities in the study used the MSR tool, however to a varying degree. Seven out of the eight municipalities have a purchasing policy and five of them have an environmental policy (Frank and Karlsson, 2007).

Yet another way to view the progress in green procurement is the range of product groups that are considered to be typical subjects in green procurement. The Swedish EPA identified them as: mobile adapters, audio/video products, lighting, construction material, fuel, electrical installations, vehicles, IT products, chemical cleaning products, officer material, food products, medical equipment, furniture, paper products, hospital consumables, textiles and white goods. for services, the list includes waste management, security, freight, park management, hotel services, installations and reparation work, transportation (person), cleaning services, printing services and finally laundry services (Sjöholm and Sunnermalm, 2008).

In terms of the preconditions for green procurement, including internal policies and staff awareness, one case study on the procurement practices in Stockholm city found the following "state of the art" (Utredings- och Statistikkontoret, 2005):

- A number of different staff are involved in green procurement
- Half of the organisations lack policies pointing to green procurement
- Fifteen percent of the organisations have a time plan for energy efficiency in accordance with the city target
- Half of the organisations claim to consider energy efficiency when purchasing energy relevant products
- Approximately one third of the organisations have plans for green procurement when it comes to chemicals and products
- Half of the organisations that buy catering services impose some type of demand on ecological food products
- Close to half of the organisations claim to need education in green procurement, energy and chemicals.

The problems and barriers is another area that has been examined case studies. In most cases, the problems concern a lack of knowledge and information, and increased costs and time (Sylvén and Lind, 2000; Söderström, 2008), among which knowledge and information were found to be the main obstacles (Söderström, 2008). Problems related to knowledge and information go hand in hand and can negatively affect the formulation of the environmental criteria and evaluation of supplier performance with regard to the criteria.

Finally, the impact of green public procurement on the suppliers has also been studied. Research has been conducted based on interviews with thirteen suppliers from three sectors: chemical products, food and transportation (Ståhl, 2008). The study found that:

- The environmental criteria imposed by the buyer were relatively easy to fulfil

- The interest for social responsibility is large however no criteria are imposed
- The environmental criteria should be higher ranked in the tenders
- The environmental criteria are not experienced as inhibiting the competitive advantage
- The quality of the posed environmental criteria and following actions are greatly varied
- There is a need for higher degree of follow-up
- The environmental criteria have contributed to making the supplier's goods and services more "sustainable"

In particular, the follow-up activity is a frequently mentioned in terms of potential bottle-necks in green procurement practices. Research both in Sweden and internationally demonstrated the lack of systematic follow-up of the environmental criteria used in procurement. It was recommended that this activity needed to be improved so that the full potential of green procurement was realised (Faith-Ell, Balfors et al., 2006; Günther and Scheibe, 2006).

Reports on Activities in the Swedish Private Sector

In contrast to the public sector, studies of procurement practices in the private sector has predominantly targeted operational experiences and management aspects, including drivers and barriers (Walker, Di Sisto et al., 2008). The overarching, statistics-based research on the application of environmental criteria in the private sector is greatly lacking in available literature. One explanation can be that the private sector is more difficult to describe in quantitative terms due to the great diversity of products and types of purchasing situations. In addition, apart from practical variations in green procurement practices, there is also a significant heterogeneity in ways that both researchers and practitioners define and conceptualise green procurement in the private sector (Kogg, 2009). As a result, current studies of green procurement in the private sector mainly include anecdotal evidence and case studies, rather than sector-wide or national/EU studies.

In Sweden, there is a belief that environmental issues are fairly high on the business agenda in Swedish companies (Falk, Frenander et al., 2004a). Some evidence indicates that Swedish companies have increased their efforts in green purchasing. A recent study of 272 Swedish manufacturing firms found that most firms with more than 100 employees practiced green purchasing, though there was no information regarding what types of requirements that are currently employed in corporate procurement (Arnfalk, Thidell et al., 2008).

Considering the heterogeneous landscape of private sector purchasing, as well as the wide range of products, including raw materials and components, a natural tendency in research so far has been to study green procurement in a particular industry sector or of a specific product group. To illustrate this, a study on environmental criteria in purchasing among building companies was done examining both public and private building companies in Sweden (Sterner, 2002). Among the participating private companies 60% included environmental aspects in their purchasing and 100% intended to do so in the future. These results could be compared to the situation in the public sector, where 100% of county councils and 20% of municipalities already apply environmental criteria in procurement practices. The study also demonstrated that the stipulation of procurement requirements was seen as relatively uncomplicated, while the evaluation of environmental impacts, mainly related to selection of materials, was found to be problematic due to the lack of adequate evaluation models (Sterner, 2002).

Another Swedish study has focused on chemicals. The Swedish Chemical Inspectorate examined the substitution of products among selected companies (Rodhe, Rozite et al., 2007). The study concluded that there were great variations in how actively and how well companies work with the substitution of hazardous substances. The differences were reflected in the level of knowledge, organisational structure and personal engagement (Rodhe, Rozite et al., 2007). The study also showed that the need for support varied among different companies, from, e.g., fundamental measures to more progressive work, as for example the development of a database on chemicals in use. The study demonstrated that the drivers for companies to engage in the substitution work were generally experienced as weak, whereas the energy to work with it were seen as high (Rodhe, Rozite et al., 2007).

Only a limited number of studies have attempted to make aggregate or cross-sector estimations of the green procurement efforts in the private sector. A recent interview series were conducted with companies from the following industry sectors: paint, paper products, construction, clothing, flooring, hotel, vehicle, transportation, food products and paper and pulp (Leire, 2009). The results show that some respondents claimed to take environmental considerations in all purchasing decisions, while others only in certain selected product groups. The most common green procurement activity was the initial qualification of suppliers according to some environmental criteria. When purchasing products, the following environmental issues were considered: chemicals, construction material, ecolabelled products, electrical installations (PCB), energy, fuels, maintenance services, oils (edible), packaging (carton, reduction, water-based ink), paints (water-based), pesticides and genetically modified organisms, phthalates, printed materials, product safety, production methods, proximity of the supplier, recyclable material, rubber for tires (cadmium contents), softeners and stabilizers have been phased out, tires, transport (loading, frequency, fuel, transport (fuel, mode and certification schemes for the vehicle operator), vehicles (fuel and certification schemes for the vehicle operator), and wood (type and forest management).

Another study, a survey, illustrated the green procurement activities taking place in 113 Swedish companies (Leire, 2005). It demonstrated that 22% of the companies regarded environmental properties of products in more than 75% of the purchasing situations, and 27% - in less than 25% purchasing situations. As for the instances of letting the environmental properties actually affect the product choice, 20% of the surveyed companies claimed that this typically happened in more than 75% of the purchasing situations, and 36% in less than 25% of the purchasing situations. Moreover, the study found that in the green procurement context, the most frequently listed product groups were chemicals, cleaning products, paper, transportation, and plastics and rubber products. The majority of the product groups of environmental concern among businesses were consumables; considerably fewer product groups involved materials and components (Leire, 2005). Moreover, the study identified tools that are used for applying environmental criteria in different product groups by companies (Table 6).

The results of the survey demonstrated indicate that the uptake of tools for certain product groups is quite heterogeneous. For some product groups all and any type of tool seem to be useful, such as chemicals, cleaning substances, transportation and vehicles, construction items and machinery, and electronic products. In some cases, products are purchased with green considerations, but no specific tools were mentioned in association with it. In eleven product groups, only one type of tool was applied. In 17 product group, more than one was used. In two product groups the respondents claim to conduct green procurement without the help of any tool.

Table 6. The product groups subject to green procurement and application of tools.

TYPES OF TOOLS						
PRODUCT GROUP	**Guidelines**	**Product lists**	**Eco-label/ EPDs**	**Criteria or spec.**	**Questionnaires**	**Black lists**
Appliances	—	—	—	—	—	X
Batteries	X	—	—	—	—	
Chemical products including additives and hazardous substance	X	X	X	X	X	X
Chemo-technical, cleaning products and detergents	X	X	X	X	—	—
Computers and IT products, Copy machines	X	X	X	—	X	—
Construction material and products	X	X	—	—	X	X
Eco-labelled products, products and services for which environmental requirements have been developed	—	—	X	—	—	—
Electricity and electrical equipment	X	—	X	—	—	—
Fire and safety products	—	—	—	—	—	—
Food	—	—	X	—	—	—
Fuel	X	—	—	—	—	—
Furniture	X	—	—	—	—	—
Heating	—	—	X	—	—	—
Hygiene products	—	—	—	—	—	—
Investment	—	X	—	—	—	—
Lighting and appliances	X	—	—	—	—	—
Machinery, installations, installation materials equipment and electronic products	X	—	X	—	X	X

Materials in general (unspecified)	X	X	X	—	—	X
Most or all product groups	X	X	X	—	X	—
Office materials	X	X	X	—	—	—
Oil	—	—	—	—	—	X
Packaging	X	—	X	—	X	—
Paints and lacquer	X	—	X	—	—	—
Paper	—	—	X	—	X	—
Plastics and rubber (including tires)	X	—	X	—	X	—
Protective equipment and textiles	X	—	—	—	—	—
Sealants, glues and solvents	X	—	X	X	—	—
Services and contracts	X	—	X	—	X	—
Steel, wood, raw material and other material for products	X	—	X	X	X	—
Surface treatments	X	—	—	—	—	—
Transportation and vehicles	X	X	X	X	X	—

Table 6 also demonstrates one way for assessing the progress in green private procurement by looking at the initiatives and tools developed for purchasing agents. Taking construction sector as an example, it can be illustrated that the existing purchasing tools have different functions, are used in different stages of the purchasing decision making process and have a different potential impact on the product's entire life cycle. The main tools that are used in the construction sector procurement include:

- "BASTA" —A sector cooperation for the phasing out of hazardous substances in construction materials, based on a database with qualified construction products.
- "Miljöstatus för byggnader" —An inventory of construction materials, focusing on environmental aspects of energy and natural resources used for indoors and outdoors. Materials are assessed and graded according to a five level scale, from good to unacceptable.
- "Miljömanualen" —Five fact sheets covering environmental concerns over the whole lifecycle of a building, which aims to help with the development of environmental criteria, problem solving, the development of an environmental program or with choosing better materials.
- "Tyréns" —Inventory and classification of buildings, based on some key indicators for environmental performance of buildings and recommendations for appropriate change measures.
- "Folksams Byggmiljöguide" —A guide that uses simple symbols to indicate environmental impacts of construction products and that gives concrete suggestions on "best in class" products.

- "Byggvarubedömning" —An evaluation of the construction product declarations that suppliers have submitted to the Swedish Construction Product Database, which can help the users to make their choice.

In the corporate sector, probably more so than in the public one, the implementation of green procurement is by nature a strategic task, which needs to be considered in relation to making profit. This issue alone is linked to a wide range of challenges for the corporate decision makers, e.g., how to identify negative environmental or social impacts that may be associated with a company, how to prioritise between different stakeholders' needs and wants, how to define what responsible behaviour is, and how to communicate corporate responsibility to key stakeholders (Kogg, 2009).

Some answers to the aforementioned problems can be found in case studies that investigate in-depth how green procurement process is organised and where difficulties emerge. Green procurement practices in a Swedish company producing paints and in a Swedish hotel are examples of such case studies (Baumann, Erlandsson et al., Forthcoming)

The first one, Akzo Nobel Decorative Coatings AB, shows that in their green purchasing activities, they targets the core products or products with a certain degree of risks due to their chemical contents. The company has implemented a product stewardship program that is run by all offices in Europe. The program qualifies and classifies suppliers and analyses various chemical products. The information for both types of considerations is collected at the time of supplier qualification. The program makes use of indicators based on the number of certified suppliers. No indicators are currently used for products. The implementation of green purchasing at AKZO is not yet an entirely smooth process. When it comes to the supplier focus, it takes time to integrate the entire pool of suppliers in the classification system. So far the company focuses on large and/or international suppliers. AKZO audits them at irregular intervals, but lacks resources or local knowledge for a more systematic approach. The company reveals that product information provided by the suppliers, mainly product declarations, is sometimes insufficient to serve as a basis for environmental assessment and therefore audits are needed. With time, as the trust between the company and its suppliers increases, it becomes easier and cheaper to request additional information. Other obstacles mentioned by the company included lack of time to search for product alternatives, and lack of knowledge and tools to apply lifecycle considerations in the purchasing.

The second example is Scandic Hotels. Their approach is predominantly centred around eco-labelled products. Since some years, Scandic strives to integrate the environmental criteria in most purchasing situations and uses indicators based on eco-labelled products to follow up on the green purchasing. For other product groups, Scandic demands environmental declarations from the suppliers, stimulating the development of their environmental management in this way. For yet other products the company applies specific criteria that are chosen to target mainly the upstream lifecycle phases. The approach to prioritize eco-labelled products is considered useful, since it helps to identify relative environmental advantages among product alternatives. Despite this "easy to count" approach, the company claims there is lack of methods to weigh the environmental criteria against other purchasing criteria. Currently, the task of choosing alternatives is given to purchasing managers, who have to justify their choices on a case to case basis.

The examples serve to illustrate that the approaches to green private procurement vary, and with them the problems that the procuring companies encounter.

CONCLUDING REMARKS

Over the last couple of decades, purchasing officers in the public and private sectors have experienced increasing pressure to integrate environmental demands into their purchasing practices, coming both from emerging policies and from other actors on the market: businesses, consumers and NGOs. Driven by this attention, green procurement practices have started to take form, hence triggering a great diversity of policies and initiatives and resulting in a wide range in the status of green purchasing among different European countries. In order to even out the level of green procurement among European countries, harmonising practices and to facilitate the progress of green procurement, it is valuable to evaluate existing experiences and outline directions for future developments. Since Sweden is one of the front-running countries in green procurement in Europe, it has been chosen as a case for studying policy development and practical implementation.

Green procurement as an instrument holds great potential in European environmental policy. One advantage is that green procurement has the possibility to alleviate environmental impacts that are caused in the home country of the suppliers, taking into consideration trade-related difficulties (Lundberg, Marklund et al., 2009).

Judging from recent EU developments, green procurement will remain an important environmental policy in the future. One interesting development would be the harmonisation of eco-label criteria with green procurement criteria. The possibility for suppliers to offer eco-labels as a proof of compliance is one way to overcome the hurdles with follow-up of stringent product requirements. Other promising developments are the European National Action Plans and related issues of target setting and potential consequences for national procurement policies. The discussions on target setting that would apply to each public organisation are already taking place in Sweden.

Together with the identified potential future developments, the overview of European and Swedish policies for green procurement and the status of green procurement practices in public and private sectors in one of the leading countries in the field - Sweden - demonstrated a number of challenges pertaining to the promotion of implementing green procurement practices. By discussing the identified challenges, insights can be gained as to how the effectiveness and efficiency of procurement policy and the dissemination of successful procurement practices in Europe can be promoted.

The case of Sweden illustrates some of these challenges. First, when it comes to policy, the conclusion is that even though there is a substantial policy framework for green procurement in Sweden, there are still quite a few questions that are left unanswered for practitioners. Perhaps the main problem both at policy and practice level is lack of clear information to be used either for interpreting the existing policy framework or for developing purchasing criteria by practitioners. There is a clear need for more support tools and guidance that would help translate the text of law into clear messages for purchasing officers, assisting them with, e.g., identifying the scope for environmental criteria in public tenders and business purchasing contracts or with auditing and monitoring supplier performance. Clear messages would not only confirm political commitment, but would provide long-term planning reliability for public and private purchasing officers. The few mentioned cases, in which the ultimate interpretation of EU law was left to the European Court of Justice, have helped to clarify how environmental criteria can be included in the tenders and contracts and how they can be balanced with the economic criteria.

Besides policy interpretation, there is lack of information regarding existing product and service alternatives on the market that can be used by purchasing officers when developing criteria or when searching for substitutes. Ecolabelling is one instrument that has been widely used in procurement both for setting environmental criteria, but also for identifying alternative products and services. Green purchasing practitioners in Sweden allegedly use eco-labels more frequently as source of information than in other EU member states (Ochoa and Erdmenger, 2003). Thus, promoting ecolabels in other countries could facilitate the green procurement practices by providing verified information about environmental features of products and services (Leire and Thidell, 2009).

In addition to the lack of information, another problem for green procurement is the difficulty for purchasers in both private and public sectors to collect, process and verify data from suppliers. It is typically a very resource and time intensive activity. Therefore support tools are needed that could assist organisations with this task. Some examples already exist, e.g., SEDEX – an information clearing house and exchange data centre, which is a cost-effective tool for suppliers to communicate environmental and social information to their customers and for customers to monitor and manage information and if needed to improve supplier standards (SEDEX, 2008). The information clearing house has built-in scoring system that helps customers to see whether suppliers conform to all the requirements of the SEDEX system or there are some aspects that need to be improved.

Yet another problematic area for practitioners, especially in the private sector, is the lack of international, national or organisations policies addressing or stimulating green procurement. While most policy efforts have been directed to public purchasing, the role of corporate purchasing is often neglected in many countries. Moreover, whereas in the public sector there are signs of increased policy pressure ahead, there are few policies that stimulate any wider dissemination practices of green procurement in the corporate sector.

A different problem related to procurement of more environmentally sound products is their perceived high price. It has however been demonstrated that for many products the higher purchasing price is compensated by lower lifecycle costs. Therefore, in order to promote green procurement, it is vital to increase general awareness about lifecycle impacts and lifecycle costs and to develop tools that would support purchasing decisions taking into account lifecycle costs and not merely the purchasing price. By creating demand, the markets of green products and services would be stimulated and the environmental and economic criteria in purchasing would be brought into a better balance.

Opportunities exist for further developing green procurement as a policy instrument and for setting realistic targets. One example is clear when comparing the policy development approaches in different countries. Sweden has so far ranked high in the investigations that measure the frequency of the use of environmental criteria. However, Sweden might not be the frontrunner in all regards when it comes to green procurement. The findings on the green procurement practices demonstrate that progress can be measured in more than one way, and that there is a risk that the environmental criteria that are posed on the products are too loose to have a real effect on, e.g., environmental impacts of products. In this regard, Sweden can take inspiration from approaches in other countries, such as Denmark, where voluntary agreements have been signed by public authorities to completely "green" all purchasing that takes place in eleven selected product groups. In light of these pragmatic efforts in Denmark, it is therefore tempting to suggest that Sweden can do more to put green procurement on the political agenda and into operational mode.

In addition to current undertakings from the Swedish Environmental Management Council, additional areas can benefit from central support. One such area is the work of clarifying how to formulate effective environmental criteria. Whereas the use of environmental criteria in the product specification has been given a green light in the policy context, there still seems to be a lack of direction in how to apply the criteria on a case-by-case basis to reach the desired environmental outcomes. Therefore, improved dissemination of best practices can become increasingly needed, along with a stronger focus on a tender-related technicalities. For this reason, a national knowledge base could be considered, which would collect case specific and technical information on the legal boundaries of green procurement.

Moreover, policy incentives for green procurement in the public sector are of great importance as they can also pave the way for green procurement practices in the corporate sector. They can stimulate the entire private sector to increasingly develop greener product alternatives and produce higher quality and clearer environmental product information. As such, green procurement activities among public buyers can have a pull effect on the corporate world, triggering positive changes upstream to the supply chains.

Finally, since corporate purchasing accounts for a greater percentage of GDP than public purchasing in almost all countries, businesses and industry deserve attention from policy makers. As of now, corporate purchasers are not bound by the directives on public procurement and there is therefore no direct political drive to improve corporate purchasing practices. Some Swedish initiatives that were reviewed in this chapter, including the Swan Club and Future Trade, are quite interesting as they can provide large retailers and producers with incentives for further efforts in sustainable purchasing. Thus, by targeting influential actors, policy makers may stimulate corporate procurement, as well as enable that these efforts can be implemented at a reasonably low cost for the practitioners.

REFERENCES

Arnfalk, P., Thidell, Å., Brorson, T. and Nawrocka, D. (2008). Miljöarbete inom teknikföretag. Lund, IIIEE.

Baumann, H., Erlandsson, J., Kogg, B., Dalhammar, C., Leire, C., Lindhqvist, T., Rex, E., Rossem, C. v. and Tillman, A.-M., Eds. (Forthcoming). Gearing up for life cycle action: Making sense of life cycle thinking in policy and business, Environmental Systems Analysis, Chalmers University of Technology, and The International Institute for Industrial Environmental Economics, Lund University.

Bouwer, M., de Jong, K., Jonk, M., Berman, T., Bersani, R., Lusser, H., Nissinen, A., Parikka, K. and Szuppinger, P. (2006). Green Public Procurement in Europe 2005 - Status overview. Haarlem, Virage Milieu & Management bv: 107.

Bowen, F. E., Cousins, P. D., Lamming, R. C. and Faruk, A. C. (2001a). "Horses for courses: Explaining the gap between the gap between theory and practice of green supply." Greener Management International 35: 41-59.

Bowen, F. E., Cousins, P. D., Lamming, R. C. and Faruk, A. C. (2001b). "The role of supply management capabilities in green supply." Production and Operations Management 10(2): 174-189.

Christensen, L. and Staalgaard, P. (2004). Støtte til indkøb og efterspørgsel af miljøvenlige tekstiler business-to-business, Miljöstyrelsen, DK.

Clarkson, M. B. E. (1995). "A Stakeholder Framework for Analyzing and Evaluating Corporate Social Performance." Academy of Management Review 20(1): 92-117.

Commission Communication [SEC(2007) 413] Report of the Environmental Technologies Action Plan (2005-2006). Brussels.

Commission Communication COM(2003) 302 final Integrated Product Policy. Building on Environmental Life-Cycle Thinking. Brussels: 30.

Commission Proposal COM(2002) 274 final Commission Interpretative Communication on the Community law applicable to public procurement and the possibilities for integrating environmental considerations into public procurement. Brussels: 28.

Commission Proposal COM(2008) 400 final Public procurement for a better environment.

Council Directive 2004/17/EC of 31 March 2004 coordinating the procurement procedures of entities operating in the water, energy, transport and postal services sectors. OJ L 134 30.04.2004: 114.

Council Directive 2004/18/EC of 31 March 2004 on the coordination of procedures for the award of public works contracts, public supply contracts and public service contracts. OJEU 134 30.04.2004: 114.

EFTA (2007). Fair Trade in Public Procurement, EFTA: 6.

Enander, G. (2000). Miljökrav vid offentlig upphandling - en oförlöst naturkraft. Miljökrav vid offentlig upphandling - så gör man.

Erdmenger, C., Ed. (2003). Buying into the environment : experiences, opportunities and potential for eco-procurement, Greenleaf Publishing.

Ernst & Young (1998). Integrated Product Policy: A study analysing national and international developments with regard to Integrated Product Policy in the environment field and providing elements for an EC policy in this area, European Commission: DGXI.

EuroFutures (2004). Miljöanpassning vid offentlig upphandling - En enkätstudie. Stockholm.

European Commission (2000). The Green Paper on Public Procurement in the European Union: Exploring the Way Forward. Brussels, EC: 59.

European Commission (2004). Buying green! A handbook on environmental public procurement. Brussels, Commission of the European Communities: 39.

European Commission (2008a). Communication from the commission to the Council and the European Parliament on the Sustainable Production and Consumption and Sustainable Industrial Policy Action Plan. Brussels, EC: 15.

European Commission (2008b). Draft commission staff working document accompanying the communication from the commission on the action plans "Sustainable Consumption And Production" and "Towards A Sustainable Industrial Policy" impact assessment. Brussels, Commission of the European Communities: 43.

European Commission (2008c). National GPP policies and guidelines. Brussels, European Commission: 6.

European Court of Justice (2002). Case C-513/99 of 17 September 2002 ("Concordia Bus Finland Oy Ab, formerly Stagecoach Finland Oy Ab v Helsinging kaupunki and HKL-Bussiliikenne"). Brussels.

European Eco-labelling Organisation (2008). The European Eco-label catalogue, European Eco-labelling Organisation.

Faith-Ell, C., Balfors, B. and Folkeson, L. (2006). "The application of environmental requirements in Swedish road maintenance contracts." Journal of Cleaner Production 14(2): 163-171.

Falk, J.-E., Ed. (2001). Miljöanpassad upphandling - offentlig och privat, Ecocompetence AB.

Falk, J.-E., Frenander, C., Nohrstedt, P. and Ryding, S.-O., Eds. (2004a). Miljöledning vid upphandling och inköp. Stockholm, Miljöstyrningsrådet.

Falk, J.-E., Frenander, C., Norhstedt, P. and Ryding, S.-O. (2004b). Miljöledning vid upphandling & inköp. Stockholm, Jure Förlag.

FOEN (2008). Marrakech Task Force on Sustainable Public Procurement, Federal Office for the Environment (FOEN).

Frank, L. and Karlsson, M. (2007). Miljöhänsyn vid offentlig upphandling. Uppsala, SLU, Institutionen för Ekonomi.

Green Electronics Council (2008). Electronics Products Environmental Assessment Tool, Green Electronics Council.

GRIP (1997). Miljøkrav og offentlige innkjøp - et studie av 6 offentlige virksomheter i Norge, Sverige og Danmark, GRIP and NIMA Rådgivning AS.

Günther, E. and Scheibe, L. (2006). "The hurdle analysis. A self-evaluation tool for municipalities to identify, analyse and overcome hurdles to green procurement." Corporate Social Responsibility and Environmental Management 13(2): 61-77.

Handfield, R., Walton, S. V., Sroufe, R. and Melnyk, S. A. (2002). "Applying environmental criteria to supplier assessment: A study in the application of the Analytical Hierarchy Process." European Journal of Operational Research 141: 70-87.

ICLEI (2008). The Procura+ Exchange, Local Governments for Sustainability.

IEFE (2005). EVER: Evaluation of EMAS and Eco-label for their Revision, European Commission.

Jonsson, A. (2004). Miljöhänsyn i statliga ramavtal, Naturvårdsverket.

Katz, J., Rifer, W. and Wilson, A. R. (2005). "EPEAT: Electronic Product Environmental Tool - development of an environmental rating system of electronic products for governmental/institutional procurement." Electronics and the Environment, 2005. Proceedings of the 2005 IEEE International Symposium on: 1-6.

Kippo-Edlund, P., Hauta-Heikkilä, H., Miettinen, H. and Nissinen, A. (2005). Measuring the Environmental Soundness of Public Procurement in Nordic Countries. Copenhagen, Nordic Council of Ministers.

Kogg, B. (2009). Responsibility in the supply chain: Interorganisational management of environmental and social aspects in the supply chain. IIIEE. Lund, Lund University.

Kok, W., Ed. (2004). Facing the Challenge. The Lisbon strategy for growth and employment. Luxembourg, European Communities.

Leire, C. (2005). The role of business procurement in sustainable consumption. 10th European Roundtabel on Sustainable Consumption and Production, Antwerp, Belgium.

Leire, C. (2006). "The application of green purchasing tools in the corporate sector." Environmental Engineering and Management Journal 5(5): 1159-1178.

Leire, C. (2009). "Green purchasing in practice: a comparison of two product categories." Forthcoming.

Leire, C. and Thidell, Å. (2005). "Product-related environmental information to guide consumer purchases - a review and analysis of research on perceptions, understanding and use among Nordic consumers." Journal of Cleaner Production 13(10-11): 1061-1070.

Leire, C. and Thidell, Å. (2009). "Indirect effects of ecolabelling in the case of GPP." Submitted to Journal of Cleaner Production.

Lobel, O. (2006). "Sustainable capitalism or ethical transnationalism: Offshore production and economic development." Journal of Asian Economics 17(1): 56-62.

London Development Agency (2008). London's Green Procurement, London Development Agency,.

Lundberg, S., Marklund, P.-O. and Brännlund, R. (2009). Miljöhänsyn i offentlig upphandling: samhällsekonomisk effektivitet och konkurrensbegränsande överväganden. Stockholm, Konkurrensverket.

Marron, D. B. (1997). "Buying green: Government procurement as an instrument of environmental policy." Public Finance Review 25(3).

McCrudden, C. (2004). "Using public procurement to achieve social outcomes." Natural Resources Forum 28(4): 257-267.

Mielisch, A. (2000). Sustainable Procurement - Adding Value, Buying Green

Green procurement at the municipal level - the local and the European dimension, ICLEI.

Min, H. and Galle, W. P. (2001). "Green purchasing practices of US firms." International Journal of Operations & Production Management 21(9): 1222-1238.

Mont, O. and Leire, C. (2009). "Socially responsible purchasing in supply chains: Drivers and barriers in Sweden." Social Responsibility Journal.

MSR (2009). Homepage of the Swedish Environmental Management Council, The Swedish Environmental Management Council.

MTF-SPP (2006). Discussion Note. Sustainable Public Procurement: Issues Facing the Marrakech Process Task Force on Sustainable Public Procurement. 1st SPP Task Force Meeting, Jongny sur Vevey, Switzerland.

Murray, G. J. (1999). "Local government demands more from purchasing." European Journal of Purchasing and Supply Management 5(1): 33-42.

Nader, R., Lewis, E. J. and Weltman, E. (1992). "Shopping for innovation: The government as smart consumer." The American Prospect 11: 71-78.

Neill, P. and Batchelor, B. (1999). "Bidding for recognition." Supply Management 4(24): 36-38.

New, S., Green, K. and Morton, B. (2002). "An analysis of private versus public sector responses to the environmental challenges of the supply chain." Journal of Public Procurement 2(1): 93-105.

Norden (2008). The Nordic Swan: From past experiences to future possibilities. The third evaluation of the Nordic ecolabelling scheme. Copenhagen, Nordic Council of Ministers: 168.

Näringsdepartementet (2009). Regeringen skärper kraven på myndigheters bilköp. Pressmeddelande.

Ochoa, A. and Erdmenger, C. (2003). Study Contract to survey the state of play of green public procurement in the European Union, ICLEI.

OECD (2002). Recommendation of the Council on Improving the Environmental Performance of Public Procurement C(2002)3. Paris, OECD.

Otto-Zimmerman, K. (2001). The World Buys Green - International Survey on National Green Procurement Activities, ICLEI.

Preuss, L. (2005). "Rhetoric and reality of corporate greening: a view from the supply chain management function." Business Strategy and the Environment 14(2): 123-139.

Rodhe, H., Rozite, V., Thidell, Å. and Trolle, A. (2007). Företags arbete med produktval och utbyte av farliga ämnen samt hur olika aktörer kan verka pådrivande i detta arbete, Kemikalieinspektionen.

Russel, T., Ed. (1998). Greener Purchasing - Opportunities and Innovation, Greenleaf Publishing.

Rüdenauer, I., Quack, D., Dross, M., Seebach, D., Eberle, U., Zimmer, W., Gensch, C.-O., Graulich, K., Hünecke, K., Hidson, M., Koch, Y., Defranceschi, P., Möller, M. and Tepper, P. (2007). Costs and Benefits of Green Public Procurement in Europe. Freiburg, Öko-Institut and ICLEI: 433.

SEDEX (2008). Supplier Ethical Data Exchange, SEDEX.

SFTI (2001). Elektronisk offentlig upphandling. Accessed on http://ehandel.skl.se/ arbeten/rap2.pdf, Arbetsgruppen för elektronisk offentlig upphandling. **2008**.

Sjöholm, U. and Sunnermalm, A. (2008). Miljöanpassad upphandling i praktiken: En genomgång av offentliga upphandlingar 2007. Stockholm, Naturvårdsverket.

Stern, P. (1999). "Information, incentives, and proenvironmental consumer behaviour." Journal of Consumer Policy 22(4): 461-478.

Sterner, E. (2002). "'Green procurement' of buildings: A study of Swedish clients' considerations." Construction Management and Economics 20(1): 21-30.

Ståhl, M. (2008). Är miljökrav vid upphandling konkurrenshämmande? Stockholm.

Swedish Energy Agency (2008). Våga vara bäst! Upphandling av energieffektiva produkter i svensk offentlig sektor. Stockholm, Statens energimyndighet: 88.

Swedish EPA (2005). En mer miljöanpassad offentlig upphandling – förslag till handlingsplan. Stockhom, Naturvårdsverket: 54.

Swedish EPA (2007). Miljöanpassad upphandling i praktiken. En genomgång av offentliga upphandlingar 2007. Stockholm, Naturvårdsverket: 183.

Sylvén, L. and Lind, L. (2000). Redovisning av Miljösamverkan Västra Götalands enkät om Miljöanpassad Upphandling, Miljösamverkan.

Söderström, M. (2008). Tar den offentliga sektorn miljöhänsyn vid upphandling? En enkätstudie 2007, Naturvårdsverket.

UN (2002). Plan of Implementation of the World Summit on Sustainable Development. New York, United Nations, Department of Economic and Social Affairs: 62.

UNEP (2003). Greener Purchasing Strategy for Local Governments Towards a Sustainable Purchasing Strategy at the Local Level, The UNEP-International Environment Technology Centre (IETC).

Utredings- och Statistikkontoret (2005). En enkät om miljöanpassad upphandling i Stockholm stad 2005. Stockholm, Utrednings- och statistikkontoret.

Walker, H., Di Sisto, L. and McBain, D. (2008). "Drivers and barriers to environmental supply chain management practices: Lessons from the public and private sectors." Journal of Purchasing and Supply Management 14(1): 69-85.

In: Handbook of Environmental Policy
Editors: Johannes Meijer and Arjan der Berg

ISBN 978-1-60741-635-7
© 2010 Nova Science Publishers, Inc.

Chapter 3

ENVIRONMENTAL AND SOCIO-ECONOMIC ASPECTS OF POSSIBLE DEVELOPMENT IN RENEWABLE ENERGY USE

Abdeen Mustafa Omer*
Nottingham, United Kingdom

ABSTRACT

The use of renewable energy sources is a fundamental factor for a possible energy policy in the future. Taking into account the sustainable character of the majority of renewable energy technologies, they are able to preserve resources and to provide security, diversity of energy supply and services, virtually without environmental impact. Sustainability has acquired great importance due to the negative impact of various developments on environment. The rapid growth during the last decade has been accompanied by active construction, which in some instances neglected the impact on the environment and human activities. Policies to promote the rational use of electric energy and to preserve natural non-renewable resources are of paramount importance. Low energy design of urban environment and buildings in densely populated areas requires consideration of wide range of factors, including urban setting, transport planning, energy system design and architectural and engineering details. The focus of the world's attention on environmental issues in recent years has stimulated response in many countries, which have led to a closer examination of energy conservation strategies for conventional fossil fuels. One way of reducing building energy consumption is to design buildings, which are more economical in their use of energy for heating, lighting, cooling, ventilation and hot water supply. Passive measures, particularly natural or hybrid ventilation rather than air-conditioning, can dramatically reduce primary energy consumption. However, exploitation of renewable energy in buildings and agricultural greenhouses can, also, significantly contribute towards reducing dependency on fossil fuels. Therefore, promoting innovative renewable applications and reinforcing the renewable energy market will contribute to preservation of the ecosystem by reducing emissions at local and global levels. This will also contribute to the amelioration of environmental conditions by replacing conventional fuels with renewable energies that

---

*17 Juniper Court, Forest Road West, Nottingham NG7 4EU, United Kingdom

produce no air pollution or greenhouse gases. This article presents review of energy sources, environment and sustainable development. This includes all the renewable energy technologies, energy savings, energy efficiency systems and measures necessary to reduce climate change.

Keywords: Renewable energies, wind power, solar energy, biomass energy, geothermal power, environment, sustainable development.

1. INTRODUCTION

Spaces without northerly orientations have an impact on the energy behaviour of a building. For sustainable development, the adverse impacts of energy production and consumption can be mitigated either by reducing consumption or by increasing the use of renewable or clean energy sources [1]. Bioclimatic design of buildings is one strategy for sustainable development as it contributes to reducing energy consumption and therefore, ultimately, air pollution and greenhouse gas emissions (GHG) from conventional energy generation. Bioclimatic design involves the application of energy conservation techniques in building construction and the use of renewable energy such as solar energy and the utilisation of clean fossil fuel technologies. Commercial buildings and institutions are generally cooling-dominated and therefore reject more heat than they extract over the annual cycle. In order to adequately dissipate the imbalanced annual loads, the required ground-loop heat exchanger lengths are significantly greater than the required length if the thermal loads were balanced.

Most businesses could make savings of up to 20% by introducing basic improvements in energy efficiency. Meeting the target of a 60% reduction in carbon dioxide (CO_2) emissions on environmental pollution is both technologically feasible and financially viable. A genuine investment of energy and resources to meet the environmental challenges the world at equity for a small planet. One compelling reason why businesses should reduce emissions:

- It is right to reduce emissions and to use energy efficiency. There are inevitably concerned about costs. They want to provide goods and services at prices their customers can afford and without a competitive detriment.
- To reduce emissions in businesses is that customers care about the environment and would give a choice and support environmentally conscious business.

Renewable energy markets, industry and investment have never grown faster than they did in 2007. Countries with the largest amounts of new capacity investment were Germany, China, the United States, Spain, Japan and India. Sources of finance and investment now come from a diverse array of private and public institutions. From private sources, mainstream and venture capital investment is accelerating, for both proven and developing technologies.

Between 1980 and 2000 governmental awareness of wind energy mainly concentrated in Denmark and Germany, where a large number of wind turbines were manufactured and installed. Nowadays, most European governments are well aware of the potential of wind energy. Generally, the development and operation of a wind farm can be subdivided into four phases:

- Initiation and feasibility.
- Pre-building (conducted by go/no-go).
- Building.
- Operation and maintenance.

2. RENEWABLE ENERGY

In the majority of cities that have installed significant amounts of renewable energy over the last 10 years. The local municipal government has played a key role in stimulating projects. When it comes to the installation of large amounts of renewables, these cities have several important factors in common. These factors include:

- Information provision about the possibilities of renewables.
- The presence of municipal departments or offices dedicated to the environment, sustainability or renewable energy.
- A strong local political commitment to the environment and sustainability.
- Obligations that some or all buildings include renewable energy.

Energy efficiency is the most cost-effective way of cutting carbon dioxide emissions and improvements to households and businesses. It can also have many other additional social, economic and health benefits, such as warmer and healthier homes, lower fuel bills and company running costs and, indirectly, jobs. Britain wastes 20 per cent of its fossil fuel and electricity use. This implies that it would be cost-effective to cut £10 billion a year off the collective fuel bill and reduce CO_2 emissions by some 120 million tones. Yet, due to lack of good information and advice on energy saving, along with the capital to finance energy efficiency improvements, this huge potential for reducing energy demand is not being realised. Traditionally, energy utilities have been essentially fuel providers and the industry has pursued profits from increased volume of sales. Institutional and market arrangements have favoured energy consumption rather than conservation. However, energy is at the centre of the sustainable development paradigm as few activities affect the environment as much as the continually increasing use of energy. Most of the used energy depends on finite resources, such as coal, oil, gas and uranium. In addition, more than three quarters of the world's consumption of these fuels is used, often inefficiently, by only one quarter of the world's population. Without even addressing these inequities or the precious, finite nature of these resources, the scale of environmental damage will force the reduction of the usage of these fuels long before they run out.

Throughout the energy generation process there are impacts on the environment on local, national and international levels, from opencast mining and oil exploration to emissions of the potent greenhouse gas carbon dioxide in ever increasing concentration. Recently, the world's leading climate scientists reached an agreement that human activities, such as burning fossil fuels for energy and transport, are causing the world's temperature to rise. The Intergovernmental Panel on Climate Change has concluded that "the balance of evidence suggests a discernible human influence on global climate". It predicts a rate of warming greater than any one seen in the last 10,000 years, in other words, throughout human history.

The exact impact of climate change is difficult to predict and will vary regionally. It could, however, include sea level rise, disrupted agriculture and food supplies and the possibility of more freak weather events such as hurricanes and droughts. Indeed, people already are waking up to the financial and social, as well as the environmental, risks of unsustainable energy generation methods that represent the costs of the impacts of climate change, acid rain and oil spills. The insurance industry, for example, concerned about the billion dollar costs of hurricanes and floods, has joined sides with environmentalists to lobby for greenhouse gas emissions reduction. Friends of the earth are campaigning for a more sustainable energy policy, guided by the principal of environmental protection and with the objectives of sound natural resource management and long-term energy security. The key priorities of such an energy policy must be to reduce fossil fuel use, move away from nuclear power, improve the efficiency with which energy is used and increase the amount of energy obtainable from sustainable, renewable sources. Efficient energy use has never been more crucial than it is today, particularly with the prospect of the imminent introduction of the climate change levy (CCL). Energy is an essential factor in development since it stimulates, and supports economic growth and development. Establishing an energy use action plan is the essential foundation to the elimination of energy waste. Alternatively energy sources can potentially help fulfil the acute energy demand and sustain economic growth in many regions of the world. Renewables are gaining widespread support, notably in the developing world.

3. WATER RESOURCES

The world was blank, white and unformed. In the system of water, clouds are a bucket brigade, not storage. The atmosphere around the planet carries only about a ten day supply of fresh water – about one inch of rain. Each day on earth almost 250 cubic miles of water evaporates from the sea and the land. Its stay in the air is short; it is always seeking particles to stick to and fall with as rain or snow.

Climatic and environmental changes and a rising water demand have increased the competition over water resources and have made cooperation between countries that share a transboundary river an important issue in water resources management and hydropolitics. Yet, in river basins around the world, international conflict and cooperation are influenced by different factors and general conclusions about forces driving conflict and cooperation have been difficult to draw. Rivers are an essential natural resource closely linked to a country's wellbeing and economic success. But rivers ignore political boundaries and competition over the water resources has lead to political tension between countries with transboundary rivers. Integrating international cooperation and conflict resolution into the water management of transboundary rivers has therefore become an important issue in water resources management and hydropolitics. The problem requires a good understanding of the history and patterns of conflict and cooperation among nations sharing international basins worldwide and of the different factors that have influenced their international relations.

Increasing water scarcity in the downstream areas of several river basins demands improved water management and conservation in upper reaches. Improved management is impossible without proper monitoring at various levels. It is well known that all existing sewage treatment plants are overloaded. Hence, the treated effluents do not comply with international effluent quality guidelines. The main reasons behind this are:

- Weak management and absence of environmental awareness.
- Public-sector institutional problems.
- Failure in process design, construction and operation.
- Lack of skilled operating staff and insufficient monitoring programmes.
- Poor maintenance and weak financial resources.
- Low level of public involvement and lack of financial commitment.

4. WIND ENERGY

Wind energy is one of the fastest growing industries nowadays. The development in wind turbine (WT) technology is not limited to the significant increase in the size of the modern units, but also includes the high reliability and availability of the current machine. Therefore, a great competition among the manufactures established on the market and newcomers in the field is witnessed nowadays. A rapid development in the wind energy technology has made it alternative to conventional energy systems in recent years. Parallel to this development, wind energy systems (WES) have made a significant contribution to daily life in developing countries, where one third of the world's people live without electricity [2].

Many developing nations need to expand their power systems to meet the demand in rural areas. However, extending central power systems to remote locations is too costly an option in most cases. Then, autonomous small-scale energy systems can meet the electricity demand in remote locations, even though they generate relatively little power. However, even little electricity would contribute greatly to the quality of life in some places of developing countries. Being one of the most promising autonomous power technologies, wind energy applications, in the power range from tens of Watts to kilowatts, are increasingly growing in rural areas of developing countries. Technical and economical aspects of WESs should further be improved to sustain this growth. Techno-economically optimal designs are crucial for wind systems in competing with the conventional and more reliable power systems. High performance at the lowest possible cost will encourage the use of such systems and lead to more cost effective systems gradually (Figure 1). Design tools, allowing system performance assessment over a certain period of time, are therefore of great importance for sizing and optimisation purposes.

Remote rural electrification projects in poorer parts of the world used to be achieved with the use of diesel engine generators. These are increasingly being replaced with decentralised, on-site stand alone and renewable energy-based hybrid power systems. Roughly 1.4 billion people in rural areas, mainly within developing countries, live without electricity. Rural electrification is therefore an issue that should be high on rural development agendas. Renewable energy technologies (RETs) have an important rule to play in rural areas in terms of the suitability and cost competitiveness of the existing technological solutions and from an environmental point of view. Climate change will affect everyone, but it is expected to have a grater impact on those living in poverty in developing countries as a result of changes in rainfall patterns, increased frequency and severity of floods, droughts, storms, heat waves, changes in water-quality and quantity, sea level rise and glacial melt.

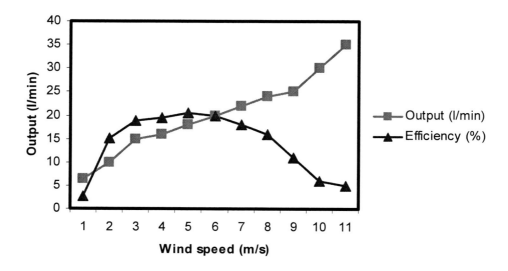

Figure 1. Performance of the wind pump.

Wind power now accounts for the dominant share of global investment in renewable energy. Total wind power capacity grew by 28% worldwide in 2007 to reach an estimated 95 GW. Annual capacity additions by market size increased even more: 40% higher in 2007 compared to 2006. Wind markets have also become geographically broad, with capacity in over 70 countries. Even as turbine prices remained high, due in part to materials costs and supply-chain troubles, the industry saw an increase in manufacturing facilities in the United States, India and China, broadening the manufacturing base away from Europe with the growth of more localised supply chains. India has been exporting components and turbines for many years and it appeared that 2006 and 2007 marked a turning point for China as well, with deals announced for the export of Chinese turbines and components. The annual energy yield is calculated by multiplying the wind turbine power curve with the wind distribution function at the site:

$$E_y = \sum_{i=1}^{i=n} f_{wi} P_{wi} \qquad\qquad (1)$$

Where:

E_y is annual energy yield in kWh.

w is the wind speed in m/s.

n is the number of data bins converting the wind speed range of the turbine (0.5 or 1 m/s intervals).

f_{wi} is the number of hours per year for which wind speed is w m/s.

P_{wi} is the power resulting from a wind speed of w m/s.

Based on power curve from Figure 2 and the Weibull wind speed distribution, with a shape factor of 2, and the gross energy yield corresponding to 7-8.5 m/s is 10 MW.

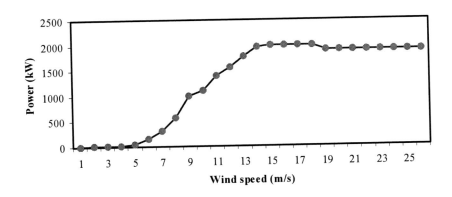

Figure 2. A power/wind speed curve.

Unchanging for all wind turbines- big or small- is a number of crucial factors that together determine the annual energy-generating potential in kWh/m^2 of rotor swept area. Key factors that impact potential energy yield and their physical relationships are expressed in the formula:

$$P = \tfrac{1}{2}\,\rho\,C_p\,\eta_{me}\,\eta_{el}\,V^3\,A \qquad (2)$$

Where:

P is the wind turbine power performance fed into the grid (Watts).

C_p is an aerodynamic efficiency of conversion of wind power into mechanical power, often called the power coefficient.

η_{me} is the conversion efficiency of mechanical power in the rotor axis into mechanical power in the generator axis. Encompasses all combined losses in the bearings, gearbox and so on.

η_{el} is the conversion efficiency of mechanical power into electric power fed into the grid, encompassing all combined losses in the generator, frequency converter, transformer, switches, etc.

ρ is the air density in kg/m^3 depends on environmental conditions.

V is the wind speed some three rotor diameters upwind from the rotor plane in m/s.

A is the rotor swept area in m^2.

Each of the elements of the performance formula has its own distinct contribution to total wind turbine power output and resulting yearly energy yield. Traditionally wind turbines applied in an open field are horizontal-axis designs fitted with an upwind rotor. In the operational output range, wind power generated increases with wind speed cubed. Rotor swept area is a function of the rotor diameter squared and is the second key wind turbine output variable. The Boyle-Gay-Laussac Law shows the impact of temperature and pressure on density, whereby density is proportional to pressure divided by temperature. The influence of air density on wind turbine performance is therefore limited.

5. ENERGY EFFICIENCY IN BUILDINGS

The world population is rising rapidly, notably in the developing countries. Historical trends suggest that increased annual energy use per capita is a good surrogate for the standard of living factors which promote a decrease in population growth rate. The term 'low energy' means achieving 'zero energy' requirements for a house or reduced energy consumption in an office (Figure 3). The main elements of energy concept are typical passive house components:

- Thermal bridges reduced to a minimum.
- Triple glazed windows with adequate frame and an optimised installation.
- A ventilation system with highly efficient heat recovery installed.
- Thermal solar collectors installed covering up 60% of the annual energy demand for domestic hot water.
- Excellent insulation level of opaque building elements: u-values range from 0.10 W/m^2K for walls and roof to 0.18 W/m^2K for basement ceilings.
- Highly efficient condensing gas boilers were installed; where possible, ducts have been insulated to a very good level; in other projects biomass boilers have been successfully tested.
- The air-tightness was improved by a factor of 6-10; the limiting value for new passive houses was achieved.

Compact development patterns can reduce infrastructure demands and the need to travel by car. As population density increases, transportation options multiply and dependence areas, per capita fuel consumption is much lower in densely populated areas because people drive so much less. Few roads and commercially viable public transport are the major merits. On the other hand, urban density is a major factor that determines the urban ventilation conditions, as well as the urban temperature. Under given circumstances, an urban area with a high density of buildings can experience poor ventilation and strong heat island effect. In warm-humid regions these features would lead to a high level of thermal stress of the inhabitants and increased use of energy in air-conditioned buildings.

The admission of daylight into buildings alone does not guarantee that the design will be energy efficient in terms of lighting. In fact, the design for increased daylight can often raise concerns relating to visual comfort (glare) and thermal comfort (increased solar gain in the summer and heat losses in the winter from larger apertures). Such issues will clearly need to be addressed in the design of the window openings, blinds, shading devices, heating system, etc. In order for a building to benefit from daylight energy terms, it is a prerequisite that lights are switched off when sufficient daylight is available. The nature of the switching regime; manual or automated, centralised or local, switched, stepped or dimmed, will determine the energy performance. In summary, achieving low energy building requires comprehensive strategy that covers; not only building designs, but also considers the environment around them in an integral manner.

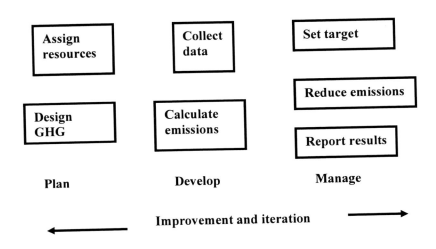

Figure 3. Shows office buildings can fight global warming.

However, it is also possible that a high-density urban area, obtained by a mixture of high and low buildings, could have better ventilation conditions than an area with lower density but with buildings of the same height. Closely spaced or high-rise buildings are also affected by the use of natural lighting, natural ventilation and solar energy. If not properly planned, energy for electric lighting and mechanical cooling/ventilation may be increased and application of solar energy systems will be greatly limited. Table 1 gives a summary of the positive and negative effects of urban density. All in all, denser city models require more careful design in order to maximise energy efficiency and satisfy other social and development requirements. Low energy design should not be considered in isolation, and in fact, it is a measure, which should work in harmony with other environmental objectives. Hence, building energy study provides opportunities not only for identifying energy and cost savings, but also for examining the indoor and outdoor environment.

Energy efficiency and renewable energy programmes could be more sustainable and pilot studies more effective and pulse releasing if the entire policy and implementation process was considered and redesigned from the outset. New financing and implementation processes are needed which allow reallocating financial resources and thus enabling countries themselves to achieve a sustainable energy infrastructure. The links between the energy policy framework, financing and implementation of renewable energy and energy efficiency projects have to be strengthened and capacity building efforts are required.

Today, the challenge before many cities is to support large numbers of people while limiting their impact on the natural environment. There are also a number of methods, which help reduce the lighting energy use, which, in turn, relate to the type of occupancy pattern of the building [6]. The light switching options include:

- Centralised timed off (or stepped)/manual on.
- Photoelectric off (or stepped)/manual on.
- Photoelectric and on (or stepped), photoelectric dimming.
- Occupant sensor (stepped) on/off (movement or noise sensor).
-

Table 1. Effects of urban density on city's energy demand.

Positive effects	Negative effects
Transport: Promote public transport and reduce the need for, and length of, trips by private cars. Infrastructure: Reduce street length needed to accommodate a given number of inhabitants. Shorten the length of infrastructure facilities such as water supply and sewage lines, reducing the energy needed for pumping. Thermal performance: Multi-story, multiunit buildings could reduce the overall area of the building's envelope and heat loss from the buildings. Shading among buildings could reduce solar exposure of buildings during the summer period. Energy systems: District cooling and heating system, which is usually more energy efficiency, is more feasible as density is higher. Ventilation: A desirable flow pattern around buildings may be obtained by proper arrangement of high-rise building blocks.	Transport: Congestion in urban areas reduces fuel efficiency of vehicles. Vertical transportation: High-rise buildings involve lifts, thus increasing the need for electricity for the vertical transportation. Ventilation: A concentration of high-rise and large buildings may impede the urban ventilation conditions. Urban heat island: Heat released and trapped in the urban areas may increase the need for air conditioning. The potential for natural lighting is generally reduced in high-density areas, increasing the need for electric lighting and the load on air conditioning to remove the heat resulting from the electric lighting. Use of solar energy: Roof and exposed areas for collection of solar energy are limited.

6. SOLAR ENERGY

Policies for solar hot water have grown substantially in recent years. In particular, mandates for solar hot water in new construction represent a recent trend at both national and local levels. Large-scale, conventional, power plant such as hydropower has an important part to play in development. It does not, however, provide a complete solution. There is an important complementary role for the greater use of small-scale, rural based, power plant. Such plant can be used to assist development since it can be made locally using local resources, enabling a rapid built-up in total equipment to be made without a corresponding and unacceptably large demand on central funds. Renewable resources are particularly suitable for providing the energy for such equipment and its use is also compatible with the long-term aims. It is possible with relatively simple flat plate solar collectors (Figure 4) to provide warmed water and enable some space heating for homes and offices which is particularly useful when the buildings are well insulated and thermal capacity sufficient for the carry over of energy from day to night is arranged. Energy efficiency is related to the provision of the desired environmental conditions while consuming the minimal quantity of energy. The sources to alleviate the energy situation in the world are sufficient to supply all foreseeable needs. Conservation of energy and rationing in some form will however have to be practised by most countries, to reduce oil imports and redress balance of payments positions. Meanwhile development and application of nuclear power and some of the traditional solar, wind and water energy alternatives must be set in hand to supplement what remains of the fossil fuels. The encouragement of greater energy use is an essential component of development.

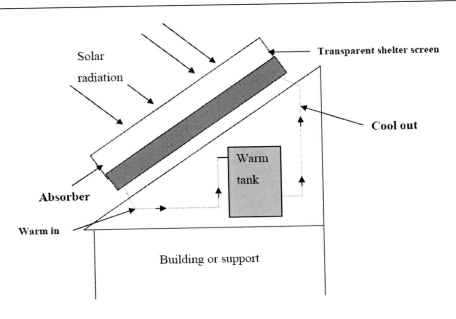

Figure 4. Solar water warmer.

There are two key elements to the fulfilling of renewable energy technology potential within the field of building design; first the installation of appropriate skills and attitudes in building design professionals and second the provision of the opportunity for such people to demonstrate their skills. This second element may only be created when the population at large and clients commissioning building design in particular, become more aware of what can be achieved and what resources are required. Terms like passive cooling or passive solar use mean that the cooling of a building or the exploitation of the energy of the sun is achieved not by machines but by the building's particular morphological organisation. Hence, the passive approach to themes of energy savings is essentially based on the morphological articulations of the constructions. Passive solar design, in particular, can realise significant energy and cost savings.

The encouragement of greater energy use is an essential component of development. In the short term it requires mechanisms to enable the rapid increase in energy/capita, and in the long term we should be working towards a way of life, which makes use of energy efficiency and without the impairment of the environment or of causing safety problems. Such a programme should as far as possible be based on renewable energy resources. As with any market, the benefit of experience is invaluable in the successful implementation of a growth strategy (Figure 5). The large-scale uptake of photovoltaics in urban areas potentially represents a vast market area that could be developed under the right conditions. A wide range of countries, project stages and stakeholders have been involved and this has led to the collection of a comprehensive set of lessons learnt and successful methods of promoting the implementation of photovoltaic (PV) within the urban planning process:

- Setting the stage- the impact of planning policy on renewables within urban areas.
- Implementation- from financing to design to construction.
- Occupation- when the real success of otherwise of a project can be seen.

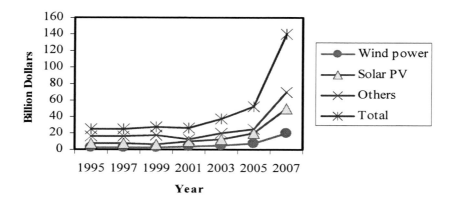

Figure 5. Annual investments in new renewable energy capacity 1995-2007 [3].

In many central European countries energy consumption for heating and domestic hot water causes around one third of national CO_2 emissions. For this reason the reduction of energy demand from buildings play an important role in efforts to control anthropogenic greenhouse gas emissions.

7. BIOMASS ENERGY

Biofuels are emerging in a world increasingly concerned by the converging global problems of rising energy demands, accelerating climate change, high priced fossil fuels, soil degradation, water scarcity and loss of biodiversity. For instance, the Intergovernmental Panel on Climate Change (IPCC) identified that in order to avoid more than an acceptable maximum 2-2.4°C rise in mean global temperature, greenhouse gas emissions will need to peak around 2015 and be reduced well below 50% of 2000 level by 2050. A lower figure is needed, which cannot be achieved by emissions reduction alone. Hence, there is a need for carbon removals, giving rise to enhanced supplies of biomass raw material and the potential of biofuels-related investments to show a profit from biofuels sales revenues.

Many nations have the ability to produce their own biofuels derived both from agricultural and forest biomass and from urban wastes, subject to adequate capacity building, technology transfer and access to finance. Trade in biofuels surplus to local requirements can thus open up new markets and stimulate the investment needed to promote the full potential of many impoverished countries. Such a development also responds to the growing threat of passing a tipping point in climate system dynamics. The urgency and the scale of the problem are such that the capital investment requirements are massive and more typical of the energy sector than the land use sectors. The time line for action is decades, not centuries, to partially shift from fossil carbon to sustainable biomass. Figure 6 shows the contribution of biomass to global primary and consumer energy supplies. Food and fodder availability is very closely related to energy availability.

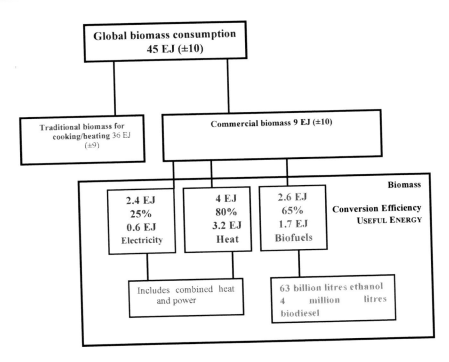

Figure 6. Contribution of biomass to global primary and consumer energy supplies.

8. APPLICATIONS

There are various successful applications of renewable technologies:

8.1. Water Management

The current global usage of water identified to be 75% of an overall water consumption for agricultural purposes [4]. Excess heat and scarcity of water are the two major problems, which are usually encountered in irrigating land especially in the arid and semi-arid regions. It is predicted that a higher future demand on crops would take place as a result of a rapid increase in the world's population leading to a serious dilemma on the global water resources in the future. Furthermore, with the current environmental problems (i.e., global warming), many plants and trees started to be exposed to additional amount of heat, which resulted in hindering their growth. The current global requirements for optimising water usage for the agricultural purposes especially in the arid and semi-arid regions, considered being the most demanding task. Irrigation in these lands, encounters two major obstacles, which are lack of water and access heat that could damage plants and prevent them from a healthy growth and attainment of a maximum production of crops. There are numerous advantages that could be associated with proper water and heat management of agricultural lands especially in the arid and semi-arid regions. Beside water is being conserved; optimum water usage for irrigation at different meteorological conditions could result in a healthy growth and optimum production of crops. The amount of heat the plant is exposed to, could play a major role in the plant's growth and sometimes proper shading could be necessary to maintain a productive land. In general, a greater economical saving could be achieved if appropriate steps are taken to manage heat and water in these lands in order to attain a better production of crops.

8.2. Greenhouses Environment

Greenhouse cultivation is one of the most absorbing and rewarding forms of gardening for anyone who enjoys growing plants. The enthusiastic gardener can adapt the greenhouse climate to suit a particular group of plants, or raise flowers, fruit and vegetables out of their natural season. The greenhouse can also be used as an essential garden tool, enabling the keen amateur to expand the scope of plants grown in the garden, as well as save money by raising their own plants and vegetables. There was a decline in large private greenhouses during the two world wars due to a shortage of materials for their construction and fuel to heat them. However, in the 1950s mass-produced, small greenhouses became widely available at affordable prices and were used mainly for raising plants [5]. Also, in recent years, the popularity of conservatories attached to the house has soared. Modern double-glazing panels can provide as much insulation as a brick wall to create a comfortable living space, as well as provide an ideal environment in which to grow and display tender plants.

The comfort in a greenhouse depends on many environmental parameters. These include temperature, relative humidity, air quality and lighting. Although greenhouse and conservatory originally both meant a place to house or conserve greens (variegated hollies, cirrus, myrtles and oleanders), a greenhouse today implies a place in which plants are raised while conservatory usually describes a glazed room where plants may or may not play a significant role. Indeed, a greenhouse can be used for so many different purposes. It is, therefore, difficult to decide how to group the information about the plants that can be grown inside it.

Population growth and less availability of food material have become global concerns. The world population increases exponentially whereas food production has increased only arithmetically, meaning that the availability of food per capita has decreased. Throughout the world urban areas have increased in size during recent decades. About 50% of the world's population and approximately 76% in the more developed countries are urban dwellers [6]. Even though there is an evidence to suggest that in many 'advanced' industrialised countries there has been a reversal in the rural-to-urban shift of populations. Virtually all population growth expected between 2000 and 2030 will be concentrated in urban areas of the world. With an expected annual growth of 1.8%, the world's urban population will double in 38 years [6]. This represents a serious contributing to the potential problem of maintaining the required food supply. Inappropriate land use and management, often driven by intensification resulting from high population pressure and market forces, is also a threat to food availability for domestic, livestock and wildlife use. Conversion to cropland and urban-industrial establishments is threatening their integrity. Improved productivity of peri-urban agriculture can, therefore, make a very large contribution to meeting food security needs of cities as well as providing income to the peri-urban farmers. Hence, greenhouses agriculture can become an engine of pro-poor 'trickle-up' growth because of the synergistic effects of agricultural growth such as [6]:

- Increased productivity increases wealth.
- Intensification by small farmers raises the demand for wage labour more than by larger farmers.
- Intensification drives rural non-farm enterprise and employment.
- Alleviation of rural and peri-urban poverty is likely to have a knock-on decrease of urban poverty.

Despite arguments for continued large-scale collective schemes there is now an increasingly compelling argument in favour of individual technologies for the development of controlled greenhouses. The main points constituting this argument are summarised by [6] as follows:

- Individual technologies enable the poorest of the poor to engage in intensified agricultural production and to reduce their vulnerability.
- Development is encouraged where it is needed most and reaches many more poor households more quickly and at a lower cost.
- Farmer-controlled greenhouses enable farmers to avoid the difficulties of joint management.

Such development brings the following challenges:

- The need to provide farmers with ready access to these individual technologies, repair services and technical assistance.
- Access to markets with worthwhile commodity prices, so that sufficient profitability is realised.
- This type of technology could be a solution to food security problems. For example, in greenhouses, advances in biotechnology like the genetic engineering, tissue culture and market-aided selection have the potential to be applied for raising yields, reducing pesticide excesses and increasing the nutrient value of basic foods.

However, the overall goal is to improve the cities in accordance with the Brundtland Report [7] and the investigation into how urban green could be protected. Indeed, greenhouses can improve the urban environment in multitude of ways. They shape the character of the town and its neighborhoods, provide places for outdoor recreation, and have important environmental functions such as mitigating the heat island effect, reduce surface water runoff, and creating habitats for wildlife. Following analysis of social, cultural and ecological values of urban green, six criteria in order to evaluate the role of green urban in towns and cities were prescribed [7]. These are as follows:

- Recreation, everyday life and public health.
- Maintenance of biodiversity - preserving diversity within species, between species, ecosystems, and of landscape types in the surrounding countryside.
- City structure - as an important element of urban structure and urban life.

- Cultural identity - enhancing awareness of the history of the city and its cultural traditions.
- Environmental quality of the urban sites - improvement of the local climate, air quality and noise reduction.
- Biological solutions to technical problems in urban areas - establishing close links between technical infrastructure and green-spaces of a city.

The main reasons why it is vital for greenhouses planners and designers to develop a better understanding of greenhouses in high-density housing can be summarised as follows [7]:

- Pressures to return to a higher density form of housing.
- The requirement to provide more sustainable food.
- The urgent need to regenerate the existing, and often decaying, houses built in the higher density, high-rise form, much of which is now suffering from technical problems.

The connection between technical change, economic policies and the environment is of primary importance as observed by most governments in developing countries, whose attempts to attain food self-sufficiency have led them to take the measures that provide incentives for adoption of the Green Revolution Technology [8]. Since, the Green Revolution Technologies were introduced in many countries actively supported by irrigation development, subsidised credit, fertiliser programmes, self-sufficiency was found to be not economically efficient and often adopted for political reasons creating excessive damage to natural resources. Also, many developing countries governments provided direct assistance to farmers to adopt soil conservation measures. They found that high costs of establishment and maintenance and the loss of land to hedgerows are the major constraints to adoption [8]. The soil erosion problem in developing countries reveals that a dynamic view of the problem is necessary to ensure that the important elements of the problem are understood for any remedial measures to be undertaken. The policy environment has, in the past, encouraged unsustainable use of land [8]. In many regions, government policies such as provision of credit facilities, subsidies, price support for certain crops, subsidies for erosion control and tariff protection, have exacerbated the erosion problem. This is because technological approaches to control soil erosion have often been promoted to the exclusion of other effective approaches. However, adoption of conservation measures and the return to conservation depend on the specific agro-ecological conditions, the technologies used and the prices of inputs and outputs of production.

8.2.1. Heat Balance

The greenhouse effect is one result of the differing properties of heat radiation generated at different temperatures. The high temperature sun emits radiation of short wavelength, which can pass through the atmosphere and through glass. Objects inside the greenhouse (or any other building), such as plants, absorb and then re-radiate this heat. Because the objects inside the greenhouse are at a lower temperature than the sun the radiated heat is of longer wavelengths and, hence, cannot re-penetrate the glass. This re-radiated heat is therefore trapped and causes the temperature inside the greenhouse to rise [9]. It has been observed that there is a significant rise of between 2.5 and 15°C in the enclosed room air temperature of a controlled environment greenhouse as reported by various authors [10-21]. This leads to a thermal energy saving for heating a greenhouse and maintaining a favourable environment for crop production during the off-season of up to 75%. At a certain prescribed depth an imposed temperature equal to a seasonal average ambient temperature, it is possible to identify the main parameters influencing the thermal behaviour of a greenhouse, which lead to a better understanding of the different processes inside the greenhouse.

Using the above concept with synthetic meteorological data (e.g., solar radiation, ambient temperature, wind speed, relative humidity, ground temperature, thermal and physical properties of the ground) will allow the prediction of energy consumption during a whole season in order to maintain the required inside air conditions for plant growth. Further modifications and introducing the more complex problem of plant growth, will allow programming the greenhouse for specific plants, thus improving the economic rentability of very specific types of solar collector.

8.2.2. Ventilation

Whereas heat loss in winter is a problem, it can be a positive advantage when greenhouse temperatures soar considerably above outside temperatures in summer. Table 2 illustrates typical greenhouse temperatures and gives an idea of temperature variation in a greenhouse. Therefore, ventilation is required to maintain the temperature at a level that plants can thrive, and to remove the still, humid air, which encourages the development of diseases. This can be achieved by natural ventilation, where warm air escapes and cool air enters through vents in the greenhouse roof (rigid vents), or by forced air ventilation, where a motorised fan designed specifically for greenhouses sucks warm air out of the greenhouse and pulling cool air in through openings on the opposite side.

The addition of side vents, which are optional extras on many small greenhouses, will, also, provide a more rapid movement of air. Cool air is drawn in from the side vents. As it heats up, it rises until it drawn out of the ridge vents. If only top ventilation is fitted, care needs to be taken not to open the vents too rapidly as rising warm air, particularly on cold days, will be quickly replaced by a block of cold air, which can prove a shock to plants. Where additional vents are available, these are best positioned on sidewalls. Table 3 gives typical ventilation space and vent numbers required for different greenhouses. Louvers will give adequate side ventilation and are less likely to cause draughts than standard ventilations [22]. Further, the door opening can provide additional ventilation, which is particularly valuable in small greenhouses. However, this is not always desirable where security is a problem. Generally, ventilators should be provided on all sides of the greenhouse, so that on windy days the windward side can be closed to prevent draughts, while ventilation is opened on the leeward end side. The temperature in small greenhouses can rise rapidly, and, hence, requires continuous monitoring and control. Note that manual control may lead to wide fluctuations in temperature. However, sufficient ventilation in greenhouse is essential for healthy plant growth. Hence, warm air gathering inside the roof may need to be released through a mop fan and replaced by cooler air. To maximise production and meet the global demand on food, vegetables, flowers and horticultural crops, it is necessary to increase the effective production span of crops. The sun is the source of energy for plants and animals.

This energy is converted into food (i.e., carbohydrates) by plants through a process called photosynthesis. This process is accomplished at suitable atmospheric conditions. These conditions are provided by nature in different seasons and artificially by a greenhouse.

8.2.3. Relative Humidity

Air humidity is measured as a percentage of water vapour in the air on a scale from 0% to 100%, where 0% being dry and 100% being full saturation level. The main environmental control factor for dust mites is relative humidity. The followings are the practical methods of controlling measures available for reducing dust mite populations:

- Chemical control.
- Cleaning and vacuuming.
- Use of electric blankets, and
- Indoor humidity.

Table 2. Typical greenhouse temperatures

Minimum winter temperature	Conditions
4°C	Frost-free
10°C	Temperate
15°C	Tropical

Table 3. Ventilation needs

Greenhouse size	Ground area	Ridge ventilators space required	Number of 0.6 m x 0.6 m ridge-vents required
1.8 m x 2.5 m	4.5 m^2	0.9 m^2	3
2.5 m x 3 m	7.5 m^2	1.5 m^2	4
3 m x 3.7 m	11.1 m^2	2.2 m^2	6
3.7 m x 4.3 m	15.9 m^2	3.2 m^2	9

Indoor relative humidity control is one of the most effective long-term mite control measures. There are many ways in which the internal relative humidity can be controlled including the use of appropriate ventilation, the reduction of internal moisture production and maintenance of adequate internal temperatures through the use of efficient heating and insulation [22]. Plants, usually, require a humidity level of 50%-60%, and succulents require 35%-40% [23]. For small greenhouses, simple, automatic controls would create ideal conditions for plants and ensure that ventilation is provided only when required. This prevents the rapid fluctuations in temperature and higher fuel bills that can occur with manual control. For larger ones, however, more sophisticated controls may need to be implemented.

As the sun comes through the glass and gives the plants the energy they need to grow, it will, also, quickly raise the temperature. With no shade and inadequate ventilation, temperatures up to 35°C can be reached very quickly. This is uncomfortable for both plants and people, although plants can tolerate high temperatures, if provided with good ventilation. Also, to allow the carbon dioxide needed for photosynthesis into the plant, the small pores in the leaf, called stomata, must be open and so water will be lost into the air. Keeping the atmosphere humid will help, but eventually the plant will have to shut its stomata, and then growth will come to a stop. All plants have a minimum, optimum and maximum temperature for growth. During the time of the year when they are actively growing, a temperature as close as possible to the optimum is preferred.

8.2.4. The Ideal Living Area

Most people would maintain a steady temperature between 15.5°C and 21°C. For comfort, the level of atmospheric humidity should be low and indeed living areas should, generally, have a dry atmosphere [23]. Hence, artificial heating in autumn, winter and spring will, usually, be required. Plants cannot live comfortably in an area of high humidity like human beings. Also, most plants do not like the combination of high temperatures and dry air. Tropical and sub-tropical plants, for instance, would make very poor growth in such environment and the leaves may shrive and dry up, or turn brown at the edges. Table 4, which is reproduced from [23], classifies plants according to their climate preferences. The table, also, indicates that there are many plants that can be grown in a humid 'microclimate'. During the summer the greenhouse is generally warm and plants are sizzling. Good thermometers are good troubleshooters and enable the greenhouse operator to monitor the shifts in temperature and take appropriate action when necessary. Also, greenhouses can be very arid places in summer. Most greenhouse plants prefer a slightly humid atmosphere. A group of plants growing together will create heir own, more humid microclimate. Plants from the Mediterranean are a good choice as they are naturally adapted to hot and dry summer.

8.2.5. Transpiration

Transpiration is a process used by plants to maintain their temperature when the environment is too warm. It serves the same functions as sweating in some animals. In order to cool the area around it, plants release water through the leaves. High rates of transpiration will occur if temperature of the growing environment is too high, and this can seriously affect the nutrient solution. As a plant transpires, it will draw up more water from the solution to replace what has been lost. This will have an impact on the water ratio of the nutrient solution, which will affect both pH (potential of Hydrogen) and the level of conductivity of these nutrients from soil to the plant. Maintaining a basic nutrient solution of pH, which, in turn, affects how easy the plant, can take in the nutrient solution. In a good growing environment, the quantity of growth and yield primarily depends on the quality of the nutrient solution. A nutrient solution contains [24]:

- Water.
- A nutrient concentrate, which provides the basic food for plants.
- One or more nutrient additives, which provide supplementary nutrients, required to promoting specific processes within the plants development.
- Possibly pH solution.

However, the effective utilisation of greenhouses has to deal with some specific climate problems like frost, during winter and overheating in summer days. These problems show the necessity of having a tool capable of predicting the thermal behaviour of a greenhouse under specific exterior conditions. Also, greenhouse industry has to deal with some problems related to a poor design of a great number of greenhouses. Such problems are mostly related to, on the one hand, its incapacity to deal with the problem of frost, which in the cold clear sky days of winter can destroy the whole work of a season, and, on the other hand, the question of overheating in the summer days.

Table 4. Atmospheric conditions for plants [23].

Types/conditions	Warm conditions and high humidity	Warm conditions and thrive in a dry atmosphere	Cool
Shrubs	Aphelandra, Gardenia	Lantana, Cestrum	Acacia, Lantana
Climbers	Allamanda, Hoya	Plumbago	Jasminum, Plumbago
Flowering pot plants short-term	Capsicum, Celosia	Ipomoea	Solanum, Primula
Flowering pot plants long-term	Begonia, Euphorbia	Fuchsis, Cacti	Clivia, Erica
Foliage pot plants	Ferns, Ficus	Sansevieria	Coleus, Hedera
Bulbous and tuberous plants	Canna	Vallota, Gloriosa	Cyclamen, Lilium
Fruits	Citrus	-	Vitis Vinifera, Prunus Persica

Plants, like human beings, need tender loving care in the form of optimum settings of light, sunshine, nourishment, and water. Hence, the control of sunlight, air humidity and temperatures in greenhouses are the key to successful greenhouse gardening. The mop fan is a simple and novel air humidifier; which is capable of removing particulate and gaseous pollutants while providing ventilation. It is a device ideally suited to greenhouse applications, which require robustness, low cost, minimum maintenance and high efficiency. A device meeting these requirements is not yet available to the farming community. Hence, implementing mop fans aids sustainable development through using a clean, environmentally friendly device that decreases load in the greenhouse and reduces energy consumption. The effect of indoor (greenhouse) conditions and outdoor (ambient) conditions (temperature and relative humidity) on system performance is illustrated in Figure 7.

The off-season production of flowers and vegetables is the unique feature of the controlled environment greenhouse. Hence, greenhouse technology has evolved to create the favourable environment, or maintaining the climate, in order to cultivate the desirable crop the year round. The use of "maintaining the climate" concept may be extended for crop drying, distillation, biogas plant heating and space conditioning. The use of greenhouses is widespread. During the last 10 years, the amount of greenhouses has increased considerably to cover up to several hundred hectares at present. Most of the production is commercialised locally or exported. In India, about 300 ha of land are under greenhouse cultivation. On the higher side, however, it is 98600 ha in Netherlands, 48000 ha in China and 40000 ha in Japan [23]. This shows that there is a large scope to extend greenhouse technology for various climates.

8.2.6. Heat transfer

The total rate of heat transfer on a surface of a soil per unit area Q''_{total}, shown in Figure 8, is a combination of radiation heat transfer Q''_{rad}, convection heat transfer Q''_{conv} and heat transfer sue to evaporation Q''_{evap}, illustrated in the following equation:

$$Q''_{total} = Q''_{rad} + Q''_{conv} + Q''_{evap} \tag{3}$$

Where:

Q''_{rad} Can be expressed in terms of emissivity ε, surface temperature T_s and the surrounding temperature T_{surr}:

$$Q''_{rad} = \varepsilon \delta (T_s - T_{surr}) \tag{4}$$

For the rate of convection heat transfer Q''_{conv}, combinations of both natural and forced convection are considered especially for low air velocities. The vapour pressure of air far away from the watered-surface $P_{v,\infty}$, is a function of relative humidity φ and saturated water vapour pressure $P_{T\infty, sat}$ [25]:

$$P_{v,\infty} = \varphi \, P_{T\infty, sat} \tag{5}$$

Treating the water vapour and air an ideal gas, the densities of water vapour, dry air and their mixture at air-water interface and far from the surface are determined in the following equations:

At the surface:

$$\rho_{v,s} = P_{v,s}/R_v T_s \tag{6}$$

$$\rho_{a,s} = P_{a,s}/R_a T_s \tag{7}$$

$$\rho_s = \rho_{v,s} + \rho_{a,s} \tag{8}$$

And away from the surface:

$$\rho_{v,\infty} = P_{v,\infty}/R_v T_\infty \tag{9}$$

$$\rho_{a,\infty} = P_{a,\infty}/R_a T_\infty \tag{10}$$

$$\rho_\infty = \rho_{v,\infty} + \rho_{a,\infty} \tag{11}$$

$G_r / R_e^2 < 0.1$ forced convection

$< G_r / R_e^2 < 10$ mixed (forced + natural) convection   (12)

$10 < G_r / R_e^2$ natural convection

Where: G_r is Grashof number and R_e is Reynolds.

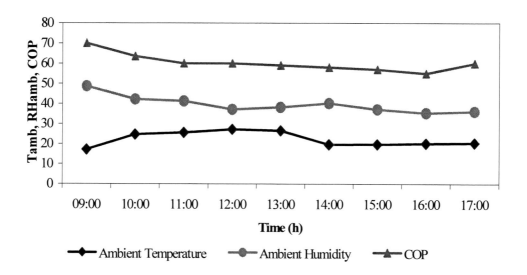

Figure 7. Ambient temperature, relative humidity and COP.

The comfort in a greenhouse depends on many environmental parameters. These include temperature, relative humidity, air quality and lighting. Although greenhouse and conservatory originally both meant a place to house or conserve greens (variegated hollies, cirrus, myrtles and oleanders), a greenhouse today implies a place in which plants are raised while conservatory usually describes a glazed room where plants may or may not play a significant role. Indeed, a greenhouse can be used for so many different purposes. It is, therefore, difficult to decide how to group the information about the plants that can be grown inside it. Whereas heat loss in winter a problem, it can be a positive advantage when greenhouse temperatures soar considerably above outside temperatures in summer. Indoor relative humidity control is one of the most effective long-term mite control measures. There are many ways in which the internal relative humidity can be controlled including the use of appropriate ventilation, the reduction of internal moisture production and maintenance of adequate internal temperatures through the use of efficient heating and insulation. The introduction of a reflecting wall at the back of a greenhouse considerably enhances the solar radiation that reaches the ground level at any particular time of the day. The energy yield of the greenhouse with any type of reflecting wall was also significantly increased.

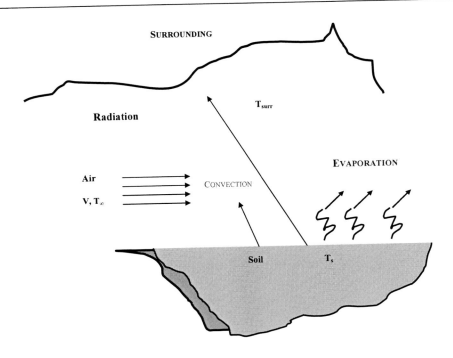

Figure 8. A combination of heat and mass transfer on the surface.

9. GROUND SOURCE HEAT PUMPS

Heat pumps function by moving (or pumping) heat from one place to another. Like a standard air-conditioner, a heat pump takes heat from inside a building and dumps it outside. The difference is that a heat pump can be reversed to take heat from a heat source outside and pump it inside. Heat pumps use electricity to operate pumps that alternately evaporate and condense a refrigerant fluid to move that heat. In the heating mode, heat pumps are far more "efficient" at converting electricity into usable heat because the electricity is used to move heat, not to generate it.

The most common type of heat pump- air-source heat pump- uses outside air as the heat source during the heating season and the heat sink during the air-conditioning season. Ground-source and water-source heat pumps work the same way, except that the heat source/sink is the ground, groundwater, or a body of surface water, such as a lake. For simplicity, water-source heat pumps are often lumped with ground-source heat pumps, as in this case (Figure 9).

The efficiency or coefficient of performance (COP) of ground-source heat pumps (GSHPs) is significantly higher than that of air-source heat pumps because the heat source is warmer during the heating season and the heat sink is cooler during the cooling season. GSHPs are also known as geothermal heat pumps.

GSHPs are environmentally attractive because they deliver so much heat or cooling energy per unit of electricity consumed. The COP is usually 3 or higher. The best GSHPs are more efficient than high-efficiency gas combustion, even when the source efficiency of the electricity is taken into account.

GSHPs are generally most appropriate for residential and small commercial buildings, such as small-town post offices. In residential and small (skin-dominated) commercial buildings, GSHPs make the most sense in mixed climates with significant heating and cooling loads because the high-cost heat pump replaces both the heating and air-conditioning system.

Because GSHPs are expensive to install in residential and small commercial buildings, it sometimes makes better economic sense to invest in energy efficiency measures that significantly reduce heating and cooling loads, then install less expensive heating and cooling equipment. The savings in equipment may be able to pay for most of the envelope improvements. If a GSHP is to be used, planning the site work and project scheduling needed so carefully that the ground loop can be installed with minimum site disturbance or in an area that will be covered by a parking lot or driveway.

GSHPs are generally classified according to the type of loop used to exchange heat with the heat source/sink. Most common are closed-loop horizontal and closed-loop vertical systems (Figure 9). Using a body of water as the heat source/sink is very effective, but seldom available as an option. Open-loop systems are less common than closed-loop systems due to performance problems (if detritus gets into the heat pump) and risk of contaminating the water source or, in the case of well water, inadequately recharging the aquifer. GSHPs are complex. Basically, water or a nontoxic antifreeze-water mix is circulated through buried polyethylene or polybutylene piping. This water is then pumped through one of two heat exchangers in the heat pump. When used in the heating mode, this circulating water is pumped through the cold heat exchanger, where its heat is absorbed by evaporation of the refrigerant. The refrigerant is then pumped to the warm heat exchanger, where the refrigerant is condensed, releasing heat in the process. This sequence is reversed for operation in the cooling mode.

Direct-exchange GSHPs use copper ground-loop coils that are charged with refrigerant. This ground loop thus serves as one of the two heat exchangers in the heat pump. The overall efficiency is higher because one of the two separate heat exchangers is eliminated, but the risk of releasing the ozone-depleting refrigerant into the environment is greater. Direct-exchange systems have a small market share.

An attractive alternative to conventional heating, cooling, and water heating equipment is the GSHP. The higher initial cost of this equipment must be justified by operating cost savings. Therefore, it is necessary to predict energy use and demand. However, there are no seasonal ratings for this type of equipment. The ratings for GSHPs calculate performance at a single fluid temperature (32°F) for heating COP and a second for cooling energy efficiency rating (EER) (77°F). These ratings reflect temperatures for an assumed location and ground heat exchanger type, and are not ideal indicators of energy use.

This problem is compounded by the nature of ratings for conventional equipment. The complexity and many assumptions used in the procedures to calculate the seasonal efficiency for air-conditioners, furnaces, and heat pumps (SEER, AFUE, and HSPF) make it difficult to compare energy use with equipment rated under different standards. The accuracy of the results is highly uncertain, even when corrected for regional weather patterns. These values are not indicators for demand since they are seasonal averages and performance at severe conditions is not heavily weighted.

The American Society of Heating, Refrigerating, and Air-Conditioning Engineers (ASHRAE) recommends a weather driven energy calculation, like the bin method, in preference to single measure methods like SEER, HSPF, EER, COP, and AFUE. The bin method permits the energy use to be calculated based on local weather data and equipment performance over a wide range of temperatures [26]. The bin method also calculates demand at the most severe conditions. This method was used to compare the energy use and demand of high efficiency equipment in Sacramento, California and Salt Lake City, Utah. The equipment considered was a high efficiency single speed air source, a variable speed air source heat pump and electric air-conditioner with a natural gas furnace, and a GSHP [26].

9.1. Heat Pump Principles

Heat flows naturally from a higher to a lower temperature. Heat pumps, however, are able to force the heat flow in the other direction, using a relatively small amount of high quality drive energy (electricity, fuel, or high-temperature waste heat). Heat pumps can transfer heat from natural heat sources such as the air, ground or water, to a building. By reversing the heat pump it can also be used for cooling. Heat is transferred in the opposite direction, from the application that is cooled, to surroundings at a higher temperature.

In order to transport heat external energy is needed to drive the heat pump. Theoretically, the total heat delivered by the heat pump is equal to the heat extracted from the heat source, plus the amount of drive energy supplied. Electrical powered heat pumps, for heating buildings, typically supply 100 kWh of heat with just 20-40 kWh of electricity. Because heat pumps consume less energy than conventional heating systems, there use will help to reduce the harmful emissions of carbon dioxide, sulpher dioxide and nitrogen oxides. However, the overall environmental impact of electric heat pumps depends very much on how the electricity is produced. Heat pumps driven by electricity generated by hydropower, wind power, photovoltaics or other renewable sources will reduce emissions more significantly than if the electricity is generated by coal, oil or gas-fired power plants.

The great majority of heat pumps work on the principle of the vapour compression cycle. The main components in such a heat pump are the compressor, the expansion valve and the two heat exchangers referred to as the evaporator and the condenser. A heat pump can take heat out of an interior space, or it can put heat into an interior space (Figures 10-11). A volatile liquid, known as the working fluid or refrigerant, circulates through the four components. In the evaporator the temperature of the refrigerant is kept lower than the temperature of the heat source. This allows heat to flow from the heat source (ground loops, air or loops in water, e.g., rivers etc.) to the refrigerant. As the refrigerant warms up it evaporates. This vapour is then compressed by the compressor to a higher pressure and temperature. The hot vapour then enters the condenser, where it condenses and gives off useful heat. Finally, the high-pressure working fluid is expanded to the evaporator pressure and temperature in the expansion valve. The refrigerant is returned to its original state and once again enters the evaporator.

The compressor is driven by an electric motor and pumps circulate the water through (ground loops, or loops in water, e.g., rivers etc.). The domestic fridge uses the same technology. When putting food and drink into fridge the low-grade heat it carries (after all it is usually warmer than the inside of the fridge) is transferred from the icebox to the refrigerant in the unit. The refrigerant is then compressed and expanded to raise the heat; this high grade heat is then expelled from the back of the fridge. This is why the inside of the fridge remains cold whilst the back of the fridge gets hot.

In the cooling mode, cool vapour arrives at the compressor after absorbing heat from the building. The compressor compresses the cool vapour into a smaller volume, increasing its heat density. The refrigerant exits the compressor as a hot vapour, which then goes into the earth loop field. The loops act as a condenser condensing the vapour until it is virtually all-liquid. The refrigerant leaves the earth loops as a warm liquid. The flow control regulates the flow from the condenser such that only liquid refrigerant passes through the control. The refrigerant expands as it exits the flow control unit and becomes a cold liquid.

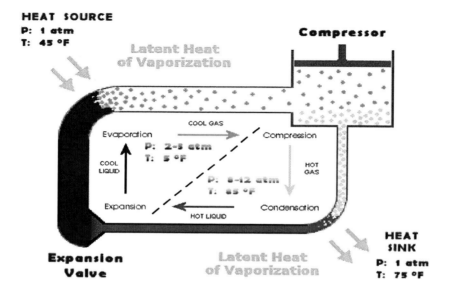

Figure 9. Heat pump works by promoting the evaporation and condensation of a refrigerant.

Ambient air is free and widely available, and it is the most common heat source for heat pumps. Air-source heat pumps, however, achieve on average 10-30% lower seasonal performance factor (SPF) than water-source heat pumps. This is mainly due to the rapid fall in capacity and performance with decreasing outdoor temperature, the relatively high temperature difference in the evaporator and the energy needed for defrosting the evaporator and to operate the fans. In mild and humid climates, frost will accumulate on the evaporator surface in the temperature range 0-6°C, leading to reduced capacity and performance of the heat pump system. Coil defrosting is achieved by reversing the heat pump cycle or by other, less energy-efficient means. Energy consumption increases and the overall coefficient of performance (COP) of the heat pump drops with increasing defrost frequency. Using demand defrosts control rather than time control can significantly improve overall efficiencies. Exhaust (ventilation) air is a common heat source for heat pumps in residential and commercial buildings. The heat pump recovers heat from the ventilation air, and provides water and/or space heating.

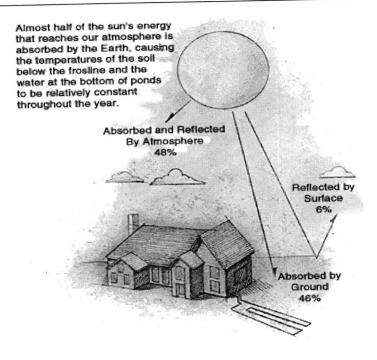

Almost half of the sun's energy that reaches our atmosphere is absorbed by the Earth, causing the temperatures of the soil below the frosline and the water at the bottom of ponds to be relatively constant throughout the year.

Absorbed and Reflected By Atmosphere 48%

Reflected by Surface 6%

Absorbed by Ground 46%

Figure 10. Earth heat.

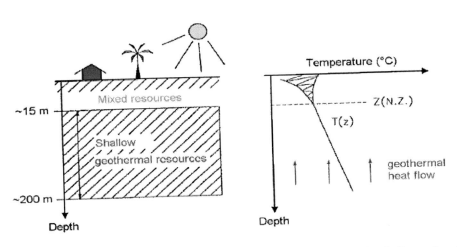

Figure 11. Geothermal energy, comprising geothermal and mixed resources in the shallow subsurface.

The term "ground source heat pump" has become an all-inclusive term to describe a heat pump system that uses the earth, ground water, or surface water as a heat source and/or sink. The GSHP systems consist of three loops or cycles as shown in Figure 12. The first loop is on the load side and is either an air/water loop or a water/water loop, depending on the application. The second loop is the refrigerant loop inside a water source heat pump. Thermodynamically, there is no difference between the well-known vapour-compression refrigeration cycle and the heat pump cycle; both systems absorb heat at a low temperature level and reject it to a higher temperature level. The difference between the two systems is that a refrigeration application is only concerned with the low temperature effect produced at the evaporator, while a heat pump may be concerned with both the cooling effect produced at the evaporator as well as the heating effect produced at the condenser. In these dual-mode GSHP systems, a reversing valve is used to switch between heating and cooling modes by reversing the refrigerant flow direction. The third loop in the system is the ground loop in which water or an antifreeze solution exchanges heat with the refrigerant and the earth.

The GSHPs utilise the thermal energy stored in the earth through either vertical or horizontal closed loop heat exchange systems buried in the ground. Many geological factors impact directly on site characterisation and subsequently the design and cost of the system. The solid geology of the United Kingdom varies significantly. Furthermore there is an extensive and variable rock head cover. The geological prognosis for a site and its anticipated rock properties influence the drilling methods and therefore system costs. Other factors important to system design include predicted subsurface temperatures and the thermal and hydrological properties of strata. GSHP technology is well established in Sweden, Germany and North America, but has had minimal impact in the United Kingdom space heating and cooling market. Perceived barriers to uptake include geological uncertainty, concerns regarding performance and reliability, high capital costs and lack of infrastructure. System performance concerns relate mostly to uncertainty in design input parameters, especially the temperature and thermal properties of the source. These in turn can impact on the capital cost, much of which is associated with the installation of the external loop in horizontal trenches or vertical boreholes. The temperate United Kingdom climate means that the potential for heating in winter and cooling in summer from a ground source is less certain owing to the temperature ranges being narrower than those encountered in continental climates. This project will develop an impartial GSHP function on the site to make available information and data on site-specific temperatures and key geotechnical characteristics.

The GSHPs are receiving increasing interest because of their potential to reduce primary energy consumption and thus reduce emissions of greenhouse gases. The technology is well established in North Americas and parts of Europe, but is at the demonstration stage in the United Kingdom. The information will be delivered from digital geoscience's themes that have been developed from observed data held in corporate records. These data will be available to GSHP installers and designers to assist the design process, therefore reducing uncertainties. The research will also be used to help inform the public as to the potential benefits of this technology.

Ground water is available with stable temperatures (4-10°C) in many regions. Open or closed systems are used to tap into this heat source. In open systems the ground water is pumped up, cooled and then reinjected in a separate well or returned to surface water. Open systems should be carefully designed to avoid problems such as freezing, corrosion and fouling. Closed systems can either be direct expansion systems, with the working fluid evaporating in underground heat exchanger pipes, or brine loop systems. Due to the extra internal temperature difference, heat pump brine systems generally have a lower performance, but are easier to maintain. A major disadvantage of ground water heat pumps is the cost of installing the heat source. Additionally, local regulations may impose severe constraints regarding interference with the water table and the possibility of soil pollution. Ground source systems are used for residential and commercial applications, and have similar advantages as (ground) water-source systems, i.e., they have relatively high annual temperatures.

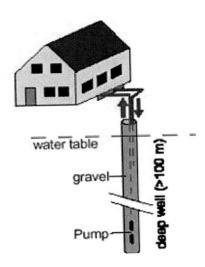

Figure 12. Standing column well.

Geothermal energy use has a net positive environmental impact. Geothermal power plants have fewer and more easily controlled atmospheric emissions than either fossil fuel or nuclear plants. Direct heat uses are even cleaner and are practically non-polluting when compared to conventional heating. Another advantage, which differentiates geothermal energy from other renewables, is its continuous availability, 24 hours a day all year round. While production costs are at times competitive and in other cases marginally higher than conventional energy, front-end investment is quite heavy and not easily funded.

9.2. Heat Pumps

A heat pump can take low temperature heat and upgrade it to a higher and more useful temperature (Figure 13). If this heat comes from an ambient source, for example outside air or the ground, the use of a heat pump can result in savings in fossil fuel consumption and thus a reduction in the emission of the GHGs and other pollutants. The GSHPs in particular are receiving increasing interest and the technology is now well established with over 550,000 units (80% of which are domestic) installed worldwide and over 66,000 installed annually [27]. Despite increasing use elsewhere, the GSHPs are a relatively unfamiliar technology in the UK although the performance of systems is now such that, properly designed and installed, they represent a very carbon-efficient form of space heating. Direct-exchange GSHPs use copper ground-loop coils that are charged with refrigerant. This ground loop thus serves as one of the two heat exchangers in the heat pump. The overall efficiency is higher because one of the two separate heat exchangers is eliminated, but the risk of releasing the ozone-depleting refrigerant into the environment is greater. Direct-exchange systems have a small market share. An attractive alternative to conventional heating, cooling, and water heating equipment is the GSHP. The higher initial cost of this equipment must be justified by operating cost savings. Therefore, it is necessary to predict energy use and demand. However, there are no seasonal ratings for this type of equipment.

Figure 13. A photograph showing the connection of heat pump to the ground source.

The GSHPs can be used to provide space and domestic water heating and, if required, space cooling to a wide range of building types and sizes. The provision of cooling, however, will result in increased energy consumption and the efficiently it is supplied. The GSHPs are particularly suitable for new build as the technology is most efficient when used to supply low temperature distribution systems such as underfloor heating. They can also be used for retrofit especially in conjunction with measures to reduce heat demand. They can be particularly cost effective in areas where mains gas is not available or for developments where there is an advantage in simplifying the infrastructure provided. This application will concentrate on the provision of space and water heating to individual dwellings but the technology can also be applied to blocks of flats or groups of houses.

10. BIOENERGY UTILISATION

The increased demand for gas and petroleum, food crops, fish and large sources of vegetative matter mean that the global harvesting of carbon has in turn intensified. It could be said that mankind is mining nearly everything except its waste piles. It is simply a matter of time until the significant carbon stream present in municipal solid waste is fully captured. In the meantime, the waste industry needs to continue on the pathway to increased awareness and better optimised biowaste resources. Optimisation of waste carbon may require widespread regulatory drivers (including strict limits on the landfilling of organic materials), public acceptance of the benefits of waste carbon products for soil improvements/crop enhancements and more investment in capital facilities. In short, a significant effort will be required in order to capture a greater portion of the carbon stream and put it to beneficial use.

From the standpoint of waste practitioners, further research and pilot programmes are necessary before the available carbon in the waste stream can be extracted in sufficient quality and quantities to create the desired end products. Other details need to be ironed out too, including measurement methods, diversion calculations, sequestration values and determination of acceptance contamination thresholds.

The internal combustion engine is a major contributor to rising CO_2 emissions worldwide and some pretty dramatic new thinking is needed if our planet is to counter the effects. With its use increasing in developing world economies, there is something to be said for the argument that the vehicles we use to help keep our inner-city environments free from waste, litter and grime should be at the forefront of developments in low-emissions technology. Materials handled by waste management companies are becoming increasingly valuable. Those responsible for the security of facilities that treat waste or manage scrap will testify to the precautions needed to fight an ongoing battle against unauthorised access by criminals and crucially, to prevent the damage they can cause through theft, vandalism or even arson. Of particular concern is the escalating level of metal theft, driven by various factors including the demand for metal in rapidly developing economies such as China and India.

10.1. Biogas Technology

Anaerobic digestion (AD) has, for some time, been considered an important technology in the treatment of waste and in the development of energy recovery solutions. Historically, many anaerobic digestion plants have tended to specialise in the treatment of manure or sludge. In today's market, the latest AD plants have to handle more complex substances and varying volume streams. As a result, the demands placed on this technology in terms of reliability, stability and robustness are significant. Also, significant is the potential contribution AD could make to solving our most pressing environmental concern- namely a reduction in the anthropogenic emission of GHGs. AD technology can reduce unwanted and uncontrolled emissions of methane by tapping the energy potential of this gas while reducing the volume of waste going to landfill. Anaerobic digestion is a biochemical process where, in the absence of oxygen, bacteria break down organic matter to produce biogas plus liquor and a fibre.

The biogas consists of 55-70% methane (CH_4) and 30-45% carbon dioxide (CO_2) and can be used to generate energy through a generator. The energy content of biogas is 20-25 MJm^3. Alternatively; the gas can be cleaned and then either compressed for use in vehicle transport (compressed natural gas) or injected into the gas distribution network. An average CH_4 yield per metric ton of treated waste (sludge, manure) ranges from 50-90 Nm^3 per ton and for municipal solid waste (MSW) the yield increases to 75-120 Nm^3 per ton. The liquid fraction, with a high nutrient content and the fibre fraction can be used as a soil improver. More modern plants have been developed to process MSW, industrial solid wastes and industrial wastewaters, but impurities and the varying content of lipids, proteins and carbohydrates can cause problems. These wastes can be characterised according to their COD concentration. COD refers to the total quantity of oxygen required for oxidation to carbon dioxide and water and is a measure of the organic content of the waste. COD loading rate is the daily quantity of organic matter, expressed COD, feed per m^3 digester volume per day, i.e., kg $COD/m^3/d$.

Some systems have been invented to process substrates with a minor COD concentration (<25 gO_2/litre raw material), for example:

- Up-flow anaerobic sludge blanket (UASB).
- Expanded granular sludge blanket (EGSB).
- Internal circulation (IC).

With a loading rate of ≥15 kg COD/m³ fermenter/d, it possessed a sharp differentiation to traditional biogas plants. The advantages of the system are the following:

- Prevention of foam and floating layers- therefore high loading rates.
- No chemical requirement, no pH regulation- therefore cost savings.
- Low hydraulic retention time- therefore low demand for fermenter volume.
- Intense contact between substrate and microorganisms- therefore high degradation rates and rapid gas production.
- No accumulation of settling sediments (e.g., sand) in the system- thus supporting continuous operation.

The organic matter was biodegradable to produce biogas and the variation show a normal methanogene bacteria activity and good working biological process as shown in Figure 14. There are a number of factors that will give rise to greater interest in technologies such as AD. These include:

- Growing energy costs and import dependency within many countries.
- Decreasing capacity for landfill.
- Increasing world energy demand, in particular in China and India.
- Climate change needing urgent reactions and activities.
- 45% of European soils suffering from low organic matter content and reduced fertility.
- The most practical environmental solution will be deriving energy from waste, not only municipal solid waste but also the residues industry.

Anaerobic digestion has significant potential for industries with organic waste streams, such as food processing, the paper and textile industry, pharmaceutical industry and biofuels production. Anaerobic digestion combines several advantages. As a technology it can be regarded as being 'CO$_2$ neutral' because there is no net addition of CO$_2$ to the atmosphere. It degrades waste while producing biogas and a fertiliser product that contains a high nutrient content (nitrogen, phosphorous and potassium), but in order for the full potential of the waste/organic substrate/input to be realised, it is vital that the waste management industry is able to develop markets for all the by-products. Biogas technology can not only provide fuel, but is also important for comprehensive utilisation of biomass forestry, animal husbandry, fishery, agricultural economy, protecting the environment, realising agricultural recycling, as well as improving the sanitary conditions, in rural areas. The introduction of biogas technology on wide scale has implications for macro planning such as the allocation of government investment and effects on the balance of payments.

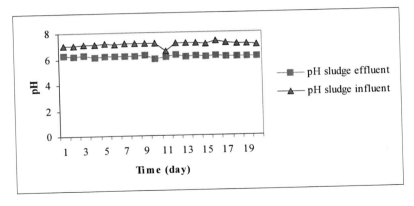

Figure 14. pH sludge before and after treatment in the digester.

10.2. Sewage Sludge

Sewage sludge is rich in nutrients such as nitrogen and phosphorous. It also contains valuable organic matter, useful for remediation of depleted or eroded soils. This is why untreated sludge has been used for many years as a soil fertiliser and for enhancing the organic matter of soil. A key concern is that treatment of sludge tends to concentrate heavy metals, poorly biodegradable trace organic compounds and potentially pathogenic organisms (viruses, bacteria and the like) present in wastewaters. These materials can pose a serious threat to the environment. When deposited in soils, heavy metals are passed through the food chain, first entering crops, and then animals that feed on the crops and eventually human beings, to whom they appear to be highly toxic. In addition they also leach from soils, getting into groundwater and further spreading contamination in an uncontrolled manner. European and American markets aiming to transform various organic wastes (animal farm wastes, industrial and municipal wastes) into two main by-products:

- A solution of humic substances (a liquid oxidate).
- A solid residue.

The key to successful future appears to lie with successful marketing of the treatment by products. There is also potential for using solid residue in the construction industry as a filling agent for concrete. Research suggests that the composition of the residue locks metals within the material, thus preventing their escape and any subsequent negative effect on the environment.

11. CONCLUSIONS

Newspapers, TV, schools, universities and politicians rant and rave about being 'green' and doing our bit for the environment, but can we as individuals change things? Energy efficiency brings health, productivity, safety, comfort and savings to homeowner, as well as local and global environmental benefits. The use of renewable energy resources could play an important role in this context, especially with regard to responsible and sustainable development. It represents an excellent opportunity to offer a higher standard of living to local people and will save local and regional resources. Implementation of greenhouses offers a chance for maintenance and repair services. It is expected that the pace of implementation will increase and the quality of work to improve in addition to building the capacity of the private and district staff in contracting procedures. The financial accountability is important and more transparent. Various passive techniques have been put in perspective, and energy saving passive strategies can be seen to reduce interior temperature and increase thermal comfort, reducing air conditioning loads. The scheme can also be employed to analyse the marginal contribution of each specific passive measure working under realistic conditions in combination with the other housing elements. In regions where heating is important during winter months, the use of top-light solar passive strategies for spaces without an equator-facing façade can efficiently reduce energy consumption for heating, lighting and ventilation.

NOMENCLATURE

E_y	Annual energy yield in kWh
w	The wind speed in m/s
n	The number of data bins converting the wind speed range of the turbine (0.5 or 1 m/s intervals)
f_{wi}	The number of hours per year for which wind speed is w m/s
P_{wi}	The power resulting from a wind speed of w m/s
Q''_{conv}	Rate of convection heat transfer per unit area, W/m^2
Q''_{evap}	Rate of heat transfer per unit area due to evaporation, W/m^2
Q''_{rad}	Rate of radiation heat transfer per unit area, W/m^2
Q''_{total}	Total rate of heat transfer per unit area, W/m^2
T_s	Soil surface temperature, °C
T_{surr}	Surrounding temperature, °C
T_∞	Airflow temperature far from the surface, °C
G_r	Grashof number, dimensionless
R_e	Reynolds number, dimensionless
DC	Direct current
HSPF	Heating season performance factor
SEER	Seasonal energy efficiency ratio
Btu	British thermal unit
EER	Energy efficiency rating
DX	Direct expansion
GS	Ground source
EPA	Environmental Protection Agency
HVAC	Heating, ventilating and air conditioning

GREEK LETTERS

ε	Surface emissivity, dimensionless
φ	Relative humidity, dimensionless
ρ	Density, kg/m^3
δ	Stefan-Boltzmann constant, 5.67×10^{-8} W/m^2K^4

Subscripts

a	Air
v	Vapour
sat	Saturated
∞	Far away from the surface
s	Surface

REFERENCES

[1]   Reddy, A., Williams, R., and Johansson, *T. Energy after Rio: prospects and challenges.* United Nations Development Programme (UNDP). 2007. http://www.undp.org/seed/energy/exec-en.html.

[2]   Cavallo, A.J., and Grubb, M.J. *Renewable energy sources for fuels and electricity.* London: Earthscan Publications. 1993.

[3]   REN21. Renewables 2007 global status report. www.ren21.net.

[4]   Gilman, K. Water of wetland areas. *In: Proceedings of a Conference on the Balance of Water – Present and Future. Dublin: Ireland.* p. 123-142. 7-9 September 1994.

[5]   John, W. The glasshouse garden. *The Royal Horticultural Society Collection.* UK. 1993.

[6]   United Nations. World Urbanisation Prospect: The 1999 Revision. New York. *The United Nations Population Division.* 2001.

[7]   WCED. *Our common future.* New York. Oxford University Press. 1987.

[8]   Herath, G. The Green Revolution in Asia: productivity, employment and the role of policies. *Oxford Agrarian Studies.* 14: 52-71. 1985.

[9]   BRECSU. Energy use in offices. *Energy Consumption Guide 19.* Watford: Building Research Energy Conservation Support Unit. 1998.

[10] Farm Energy Centre. Helping agriculture and horticulture through technology, energy efficiency and environmental protection. Warwickshire. 2000.

[11] Randall M. *Environmental Science in Building.* Third Edition. 1992.

[12] Tiwari, G.N., and Goyal, R.K. *Greenhouse technology.* New Delhi: Narosa Publishing House. 1998.

[13] Santamouris, M., Balaras, C.A., Dascalaki, E., and Vallindras, M. Passive solar agricultural greenhouse; a worldwide classification evaluation of technologies and systems used for heating purpose. *Solar Energy* 53(5): 411-26. 1994.

[14] Santamouris, M., Arigirious, A. and Vallindras, M. Design and operation of a low energy consumption passive greenhouse. *Solar Energy* 52(5): 371-8. 1993.

[15] Mercier, I. Design and operation of a solar passive greenhouse in the South West France. *In: Proceedings of the International Congress on Energy Conservation of Agriculture and Fishculture*. London. 1982.

[16] Grafiadellis, M. Greenhouse heating with solar energy. *In: Von Zabettitz C* (Editor), FAO, Rome. 1987.

[17] Fotiades, I. Energy conservation and renewable energies. *In: Von Zabettitz C* (Editor), FAO, Rome. 1987.

[18] Pacheco, M., Marreivos, S., and Rosa, M. Energy conservation and renewable energies for greenhouse heating. *In: Von Zabettitz C (Editor)*, FAO, Rome. 1987.

[19] Santamouris, M., Mihalakakow, G., Belaras, C.A., Lewis, J.O., Vallindras, M., and Argiriou, A. Energy conservation in greenhouses with buried pipes. *Energy* 21(5): 353-60. 1996.

[20] Garzoli, K.V., and Blackwell, J. An analysis of the nocturnal heat loss from a single skim plastic greenhouse. *J. Agric. Engg. Res*. 26: 203-14. 1981.

[21] Chandra, P., and Albright, L.D. Analytical determination of the effect on greenhouse heating requirements of using night curtains. *Trans ASAE* 23(4): 9994-1000. 1980.

[22] Anne, S. *A handbook of greenhouse and conservatory plants*. Paston Press. London. 1989.

[23] Lynn, B. *The pleasure of gardening*. ANAYA Publishers Limited. London. 1993.

[24] Paul, F. *Indoor hydroponics: A guide to understanding and maintaining a hydroponic nutrient solution*. UK. 2001.

[25] Cengel, Y. Heat Transfer-*A Practical Approach*. First ed. McGraw-Hill, Inc. 1998.

[26] Heinonen, E.W., Tapscott, R.E., Wildin, M.W., and Beall, A.N. *Assessment of anti-freeze solutions for ground-source heat pumps systems*. New Mexico Engineering Research Institute NMERI 96/15/32580, p. 156. 1996.

[27] Huttrer, G. The status of world geothermal power generation 1995-2000. *Geothermics* 30: 1-27. 2001.

In: Handbook of Environmental Policy
Editors: Johannes Meijer and Arjan der Berg

ISBN 978-1-60741-635-7
© 2010 Nova Science Publishers, Inc.

Chapter 4

GOVERNANCE BY THE COMMONS: EMERGING PERSPECTIVES IN GLOBAL ENVIRONMENTAL GOVERNANCE

Patrick E. Meyer[*]

Center for Energy and Environmental Policy, University of Delaware, Newark, Delaware, USA

ABSTRACT

In the realm of environmental policy, it is evident that a widespread shift is taking place from what has traditionally been known as "government" to a new form of "governance". There is a trend throughout literature which refers to government becoming one party equal in power to members of a group comprised of many other parties. That is, government as traditionally defined, is becoming an equal peer amongst nonprofit organizations, nongovernmental organizations, activist groups, citizens, and other entities, as opposed to serving as the dominant ruling body. Globalization has fueled this shift by changing the way in which the world operates – increasing complexity and intricacy of all interactions, especially in the environmental realm. Embodied in the aforementioned trends, *global environmental governance* has emerged as a new paradigm and, for some, a discipline of environmental policy. While no correct rule for global environmental governance has emerged, analysis of the concept is imperative to ensure that the world's environmental problems are addressed in an organized, effective, and mutually-beneficial manner. This chapter identifies challenges associated with the conceptualization of global environmental governance, focusing on the perspectives of authors in relevant fields. It is determined that the challenges of global environmental governance can be accurately described through a handful of overarching themes, including public procurement programs, private certification systems, minority environmental groups (women's groups, among others), and the emergence of post-sovereign environmental governance. This chapter characterizes and compares these four major themes – as well as other minor themes – and demonstrates how various authors have made contributions to the literature supporting or contending the reigning paradigm. Given the abovementioned review of the reigning paradigm, this chapter provides a

---

[*] patrickmeyer@gmail.com

concise summary of who should be involved in global environmental governance regimes, including a discussion on which parties may hold more power than others and which parties have potential to become more powerful in the future. Lastly, this chapter looks to the future and provides a discussion on potential directions of the field of global environmental governance, including an identification of the primary areas where more work is needed. This chapter concludes that considerable contradiction exists amongst various authors within the global environmental governance debate. It is further concluded that the existing body of literature should serve only as a foundation for what could become a complex discipline. That is, there is room for a tremendous amount of future work to be completed in the realm of governance and the fundamental concepts discussed in this chapter serve as compelling leverage points for future deliberation.

INTRODUCTION

Globalization is a concept which, some argue, has been rapidly pursued by nations since the 1490s [O'Rourke & Williamson, 2002]. Other scholars argue that globalized trade routes were fragmented and poorly integrated until the nineteenth century [Menard, 1991]. And others argue that there has been a *same world system* stretching back at least 5,000 years [Frank & Gills, 1993]. More recently, throughout the 1990s globalization was touted as a novel concept which would integrate world economies and bring new wealth to new people across the globe. Regardless of the exact date of conception, it is evident that globalization is far from novel and the concept dates back a few hundred years at an absolute minimum. Nevertheless there have recently emerged trends within the globalization paradigm which may have never been witnessed before in world history. That is, there has been an emergence of global *governance* and an overall diminishment of the dominance of *government* in globalization schemes.

Governance refers to the patterns and distribution of the institutionalized capacity to take and influence decisions with regard to a particular locality [Danson & Whittam, 2005]. Further, governance refers to "the ways in which activities of governing are now distributed over a much wider group of actors than they were in the past, none of whom can unilaterally control outcomes. Hence, governance implies a process of coordinating and conciliating multiple actors" [Ansell, 2002, p. 668]. Importantly, most definitions of governance do not include an explicit explanation of what entity is undertaking a governing role.

It is of utmost importance to realize that "governance" is not the same as "government" [Finkelstein, 1995; Lemos & Agrawal, 2006]. The specific lack of mention of "government" is intentional since a key characteristic of the governance concept is the notion that there has been an overall shift in scholarly discussions from government to governance which is characterized by privatization, state transformation, shared public and private authority, cooperative partnerships complementing authoritative top-down regulation, and increased citizen participation in decision making processes [Ansell, 2002; Gulbrandsen, 2004]. Indeed, global environmental governance includes not only the actions of governments themselves, but also actions of communities, businesses, nongovernmental organizations (NGOs), and other entities. For more precise definitions of governance, see Box 1 which recites the definitions of governance by various international organizations.

Box 1. Definitions of Governance by Various International Organizations.

Definitions of "governance" by various international organizations, as researched by and reported in Weiss [2000]:

World Bank. Governance is defined as the manner in which power is exercised in the management of a country's economic and social resources. The World Bank has identified three distinct aspects of governance: (i) the form of political regime; (ii) the process by which authority is exercised in the management of a country's economic and social resources for development; and (iii) the capacity of governments to design, formulate, and implement policies and discharge functions.

UNDP. Governance is viewed as the exercise of economic, political and administrative authority to manage a country's affairs at all levels. It comprises mechanisms, processes and institutions through which citizens and groups articulate their interests, exercise their legal rights, meet their obligations and mediate their differences.

Institute of Governance, Ottawa. Governance comprises the institutions, processes and conventions in a society which determine how power is exercised, how important decisions affecting society are made and how various interests are accorded a place in such decisions.

Commission on Global Governance. Governance is the sum of the many ways individuals and institutions, public and private, manage their common affairs. It is a continuing process through which conflicting or diverse interests may be accommodated and co-operative action may be taken. It includes formal institutions and regimes empowered to enforce compliance, as well as informal arrangements that people and institutions either have agreed to or perceive to be in their interest.

UN Secretary-General Kofi Annan. Good governance is ensuring respect for human rights and the rule of law; strengthening democracy; promoting transparency and capacity in public administration.

Tokyo Institute of Technology. The concept of governance refers to the complex set of values, norms, processes and institutions by which society manages its development and resolves conflict, formally and informally. It involves the state, but also the civil society (economic and social actors, community-based institutions and unstructured groups, the media, etc) at the local, national, regional and global levels.

A recently-emerged sub-category of governance is environmental governance, commonly referred to as *global environmental governance*. Global environmental governance follows the aforementioned definition of governance, but pertains to specific governance of natural systems, air and water quality, and economy- and society-environment relationships. According to Lemos and Agrawal [2006], global environmental governance refers to the "set of regulatory processes, mechanisms and organizations through which political actors influence environmental actions and outcomes" [p. 298]. Global environmental governance has become a topic of extreme interest amongst academics, business people, industry representatives, politicians, government employees, and citizens alike due to the negative impacts that may be wrought upon societies and economies by a continually degrading environmental condition. Indeed, the emergence of global environmental governance has not occurred due to any planning by any particular organization, but instead has emerged due to an overall realization of the utter inadequacy of international policy cooperation in response to transboundary problems, and the need for better management of shared natural resources [Esty, 2008].

Here, we explore the concept of global environmental governance through multiple lenses. First we define the basic premise behind global environmental governance, explaining how it has emerged with the assistance and influence of multiple other broad global trends. Then, we will analyze the points of view and specific applications of global environmental governance of four authors. These authors apply the concept of global environmental governance to the realm of public procurement programs [Arnold & Whitford, 2006], private certification systems [Bartley, 2003], networked women's environmental groups [Bretherton, 2003], and post-sovereign environmental governance [Karkkainen, 2004]. The primary arguments of each author will be applied to the existing body of literature in the respective field and points of agreement and disagreement will be identified. The chapter is concluded with a discussion of which actors should be involved in global environmental governance, according to the aforementioned authors.

THE EMERGENCE OF GLOBAL ENVIRONMENTAL GOVERNANCE

Introducing the concept of global environmental governance for Yale University's *Global Environmental Governance*, Speth [2002] wrote:

> We have been moving rapidly to a swift and pervasive deterioration of our environmental assets. In response, there has been an upsurge of international environmental law and diplomacy, a vast outpouring of impressive scientific research, and thoughtful policy analysis. What has emerged over the past two decades is the international community's first attempt at global environmental governance [p. 1].

Speth argues that the emergence of global environmental governance has been made possible only due to the relatively recent success of environmental policy at the national level of most sovereign states, and the slow, but final realization that the life-sustaining processes of the biosphere are a common concern of all people on Earth.

The concept of global environmental governance is generally thought of to include four overarching themes. These are globalization, decentralized environmental governance, market- and individual-focused instruments, and governance across scales. These four issues have been discussed in some detail by Lemos and Agrawal [2006] and are worth briefly summarizing here as a gateway to the wider discussion of this chapter.

Popular globalization discourse argues that humanity is in the midst of a new historical transition with implications no less profound than the emergence of settled agriculture and the industrial system [Harris, 1992]. Globalization is an incredibly vast concept, encompassing changes in the nature of the planetary environment, economic interdependence, a revolution in information technology, an increasing hegemony of dominant cultural paradigms, and new social and geopolitical fissures [Raskin et al., 2002]. Lemos and Agrawal explain that globalization produces both negative and positive pressures on governance. Negatives, according to the authors, are mostly related to corporate power, and include the notion that "by integrating far-flung markets and increasing demand, globalization may intensify the use and depletion of natural resources, increase waste production, and lead to a 'race to the bottom' as capital moves globally to countries and locations that have less stringent environmental standards" [p. 300]. Our later discussion on mandatory environmental management systems [Arnold & Whitford, 2006] will expand on the roles played by corporations in environmental governance in a globalized world.

Lemos and Agrawal argue that positive impacts of globalization include the potential for diffusion of "positive environmental policy initiatives" on a broad scale, creation and development of new global regimes dedicated to environmental governance, more efficient use and transfer of technology, freer flow of information, and an enhanced depth of participation [p. 300]. Whether or not globalization provides positive or negative influence on environmental governance aside, the fact remains that the increasing complexity of the world "requires more holistic or comprehensive policies to address environmental externalities and to support sustainable development" [Haas, 2004, p. 2]. Simply put, globalization has changed the way that governance works.

The second theme presented by Lemos and Agrawal is that of decentralized environmental governance. Lemos and Agrawal argue that "some of the most important contemporary changes in environmental governance are occurring at the subnational level and relate to efforts to incorporate lower-level administrative units and social groups better into formal processes of environmental governance" [p. 302]. The authors claim that there has been "a shift toward co-management, community-based natural resource management, and environmental policy decentralizations" [p 302]. Further, they contend decentralized environmental governance to be beneficial in that it can produce greater efficiencies because of competition among subnational units and it can bring decision making closer to those affected by governance.

The third theme presented by Lemos and Agrawal is market and agent-focused instruments of governance. The authors explain that since the 1970s, there has been a shift towards market and voluntary-based mechanisms of environmental governance. These mechanisms "aim to mobilize individual incentives in favor of environmentally positive outcomes through a careful calculation and modulation of costs and benefits associated with particular environmental strategies" [p. 305]. The mechanisms typically are broad in range with examples such as eco-taxes, subsidies, regulation, market incentives, voluntary agreements, certification, eco-labeling, and informational systems. Finally, the fourth theme presented by Lemos and Agrawal is that of cross-scale environmental governance. The authors argue that "the multi-scalar character of environmental problems – spatially, socio-politically, and temporally – adds complexity to their governance" [p. 308]. Because sociopolitical, cross-scale environmental problems affect and are affected by institutionalized decision making at local, subnational, national, and transnational levels, multilevel governance has emerged as a method to address environmental issues while counteracting the "fragmentation that is characteristic of sectorally based decision making" or "of decision making that is organized by territorial, social, and political divisions" [p. 308]. These notions are fortified by others. Haas [2004], for example, argues that as a result of cross-scale environmental governance, "more actors now engage in more governance functions at multiple levels of governance" [p. 13].

Yet, it must be realized that the definition of global environmental governance is not necessarily embodied *only* in the four aforementioned themes. "Global environmental governance" is a slippery slope; i.e. it means many different things to many different people. Defining global environmental governance is similarly ambiguous as defining "sustainable development". Although the concept of sustainable development – despite its many ambivalent definitions – has provided the theoretical bridge connecting the environment with economy and society [Vogler & Stephan, 2007], it remains no more well defined today than it was twenty years ago shortly after the publication of *Our Common Future*. Indeed, in one recent review of the existing definitions of sustainable development, it was found that since 1987 more than 300 alternative and variously-modified definitions of the concept have emerged [Johnston et al., 2007]. Although the concept of global environmental governance has not received as much attention as sustainable development, there exists almost as much variability in available definitions as that of sustainable development.

We must wonder why global environmental governance has been aptly named so. Argued by Finkelstein [1995], global governance "reflects inescapable ambiguity about the nature of the 'international system,' indeed about what the international system is, or what it encompasses. Does *global* mean what has been signified by *international, interstate, intergovernmental*, or even, often, *transnational?* If so, why not use one of those terms, instead of choosing a more ambiguous one?" [p. 367]. "Global" is inherently an indistinct term, implying the involvement of all of Earth, but not defining at which level such involvement should manifest. The point is that global environmental governance as a theme or discipline has not yet been satisfactorily defined. There remains tremendous ambiguity in the limits, bounds, and overall definition of the term, which then adds ambiguity to most discussions focusing on governance. Thus discussions such as those developed in this chapter seek to further refine and define the governance paradigm, working towards a goal of establishing a defined order in the years to come.

A distinction between the concepts of global environmental governance and global environmental regimes is required. A global environmental regime has been defined as "a partially integrated collection of world-level organizations, understandings, and assumptions that specify the relationship of human society to nature" [Meyer et al., 1997]. While similar to governance, regimes are distinctly different in that:

> Regime analysis tends to study governance through statist lens, focusing on the creation and operation of rules in *international affairs*. The term *global governance*, on the other hand, encompasses not only those phenomena but also situations in which the creators and operators of rules are nonstate actors of various kinds, working within and across state boundaries [Stokke, 1997, p. 28].

Along these lines, it has been shown that the governance concept is "beyond" the regime concept in that most governance advocates will confess they are studying a new, novel concept which is more progressive than global environmental regimes [Okereke & Bulkeley, 2007].

As is evident in the aforementioned discussion, much ambiguity remains in the definition of global environmental governance and how the concept compares to similar bodies of theory. Despite the ambiguity, the point remains that global environmental governance has gained tremendous footing in academic studies. In the last decade the number of scholars studying global environmental governance has skyrocketed, indicating the overall importance and permanence of the concept. Some example articles, books, and analyses include: Levy and John [2005] who provide a political economy approach to understanding the role of business in global environmental politics; Oberthür and Gehring [2006] on governing the trade of genetically modified organisms; Pattberg [2005] on the lack of understanding of the trend of rule-making by private multi-stakeholder initiatives in the global environmental governance arena; Himley [2008] on global environmental governance in the realm of nature-society geography; Mason [2008] on the failure of global environmental governance initiatives to address "accountability deficits" left by lack of state-centered accountability and responsibility; Najam et al. [2004] on the overall lack of organization in global environmental governance systems and methods by which the systems can be made more effective at achieving goals; Arnold and Whitford [2006] on making environmental self-regulation mandatory in order for governmental departments to take advantage of procurement programs; Bartley [2003] on private certification systems in the forestry and apparel industries; Bretherton [2003] on the importance of including women in environmental governance discourse; and Karkkainen [2004] on the flattening of national power and the emergence of post-sovereign environmental governance. The broadness of the aforementioned topics further demonstrate the difficulty in defining global environmental governance as is spans so many diverse issues.

The abovementioned articles pertaining to global environmental governance are by no means meant to comprise the entire field, but instead serve to shed light on the diversity and increasing interest in the governance arena. Four works listed above – Arnold and Whitford; Bartley; Bretherton; and Karkkainen – serve as particularly important pieces worthy of in-depth analysis through which we can gain a clear perspective on priority issues in the realm of global environmental governance. The following sections will analyze each of these four works in detail.

FOUR ENVIRONMENTAL GOVERNANCE TOPICS IN DEPTH

Making Environmental Self-Regulation Mandatory

Arnold and Whitford [2006] present a two-part argument with an aim to fortify the realm of global environmental governance. First, they argue that private firms' participation in public procurement programs should be contingent on their adoption of an environmental management system such as ISO 14001. Second, they argue that the role of politicians must be separated from that of the actual implementation of any policy aimed at promoting procurement requirements under an environmental management system.

Arnold and Whitford begin their discussion by pointing out that private firms and corporations have the greatest impact on quality of life at almost all levels, including the local, regional, nation-state, and international levels, and thus should act as the leading force in environmentally-friendly action. They argue that there has been much interest in "corporate social responsibility" or "corporate citizenship" – typically involving voluntary environmental self-regulation. The rise of corporate social responsibility has been discussed in-depth elsewhere. For example, Levy and Kaplan [2007] provide a review of the rise of corporate social responsibility and the concept's role in governance, international law, and environmental responsibility. What sets Arnold and Whitford's work apart is their argument that the political nature of corporate social responsibility initiatives is a hindrance to the programs' success. Indeed, while a self-regulated program may be highly successful during the moment at hand, changes in political direction can often mean the eradication of self-regulated environmental programs – even those that have been proven to be effective.

As an alternative to voluntary corporate environmental self-regulation, Arnold and Whitford argue for a *mandatory* environmental management system (EMS). They define an EMS as "a way that firms or other organizations (including public agencies) build processes that continually plan, implement, review and improve the ways the organization tries to meet its business and environmental goals" [p. 2]. Elsewhere, an EMS is similarly defined as a standard that encourages "firms to voluntarily adopt policies that go beyond the requirements of extant law and promote continuous improvement in firms' environmental performance" [Kollman & Prakash, 2001, p. 399]. EMS standards are playing an increasingly important role in the policy realm. Kollman and Prakash further argue that EMS standards can today be seen as a part of broader trends that are fundamentally changing the way business and policy-making are regulated. In other words, EMS is one of many tools being developed as part of the wider global environmental governance portfolio.

As a case example, Arnold and Whitford focus on a popular EMS: the ISO 14001. ISO (The International Organization for Standardization) located in Geneva, Switzerland, promotes the development and implementation of voluntary international standards, and has been recognized for its particular level of success in environmental management issues [EPA, 2006]. Although it is a voluntary system, it has proven successful due to its ability to "pin firms down by creating external and internal pressures without the intervention of government" [Arnold & Whitford, 2006, p. 2]. ISO 14001 is further promoted by state agencies though the use of awards and general recognition. However, the problem with this system is that "these programs are always susceptible to change when political administrations turn over or when current leaders have a change of heart about using public dollars to encourage environmentalism" [p. 3].

Arnold and Whitford argue that an effective way for an EMS to be guaranteed successful is to require adherence to the EMS if a firm or corporation wishes to participate in government procurement programs. Programs such as the ISO 14001 should not be voluntary; failure to participate in the programs should have such dire consequences that corporations envision the programs as *mandatory*. An example of this type of program can be found in "green procurement" programs, which have become increasingly popular. Green procurement is the selection of products and services that minimize environmental impacts. It usually requires an organization to carry out an assessment of the environmental consequences of a product at all the various stages of its lifecycle and the selection of less-environmentally-damaging products for purchase [IISD, 2007]. "Green procurement is simply a result of EMS-type activity within the government – but it can have important effects on the broader community of suppliers" [p. 6]. In this sense, "EMS adoption can signal technical expertise for the supplier, and deliver environmental protection benefits for society as a whole" [p. 8].

Finally, Arnold and Whitford argue that it is essential that the role of politicians be separated from that of the actual implementation of any EMS policy. In other words, they argue for the further removal of government from governance. The authors contend that "politicians have incentives to *fail* to provide sufficient leadership in implementing these sorts of programs, and even if any given politician can see their way to implement that program, that attention can change with change in political regimes" [p. 8]. Instead of political-oriented programs, the authors argue for a heavier reliance on independent regulatory agencies which would help to ensure that EMS programs survive a change in administration – again, a shift from government to governance.

Although Arnold and Whitford focus on national-level EMS standards, it is important to note that there exists a vast array of literature regarding international EMS standards as well. In international relations, many scholars argue that EMS regimes represent a new form of governance in which actual governments "play a more limited role in the establishment, monitoring, and enforcement of regulatory regimes" [Kollman & Prakash, 2001, p. 400]. That is, the adoption of EMS standards has many times lead to a situation where national-level policy making is transferred to supranational (regional or international) bodies. In doing this, individual states in effect forfeit a certain amount of sovereignty to these broader institutions. It has been argued that:

> The increase in the number of supranational environmental regimes ... has led to an extremely fragmented and decentralized form of governance, with the result that states have had to adjust their national regulation styles. Additionally, and less recognized in the literature, many of these regimes are transnational in nature and include nongovernmental actors in their negotiations and compliance structures. As a result, the unquestionable dominance of national governments over environmental policy-making seems to be eroding [Kollman & Prakash, 2001, p. 400].

The notion of decreased state power will reoccur throughout the remainder of this chapter. For example, we next discuss Bartley [2003] who provides a discussion regarding the shift from government- to private-oriented certification schemes. Later, we discuss Karkkainen [2004] who confers on the emergence of post-sovereign environmental governance.

Private Certification of Forests and Factories

Private certification systems within the environmental sector have recently emerged as a successful method to raise awareness, boost knowledge and increase discourse among a broad range of stakeholders [Gulbrandsen, 2004]. Bartley [2003] presents an interesting comparison between the simultaneous rise of private certification systems in the forestry and apparel industries, demonstrating the success of certification systems in both fields. The author presents an analysis which "identifies two types of mechanisms driving the rise of private regulation – one about social movements, the other about free trade rules and regimes" [p. 457]. By doing so, he sheds light on the process by which certification associations emerge, but more importantly, shows which specific events helped mobilize moral and material support for certification. As such, he provides an analysis and framework for the potential future emergence of certification systems in a vast array of industries.

Similar to Arnold and Whitford's argument, Bartley begins with a brief discussion regarding the decline of "traditional 'command and control' strategies," arguing that there has been a recent shift toward regulatory forms based on different social control strategies. He then discusses the application of certification as one such strategy to regulate the forestry and apparel industries. It is worth noting that elsewhere, certification in the forestry industry has been recognized as representing "the most advanced case of nonstate-driven rule making dynamics globally in the environmental field" [Gulbrandsen, 2004, p. 76]. This success case further solidifies the importance of social control strategies as a form of governance. Bartley's analysis is discussed in depth below.

First, "in both the apparel and forest products fields, certification systems emerged in a context of social movement activity and public controversy about the social or environmental dimensions of the industry" [p. 442]. To make his point, Bartley begins with a discussion regarding the history of certification systems in each field. In the forestry industry in the 1980s, environmental organizations were largely responsible for focusing attention on tropical deforestation. These organizations would often organize consumer boycotts, to which tropical timber companies responded by making claims about the supposed environmental friendliness of their forest operations and by undertaking voluntary self-monitoring. Bartley argues that these claims were heavily criticized for being improvable, unable to be monitored, and simple fabrication of the truth; and the self-monitoring process to be heavily flawed. However, eventually "the boycott strategy came under fire for hurting forest-dependent populations in developing countries" and thus the environmental groups began the construction of an alternative method: certification systems. Bartley's point is that "as social movement organizations put pressure on tropical timber producers and retailers, this created a demand for a more overarching system for evaluating claims about forest management, harvesting practices, and the use of the infamously ambiguous term 'sustainability'" [p. 445].

An even more important point, however, is that a very similar development was occurring in the seemingly unrelated apparel industry. According to Bartley, throughout the 1980s companies were faced with media exposés, protests at their stores and shareholder resolutions over the issue of child labor and the existence sweat shops. As a result, the apparel industry responded by adopting corporate codes of conduct that addressed working conditions in the factories which had come under fire. However, these corporate codes (similar to self-monitoring in the forestry industry) were criticized as ineffective. Indeed, "in attempts to verify their codes, companies became more likely to bring outsiders in to inspect or monitor factory conditions" [p. 446], but Bartley argues that this monitoring process was heavily flawed. As a result of continued social movement pressure, there became an underlying demand for a more credible system for evaluating factory conditions – leading to a certification process. This notion is explained further by Rodriguez-Garavito [2005]:

> Given the lack of credibility of first-party monitoring (in which a single manufacturer monitors compliance based on audits carried out by its own personnel or by a commercial auditor hired by it) and second-party monitoring (which involves a business association formulating a code of conduct and hiring commercial firms to report on compliance), it is with regard to third-party monitoring that the most consequential debates and experiments are taking place [p. 214].

"Third-party monitoring" could take the form of a private certification system, and as Bartley would argue, is the preferred method to overcome the shortcomings of first- and second-party monitoring.

According to Bartley, a second similarity between the two fields existed in the notion that overarching social shifts in states, social movement groups, and NGOs has helped to shift efforts and resources to a path toward private forms of regulation, as opposed to governmental or intergovernmental regulatory systems. In the forestry industry, the Austrian government attempted to ban the import of unsustainable tropical timber, only to be internationally accused of hindering free trade under the Generalized Agreement on Tariffs and Trade (GATT). As a result, the Austrian government yielded, instead allocating money to a private labeling program. The private program proved highly successful. Similarly, in the apparel industry, the US government, behind the front of the International Labor Organization (ILO), attempted to construct a "no sweat" labeling system, only to be internationally accused of protectionism and criticisms of attempting to act as a trade organization. As a result, the US government backed down, but instead allocated money to two private labeling programs which also proved to be highly successful. These examples show that private certification can be far more effective than government-initiated certification.

Barley argues that in forestry and apparel, "private certification systems emerged as state action got directed toward private forms of regulation, rather than governmental or intergovernmental systems" [p. 447]. This notion is echoed by others, such as Gulbrandsen [2004] who states that "*nonstate* forest certification schemes have emerged in the shape of powerful market-driven governance and rule-making systems" [p. 76]. In the apparel industry, Rodriguez-Garavito [2005] argues that national and international labor advocacy organizations have increasingly engaged in *nonstate* forms of regulation. However, more broadly, Bartley's argument may be contested; popular notions argue that there is only limited participation from nonstate actors in new realms of environmental governance – which may be dominated by government actors, democratic or not [Lemos & Agrawal, 2006]. Thus, it is arguable that governments still play a major role in global environmental governance; that they have not been unsuccessful to the point argued by Bartley (or Gulbrandsen and Rodriguez-Garavito). This notion deserves further analysis and will be explored further when we analyze Karkkainen's post-sovereign environmental governance.

A third similarity between the forestry and apparel industries exists in that in both cases "experiences in intergovernmental arenas led NGOs and social movement groups to put more energy into private, nongovernmental approaches to labor and environmental issues, as they became increasingly discouraged and disenchanted with governmental and intergovernmental approaches" [p. 451]. Indeed, in the forestry industry for example, environmental groups became discouraged at such events as the failure of the UK-based International Tropical Timber Organization (ITTO) to set certification standards and viewed the certification measures enacted under the UNCED "Earth Summit" as a complete failure. In the apparel industry, labor groups became discouraged at the North American Free Trade Agreement (NAFTA), which initially was supposed to contain labor protection, but ended up including only "weak labor 'side agreements'." "In both cases, as NGOs and social movement organization experienced failures in intergovernmental arenas, they tended to shift their energies and resources toward private alternatives" [p. 454]. These private alternatives proved to be highly successful in comparison to the governmental approaches.

By demonstrating the close parallels between the rise of private certification systems in the forestry and apparel industries, Bartley has shown how private certification systems have emerged independently in two seemingly unrelated industries. By doing so, Bartley provides a useful argument supporting the usage of private certification systems in a vast array of industries and demonstrates through specific examples how the regulatory arena is shifting from government to governance; i.e. Bartley demonstrates through a discussion of certification the emergence of global environmental governance initiatives. Although, as identified above, there are a few points in his argument which may be contested, overall, he has solidified the notion that private certification systems are functional, and widely applicable.

The Necessary Involvement of Women and Minority Groups

Around the globe there are countless minority and sub-groups of people who must cope with an unfair and unequal share of pollution and other negative environmental impacts. For example, native Indians of Brazil have been subjugated by powerful interest groups such as bankers and rubber estate owners who are interested in for-profit exploitation of the Amazon rather than forest preservation [Conca & Dabelko, 1998]. Elsewhere, African countries that are heavily dependent on the generation of foreign exchange on a limited range of primary agricultural and mineral commodities such as coffee, cocoa, cotton and copper are continually devastated by changes and depressions in world commodity markets or local climate [Potter et al., 1999]. In the United States it has been found that hazardous waste sites are most likely to be found in poor neighborhoods and neighborhoods with a large percentage of racial minorities [Smith, 2004]. Indeed, a recent General Accounting Office study of hazardous waste landfills in the Southeast US found that three out of four landfills were in areas that were predominantly black and poor. It has been argued that there is a functional link between racism, poverty, and powerlessness and industry's assault on the environment; a trend that has been called "environmental racism" [Russell, 1989; Smith, 2004].

Cultural norms can also influence the share of adverse environmental impacts imposed upon a select group within a society. It has long been argued that women, for example, and especially women in poor countries, are forced to cope with a disproportionately large share of environmental burden. For instance, biofuel stoves are used throughout many developing nations, and are used primarily by women since it is usually the woman's task to cook and maintain the home. In these situations, women are particularly impacted by environmental negativities due to their social role. In specific cases it has been shown that women have been disproportionately exposed to poor indoor air quality due to the usage of biofuel stoves in highland Guatemala [Bruce et al., 1998], in Zimbabwe [Rumchev et al., 2007], and elsewhere. Although not to discredit or downplay the importance of deliberating the prejudices imposed on other minority groups, the impact on women has been particularly profound and thus the remainder of this section will focus on women's role in global environmental governance initiatives.

Bretherton [2003] presents an analysis of the role of women in global environmental governance initiatives and presents methods for women to become empowered in an increasingly complex arena. She begins by explaining why it is necessary that the role of women be recognized, analyzed, and furthered in the realm of environmental governance. She points out that women have typically been repressed and represent a particularly challenging social group needing much work. For example, she states that 70 percent of those today living in absolute poverty are believed to be women and that women relatively poor access to economic and political resources around the globe. These notions are repeated elsewhere; for example, Gremmill and Bamidele-Izu [2002] argue that women – along with children, youth, and indigenous peoples – are a particularly important, but often overlooked group in global environmental governance. Auer [2000] identifies numerous case studies in India, the Philippines, and Japan where women played an essential and important, but relatively unanalyzed role in global environmental governance schemes. In Auer's Philippines example, local women managed policy resources in a sustainable development project which involved the sale of local handmade goods and resulted in the development of a lucrative local market, a substantial rise in the area's household incomes, and the improvement of the ecological stability of the coconut groves on which the production of the goods relied.

Bretherton argues that women find it particularly difficult to make their voices heard due to the fact that "global environmental governance is informed by two deeply embedded and mutually reinforcing sets of norms; those deriving from contemporary neo-liberalism, and those deriving from gender structures and practices associated with contemporary hegemonic masculinity" [p. 104]. She defines hegemonic masculinity as "the particular conceptualization of masculinity which is dominant in a given set of gender relations at a particular time and in relation to which other conceptualizations are seen as subordinated, marginalized, or complicit." Contemporary neo-liberalism is defined elsewhere as a restatement of classical liberalism, reasserting the liberal principles of freedom, market individualism, and small government [Shaver, 1996]. Above other components, neo-liberalism has a strong emphasis on concepts of freedom:

> Neo-liberalism has gained strength in the last two decades, taking up political ground between conservatism and the 'socialist' collectivisms of the welfare state and monopoly capitalism. Like classical liberalism, neo-liberalism gives primacy to freedom, which it understands in the narrow and negative sense of restriction of the individual by the powers of the state [Shaver, 1996, p. 14].

However, despite the fact that neo-liberalism claims to uphold freedom and alleviation from suppression by powers of the state, Bretherton argues that the hegemonic masculinity dominant in neo-liberal movements serves only to place further suppression on women.

Bretherton argues that there have been four general historical approaches by women to make their voice heard in environmental debates: efficiency arguments, equity arguments, ecofeminist arguments, and emancipatory arguments. According to Bretherton, efficiency arguments coincide well with existing global environmental governance structure and are founded in the idea that "the involvement of women, particularly at the grassroots[*] level, is a cheap, effective way of ensuring that policies are implemented. ... From the efficiency perceptive, women are viewed as objects and/or implementers, rather than formulators, of policy and a resource to be employed in defense of the environment" [Bretherton, 2003, p. 105].

The equity arguments fundamentally provide more promising results for women's rights – but tend to focus on issues associated with women's political, economic and social equity

---

[*] "Grassroots" is defined as "the common people at the local level (as distinguished from the centers of political activity)" [princeton.edu, 2008].

over the interest of the natural environment. However, such arguments have been successful, challenging the "dominant norms and values of capitalist patriarchy" [p. 106]. On the other end of the spectrum are the ecofeminist arguments, which, perhaps at the expense of the overall progression of the movement, place a great interest on the natural environment and less on political, economic, and social issues. "Ecofeminists urge that admiration for masculine traits be replaced by reverence for, and identification with, the feminine" thus returning proper value to the life-giving characteristics of women and nature [p. 106]. However, this prioritization of the interests of the natural world over those of women has been criticized by radicals, perhaps at the expense of the ecofeminist movement [p. 109]. Finally, emancipatory arguments emphasize women's positions in social structures over their affinity with nature. Bretherton argues that "women generally remain politically and economically subordinate to men of their class or ethnic group" [p. 107]; the emancipatory arguments challenge this subordination by challenging the hegemonic structures and their constitutive norms in societies.

With the four historical approaches in mind, Bretherton contends that women are faced with the choice between working alongside men within mixed environmental movements or organizing separately as women. This choice, Bretherton maintains, is not trivial: "choice of strategy and related ideological position are important determinants of cohesiveness in women's movements" [p. 109]. She points out that historically, working alongside men has proven less successful and "consequently, women seeking to advance a radical feminist critique and politics around environmental issues have preferred to organize as women" [p. 108]. Such movements have had a number of common characteristics including that they are drawn from a population which shares a sense of grievance, are intended to be disruptive, exist primarily outside conventional government, are in contention with powerholders, emphasize grassroots, and are distinct from their mass social base. In this sense, "if women's movements are to enjoy success in opposing and undermining dominant norms they will need to realize the potential of already engaged women by creating/strengthening links between movements and attracting new recruits by framing issues in ways that are widely perceived as relevant" [p. 110].

Thus, Bretherton is arguing that for women's movements to be successful, they need to engage the local populace where the issues are most prominent: they need to have a grassroots focus. For large, transnational women's movements, "it is essential that issues are framed in ways that resonate in a variety of cultural contexts" but such framing proves to be highly difficult if not impossible due to the fact that women around the globe interact with their natural environment in vastly different social and cultural contexts [p. 112]. Indeed, "it is relatively rare that a transnational social movement can promote the formulation of collective identities around common interests 'among people who live far apart, speak different languages, have varied social and ethnic background, and face different opponents'" [p. 114]. Working on a transnational level whether with or without transnational organizations such as the World Bank, International Monetary Fund (IMF) and World Trade Organization (WTO), has proven unsuccessful on most levels. Working with transnationals tends to only further "advocate the mainstreaming of gender issues within the existing neoliberal policy frames" [p. 116]. Interestingly, the World Bank has elsewhere been criticized for its operation of a reinforcing knowledge structure, which only allows knowledge to penetrate the mainstream if that knowledge coincides with the existing policy frame [St. Clair, 2006]. Thus, if the existing policy frame does not allow for the liberation or involvement of women, then only masculine-dominated policy will survive the knowledge structure and penetrate the mainstream.

However, working *without* transnationals has proven equally challenging; working outside of these organizations has proven to "fail to challenge the related norms of hegemonic masculinity" [Bretherton, 2003, p. 116]. As such, there is a need for "social movements to be rooted in local social networks facilitating development of a sense of collective identity and feelings of solidarity necessary for collective action" [pp. 113-114]. Haas [2004] agrees that taking advantage of such networks will improve global governance and the prospects for achieving positive environmental results [p. 13].

Examples of locally-oriented transnational programs can be found in literature. For example, consider the Convention to Combat Desertification, which is a transnational program aimed at protecting biodiversity and forest preservation. The program emphasizes a participatory, bottom-up strategy specifically aimed at engaging local groups of women in a common fight against dryland degradation [Jasanoff & Martello, 2004]. Programs such as this are important in solidifying Bretherton's argument and show that although she is studying this issue from the academic realm, real-world organizations are already putting similar ideas to practice.

Bretherton's article is immensely useful to the body of global environmental governance literature in that it provides an assessment of the political impact of grassroots groups in differing social and cultural contexts. It also provides an overview of the many factors within a global institutional framework which present continuous challenges when formulating, implementing, and evaluating environmental policy, whether at the local- or transnational-level. This is a particularly pressing issue because, as seen in the case of the aforementioned supranational EMS standards or private certification systems, there is an overarching trend for environmental governance to take on a transnational approach. Whereas Bretherton agrees that there is an inherent benefit in attempting to create transnational women's environmental organizations, it remains extremely important to keep the interest of these organizations rooted in locales.

The Post-Sovereign Future of Global Environmental Governance

There are numerous theories regarding the future of global environmental governance. Many scholars contend that by 2050 the world will witness the emergence of a "World Environment Organization" potentially as a stem from the United Nations Environment Program or from some other organization [worldbank.org, 2008]. Another particularly well-received theory is that raised by Karkkainen [2004], who argues that global environmental governance is transitioning to a state of "post-sovereign environmental governance." Karkkainen's argument ties in very neatly with the aforementioned arguments of Arnold and Whitford and Bartley, who argue that the power of nation-states is shifting to become one power in an arena of many. When originally presenting his argument in 2002 at the Berlin Conference on the Human Dimensions of Global Environmental Change, Karkkainen argued that "increasingly, environmental problems too complex to be resolved though fixed international rules or straightforward exercises of state sovereignty are addressed through hybrid, multi-party, collaborative governance arrangements that pool and recombine the resources and competencies of a variety of state and non-state actors" [Karkkainen, 2002, p. 206]. More recently, in his 2004 piece, he expands on his previous notions.

Karkkainen begins by presenting a background on the developments within environmental governance over the last couple decades. He argues that two contending points of view have emerged: first, the "vertical institutionalists" who emphasize the role of international rules, norms and regimes, and second, the "horizontal diffusionists" who emphasize the power of the nation state, benchmarking, and networking among similarly situated national-level decision makers. Karkkainen points out that each of these bodies of theory take state sovereignty as a bedrock principle – and therein lay the departure to his argument.

Karkkainen argues that the world is "now witnessing a bifurcation of authority in which 'the state-centric world is no longer predominant'" [p. 73]. Instead, there has emerged a "complex multi-centric world of diverse actors replete with structures, processes and decision rules of its own." As such, as a contending view to the abovementioned vertical institutionalists or horizontal diffusionists, Karkkainen presents a new body of thought: "post-sovereign environmental governance." Under this form of governance exists "a mode of hybrid problem-solving governance in which sovereign states and nonstate parties actively collaborate, roughly as equal partners, to address certain kinds of highly complex problems that appear to be beyond the capacity of sovereign states alone to serve" while understanding that "decision-making is characterized by a self-consciously experimentalist problem-solving approach, emphasizing continuous generation of new information which leads in turn to continuous adjustment, refinement, and reconfiguration" [pp. 74-75].

Karkkainen explains that post-sovereign governance has three distinct characteristics: it is non-exclusive, non-hierarchical, and post-territorial. By non-exclusive, Karkkainen means that post-sovereign governance departs from "the conventional state-centric understanding that sovereign states hold exclusive authority over environmental and natural resource policies within their territorial jurisdictions" [p. 75]. Indeed, under post-sovereign governance, "decision-making and policy implementation are understood to be the joint responsibility of both state and nonstate actors." By non-hierarchical, Karkkainen means that post-sovereign governance "does not rely exclusively, or even primarily, on traditional modalities of hierarchical authority. Post-sovereign environmental governance arrangements are founded upon ongoing, open-ended commitments by multiple parties to 'do whatever it takes' individually and jointly to restore ecological integrity in particular locales" [p. 76].

Lastly, by post-territorial, Karkkainen means that post-sovereign governance no longer adheres to traditional territorial boundaries. Indeed, in post-sovereign governance, "special and conceptual boundaries are defined not by reference to fixed, territorially delimited jurisdictional lines, but by reference to shares understanding of the nature, scale, and causes of the problem to be addressed" [p. 77]. Karkkainen is careful to explain that post-sovereign governance does not mean the elimination of traditional government. Instead, it means a shifting away from "command-and-control" techniques and towards a balanced, multi-party approach in which government is only one party among many. Haas [2004] would agree with Karkkainen. Haas states:

> The new geopolitical reality is the proliferation of new political actors and the diffusion of political authority over major governance functions, particularly in the environmental sphere. These new actors include NGOs, multinational

corporations, organized transitional scientific networks known as epistemic communities, global policy networks, and selective international institutions that are capable of exercising discretionary behavior independently of the wishes of their dominant member states [p. 2].

Yet, other authors would argue against Karkkainen and Haas' notion of government being one of many entities, each possessing an equal amount of influence. Ansell [2002], summarizing the views of Oxford University's Jon Pierre, argues:

> Governance seeks to acknowledge and call attention to the relative shift away from the state-centric model of governing toward a more pluralistic model. The end result does not fit into either the state-centric category or the opposing category of society-centered pluralism. While constrained, the state remains too active and interventionist in the governance model to be described in conventional pluralist terms [p. 668].

To Pierre, state power retains higher influence than most other parties. He does acknowledge that state power has become one power in a realm of many – but does not agree that government power has reduced itself to the point of being "equal" with the other parties. The bottom line is that although some debate may still exist regarding the true power of government in governance, governments are no longer the only player – they have indeed become one party among a multitude of other parties.

Karkkainen indicates that such a shift is not necessarily at the disliking of traditional governments. Indeed, many governments have recognized that the "complex and dynamic nature of ecosystems, coupled with the need to maintain a flexible, dynamic, continuous-learning approach" place most problems beyond the capacity of traditional governments. This notion is further enforced by Lemos and Agrawal [2006] who argue that "hybrid forms of environmental governance" – such as Karkkainen's post-sovereign governance – have emerged "based upon the recognition that no single agent possesses the capabilities to address the multiple facets, interdependencies, and scales of environmental problems that may appear at first glance to be quite simple" [p. 311]. As a result of this recognition, many of these governments have openly welcomed the involvement of additional parties.

With his definition of post-sovereign governance explained, Karkkainen continues on to provide two solid case examples of post sovereign governance in action. He shows that both case examples – the Chesapeake Bay Program and the US-Canadian Great Lakes Program – have proven to be non-exclusive, non-hierarchical, and post-territorial. The Chesapeake Bay Program, for example, is non-exclusive because it involves the participation of nongovernmental organizations, individual citizen activists, independent scientists, governments, and other entities. It is non-hierarchical because the program largely involves "procedural framework agreement within which subsequent objectives and implementation measures may be progressively specified" [p. 84], but the agreements and directives have no formal legal status, thus eliminating the hierarchical factor. Lastly, the program is post-territorial because it spans multiple states and two countries. Indeed, even from the outset, the program "involved a kind of regional pooling and coordination of the combined capabilities of multiple tiers of government, none of them matched precisely to the geographical scale or functional scope of the problem" [p. 84]. Karkkainen provides similar analysis of the US-Canadian Great Lakes Program, showing that it too is non-exclusive, non-hierarchical, and post-territorial. It is worth noting that in Karkkainen's 2002 work, he also applied the concept of post-sovereign governance to the Baltic Sea, showing similarly positive results as the two other cases.

Overall, Karkkainen has presented a strong argument showing that global environmental governance has begun to evolve to a new type of governance, dubbed post-sovereign environmental governance. In asking whether this new form of governance has broader implications past the two case studies presented, Karkkainen indicates that post-sovereign governance is an "experimentalist" form of governance. Furthermore, he points out that some environmentalists "remain skeptical of the degree to which the Chesapeake Bay and Great Lakes programs can yet claim success in actually resolving complex environmental problems" [p. 91]. However, the point remains that both programs, which fall under Karkkainen's post-sovereign governance thesis "have been widely hailed in policy circles as innovative and successful prototypes" [pp. 91-92].

THE ACTORS OF GLOBAL ENVIRONMENTAL GOVERNANCE

There are a number of common themes across the discussions of environmental governance regarding exactly who needs to be involved in global environmental governance initiatives. This section serves as a summary of these themes, with specific focus on the role of government in governance. First, Arnold and Whitford argue that in establishing mandatory EMS standards as a form of environmental governance, the process must include corporations, independent regulatory agencies, national governments, supranational bodies/governments, and policy makers at either the local, national, or supranational level. Considering that their argument revolves around governmental procurement programs as a method to enforce environmental standards, for Arnold and Whitford governments play a relatively central role in environmental governance. However, it is important to point out that one of the primary arguments of the authors is to remove politics from the EMS standards. Thus, they call for governments to remain involved, but for politicians to become uninvolved.

In discussing private certification systems, Bartley argues that certification schemes must involve private organizations, nonprofit organizations, nongovernmental organizations, activist groups, and trade unions. He does also call for the involvement of governments, but it is very important to note that for Bartley, government can act as a hindrance to the certification systems. Indeed, as previously shown, Bartley argues that NGOs and other organizations have found governments to be immobile and unable to create effective certification and thus have resorted to nonstate efforts – which have proven highly successful.

In discussing the role of women in global environmental governance, Bretherton argues for the involvement of grassroots activists, individual women, multiple diverse cultures and societies, *new* transnational organizations, and governments. Regarding governments, she argues that women's environmental groups need to work with governments to increase their success, but the groups should not be dominated by government. That is, in the spirit of Karkkainen's argument, Bretherton argues that new transnational women's environmental organization must attempt to make their own groups to be equal to governmental groups; that the women's organizations be able to compete on a level field with the government groups.

In comparison to the other authors, Bretherton provides the strongest bottom-up argument. She very specifically emphasizes the role of individuals and communities and must be given credit for doing so.

Lastly, in discussing his post-sovereign environmental governance thesis, Karkkainen argues for the involvement of nongovernmental organizations, multilateral institutions, activist groups, activist citizens, scientists, sovereign states (subnational, national, and international regimes), intergovernmental and non-governmental organizations, business interests, independent scientific community, and essentially anyone else who wants to be and has the means to be involved. In this way, Karkkainen's argument contains more actors than any of the other authors. Essentially, Karkkainen insists that the world is now a place where everyone has a fair chance at making their voice heard and participating in decision making and problem solving endeavors. Indeed, he argues that the world truly is post-sovereign in the broadest definition of the term. That is, not only do nation-states no longer hold rights to sovereignty, but no organization, person, or group of any type holds rights to sovereignty. Authority to make decisions and to carry out actions is now held in the hands of the commons.

Yet it has been elsewhere maintained that the evolving governance system is, in some parts, hierarchical rather than open and transparent as Karkkainen would envision it. That is, some scholars argue that corporate and national power have been relocated to less-obvious but equally as powerful sources, rather than reduced or lost entirely. For example, Bulkeley [2005] shows through literature review that within both traditional and constructivist accounts of environmental governance, decision-making takes place in a hierarchical manner. Bulkeley states:

> For the most part, the scope of global environmental governance is confined to an imagined global scale, either in terms of the nature of the problems to be governed or in terms of the institutional solutions, which are considered appropriate. ... the governance of global environmental issues requires global solutions, which are then 'cascaded' down through national, and implicitly, subnational arenas of governance. ... the scales of governance remain bounded, and there is little consideration of the possibilities that the governance of global environmental issues might emanate from the 'bottom up' [p. 879].

Vogler [2003] argues that governance institutions' "defining characteristic must be the collective acceptance of their authority in superintending other institutions. One can envisage a nested hierarchy of governance institutions reaching down from the inter-state to the local level" [p. 30]. Indeed, there are strong arguments that it is hierarchy itself which is necessary in order for organizations to know what they can do and where they can do it. This notion may stem from the hierarchical nature of government itself and serves as a yet-to-be-tackled challenge within governance initiatives.

Despite trends which indicate the presence of hierarchical governance, Bulkeley identifies that there has recently emerged a number of alternative approaches focusing on horizontal governance structures. These structures involve the involvement of actors and institutions operating simultaneously across multiple scales, in a similar fashion to that envisioned in Karkkainen's post-sovereign system. It is important that these new structures are allowed to emerge and maintain their non-hierarchical nature due to the fundamentals of management studies in which it has been shown that:

> A low degree of hierarchy allows for comparatively high flexibility in the actions of individual organizational units. The dynamics and complexity of problems in environmental policy appear to advise in favor of a more horizontal and flexible structure of international organizations. It appears plausible to assume that organizations with a high degree of flexibility in organizational structure are more effective than organizations with more vertical hierarchies [Biermann & Bauer, 2004, p. 192].

Thus, the success of future environmental governance initiatives may well be entirely dependent on the structure of the initiatives. Initiatives with a high degree of hierarchy are potentially too resembling of government and thus may prove to be as ineffective as past governmental attempts to solve environmental problems. Initiatives with a horizontal and transparent structure, on the other hand, may prove to be highly effective and mobile, while progressively assisting environmental preservation and remediation.

CONCLUSION

Engrained in the four major topics discussed in this chapter are numerous calls for some form of future work and direction in the field of global environmental governance. Arnold and Whitford, for example, call for a more widespread adoption of EMS standards in government policies, agencies, and procurement programs. They point out that large-scale procurement is a recent phenomenon, and although they do not state it plainly, this implies that more analysis on large-scale procurement may be warranted.

Bartley argues that "on the whole, comparing forest and labor standards certification can provide a good deal of analytical leverage for understanding the rise of private regulatory systems. While a larger set of comparisons may also be helpful, this initial two-case comparison takes us far beyond the current norm of industry-specific case studies" [p. 439]. Indeed, Bartley's concept of private certification systems begs to be analyzed further and applied to a broader range of issues both within and outside of the environmental realm.

Bretherton indicates that no grassroots organization of women has yet to be tremendously successful. However, this notion may be challenged by the aforementioned example provided by Jasanoff and Martello. The fact that contending views exist regarding the success of women's grassroots environmental organizations is a call for further research in itself. Bretherton calls for greater organization at the local level across multiple cultures in the formulation of new, women-run organizations, but does not specifically call for future research. Nevertheless, future research is not only warranted, it is absolutely necessary.

Finally, Karkkainen specifically calls on policy makers to consider his thesis for applications to environmental issues both to water-issues similar to his cases, and also to wider environmental policy agendas. At the same time, he seems to indicate that a transition to post-sovereign governance is occurring no matter what and that all parties must learn to function in a new world in which no one party has authority over another. This notion contains within it inherent desire for future research on the subject. Thus, a common theme between all four authors is the shift to "cross-scale environmental governance" and the emergence of a multitude of actors. Indeed, Haas' argument that "more actors now engage in more governance functions at multiple levels of governance" is solidly upheld in all four cases.

Outside of the arguments of the aforementioned authors, there are numerous research areas in need of further development which pertain to the realm of global environmental governance. Of utmost concern is the essentiality that new and progressive work be conducted which seeks to increase public access to environmental and governance-related information and increase public participation overall. According to Margot Wallström, Environment Commissioner of the European Commission (EC) (now Vice President of the EC in charge of Institutional Relations and Communication), the absolute priority issue in improving global environmental governance is improving society's involvement in the environmental decision-making process, including involvement in the implementation of environmental law and, more generally, in greening of policies [Wallström, 2003]. She argued in Brussels in 2003 that "the real challenge is to generate the momentum at all levels of society to work towards sustainable development."

Additionally, more work must be completed regarding the conceptual dimensions of global environmental governance. That is, who should be involved and what are the specific roles of those involved? According to Karkkainen, *everyone* should be involved in global environmental governance initiatives, while other authors maintain that certain parties' influence should be limited. Of course one must realize that not all people interested in participating in governance actually have the means to do so, and facilitating those means serve as a conceptual challenge in itself. Global environmental governance has not yet been established as a discipline with bounds, and thus conceptualizing the idea itself can be a complex matter certainly in need of further study.

Along these lines, more work must be conducted which sets global environmental governance as a unique model for successfully addressing environmental problems. That is, the debate still exists whether or not global environmental governance is a distinctive and effective model, rather than just one of a multitude of regime-like options. For example, the Wisconsin Department of Natural Resources (WDNR) has been experimenting with an alternative form of environmental governance which uses cooperative agreements to provide flexibility to firms in exchange for superior environmental performance. The program attempts to change norms of adversarial and rule driven regulation, to norms of cooperative and flexible regulation [Amengual, 2005]. What, if anything, sets global environmental governance above emerging alternatives such as that by the WDNR? Further, there has yet to be a general academic agreement that global environmental governance initiatives are superior compared to environmental regimes as discussed in the introduction of this paper. As such, new work must be conducted which seeks to solidify the governance model. Doing so will ensure that the methodology receives even greater attention and is further refined and made more efficient.

This chapter has introduced the many diverse challenges of conceptualizing global environmental governance while focusing primarily on the basic ideas, issues, and perspectives of four sub-topics. It has characterized and compared each topic to the others, but also to the reigning body of academic literature on each specific issue. In this way, it has been shown how each sub-topic of environmental governance makes a contribution to, agrees with, or contends with the existing body of literature. Furthermore, although inherent in the body of the chapter, a concise summary of who should be involved in global environmental governance regimes has been provided, according to the arguments of the primary authors. Lastly, a brief discussion regarding each author's vision for future work in the realm of global environmental governance has been offered.

It is evident that a widespread and extremely significant shift is talking place from what has been traditionally known as "government" to a new form of "governance." There is a reigning common trend throughout the literature referring to government becoming only one party among many other parties consisting of nonprofit organizations, nongovernmental organizations, activist groups, citizens, and numerous other entities. Globalization has been the fuel of this shift, and has changed the way in which the world operates, simultaneously increasing complexity and intricacy of our world. With this shift have emerged numerous global environmental governance initiatives which seek to correct environmental dilemmas through innovative forms of governance, rather than government.

Despite the many complications with governance, the future for the environmental governance field looks bright as it continually expands. Governance has become increasingly incorporated into academic curriculums of various disciplines including geography, political science, economics, agriculture, energy and environment, and resource economics. In Fall 2008 the University of Guelph in Ontario, Canada, became the first university to offer an undergraduate degree in Environmental Governance with a specific focus on teaching students how to safeguard the environment through innovative approaches to governing [uoguelph.ca, 2008]. Chances are that other similar programs will follow at other universities around the world. While no "correct" system for global environmental governance has yet to emerge, analysis and expanded discourse on the concept has become imperatively important to ensure that the world's environmental problems are addressed in an organized, effective, and mutually beneficial manner for all citizens of Planet Earth.

AUTHOR BIOGRAPHY

Patrick E. Meyer is a doctoral candidate and research associate at the University of Delaware's Center for Energy and Environmental Policy and is also a research associate with Energy and Environmental Research Associates, LLC., Pittsford, NY, specializing in energy and environmental life-cycle analysis. Meyer serves as the Energy, Environment & Sustainability Editor for *Today's Engineer Online*, a publication of the Institute of Electrical and Electronics Engineers (IEEE). He holds a B.S. in Public Policy and a M.S. in Science, Technology, and Public Policy from the Rochester Institute of Technology. His current research pertains to energy sustainability, infrastructure and transportation relationships in industrialized and developing nations, technology–society–environment relationships, and methods to incorporate social equity into energy and environmental policy initiatives.

ACKNOWLEDGMENTS

The author wishes to thank Robert Warren, Ph.D., Professor of Urban Affairs and Public Policy at the University of Delaware for providing invaluable feedback and foundational concepts for this chapter. The author also wishes to thank Erin Green and Lyndsey McGrath at the Rochester Institute of Technology for their constructive advice.

REFERENCES

Amengual, M. (2005). Flexibility with accountability: an experiment in environmental governance. *Massachusetts Institute of Technology,* Cambridge, MA.

Ansell, C. (2002). Debating Governance: Authority, Steering, and Democracy. *American Political Science Review,* 96(3), 668.

Arnold, R., & Whitford, A. B. (2006). Making Environmental Self-Regulation Mandatory. *Global Environmental Politics,* 6(4), 1-12.

Auer, M. R. (2000). Who participates in global environmental governance? Partial answers from international relations theory. *Policy Sciences,* 33, 155-180.

Bartley, T. (2003). Certifying Forests and Factories: States, Social Movements, and the Rise of Private Regulation in the Apparel and Forest Products Fields. *Politics & Society,* 31(3), 433-464.

Biermann, F., & Bauer, S. (2004). Institutions for global environmental change. *Global Environmental Change,* 14, 189-193.

Bretherton, C. (2003). Movements, Networks, Hierarchies: A Gender Perspective on Global Environmental Governance. *Global Environmental Politics,* 3(2), 103-119.

Bruce, N., Neufeld, L., Boy, E., & West, C. (1998). Indoor biofuel air pollution and respiratory health: the role of confounding factors among women in highland Guatemala. *International Journal of Epidemiology,* 27, 454-458.

Bulkeley, H. (2005). Reconfiguring environmental governance: Towards a politics of scales and networks. *Political Geography,* 24, 875-902.

Conca, K., & Dabelko, G. (1998). *Green Planet Blues* (2nd ed.). Boulder, Colorado: Westview Press.

Danson, M., & Whittam, G. (2005). *Web Book of Regional Science.* Regional Research Institute, West Virginia University Retrieved 2008, 19 December, from http://www.rri.wvu.edu/WebBook/Danson/glossaryterms.htm

EPA. (2006). *Environmental Management Systems*/ISO 14001. US Environmental Protection Agency. Retrieved 17 June, 2007, from http://www.epa.gov/ owm/iso14001/isofaq.htm

Esty, D. C. (2008). Climate Change and Global Environmental Governance. *Global Governance,* 14, 111-118.

Finkelstein, L. S. (1995). What is Global Governance. *Global Governance,* 1(1), 367-372.

Frank, A. G., & Gills, B. (Eds.). (1993). *The World System: Five Hundred Years or Five Thousand?* London: Routledge.

Gremmill, B., & Bamidele-Izu, A. (2002). The Role of NGOs and Civil Society in Global Environmental Governance. In D. C. Esty & M. H. Ivanova (Eds.), *Global Environmental Governance: Global Environmental Governance: Yale Center for Environmental Law & Policy.*

Gulbrandsen, L. H. (2004). Overlapping Public and Private Governance: Can Forest Certification Fill the Gaps in the Global Forest Regime? *Global Environmental Politics,* 4(2).

Haas, P. M. (2004). Addressing the Global Governance Deficit. *Global Environmental Politics*, 4(4).

Harris, P. (1992). *The Third Revolution*. London: Tauris.

Himley, M. (2008). Geographies of Environmental Governance: The Nexus of Nature and Neoliberalism. *Geography Compass*, 2(2), 433-451.

IISD. (2007). *Business and Sustainable Development: A Global Guide: Green Procurement. International Institute for Sustainable Development*. Retrieved 19 December, 2008, from http://www.bsdglobal.com/tools/bt_green_pro.asp

Jasanoff, S., & Martello, M. L. (2004). *Earthly Politics: Local and Global in Environmental Governance*. Cambridge: MIT Press.

Johnston, P., Everard, M., Santillo, D., & Robèrt, K.-H. (2007). Reclaiming the Definition of Sustainability. *Environ Sci Pollut Res Int.*, 14(1), 60-66.

Karkkainen, B. (2002). Post-Sovereign Environmental Governance: The Collaborative Problem-Solving Model. Paper presented at the Conference on the Human Dimensions of Global Environmental Change: "*Global Environmental Change and the Nation State*", Berlin, Germany.

Karkkainen, B. (2004). Post-Sovereign Environmental Governance. Global *Environmental Politics*, 4(1), 72-96.

Kollman, K., & Prakash, A. (2001). Green by Choice?: Cross-National Variations in Firm' Responses to EMS-Based Environmental Regimes. *World Politics*, 53, 399-430.

Lemos, M. C., & Agrawal, A. (2006). *Environmental Governance. Annual Review of Environment and Resources*, 31, 297-325.

Levy, D. L., & John, P. (Eds.). (2005). *The Business of Global Environmental Governance*. Cambridge, MA: MIT Press.

Levy, D. L., & Kaplan, R. (2007). CSR and Theories of Global Governance: Strategic Contestation in Global Issue Arenas. *In A. Crane, A. McWilliams, D. Matten, J. Moon & D. Siegel (Eds.)*, The Oxford Handbook of CSR: Oxford University Press.

Mason, M. (2008). The Governance of Transnational Environmental Harm: Addressing New Modes of Accountability/Responsibility. *Global Environmental Politics*, 8(3), 8-24.

Menard, R. (1991). Transport costs and long-range trade, 1300-1800: was there a European 'transport revolution' in the early modern era? *In J. D. Tracy (Ed.), Political Economy of Merchant Empires*. Cambridge: Cambridge University Press.

Meyer, J. W., Frank, D. J., Hironaka, A., Schofer, E., & Tuma, N. B. (1997). *The Structuring of a World Environmental Regime*, 1870–1990 International Organization, 51(4), 623-651

Najam, A., Christopoulou, I., & Moomaw, W. R. (2004). *The Emergent "System" of Global Environmental Governance. Global Environmental Politics*, 4(4), 23-35.

O'Rourke, K. H., & Williamson, J. G. (2002). *When did globalisation begin? European Review of Economic History*, 6(1), 23.

Oberthür, S., & Gehring, T. (2006). *Institutional Interaction in Global Environmental Governance: The Case of the Cartagena Protocol and the World Trade Organization*. Global Environmental Politics, 6(2), 1-31.

Okereke, C., & Bulkeley, H. (2007). *Conceptualizing climate change governance beyond the international regime: a review of four theoretical approaches*. Norwich, UK: Tyndall Centre.

Pattberg, P. (2005). What Role for Private Rule-Making in Global Environmental Governance? *Analysing the Forest Stewardship Council (FSC) International Environmental Agreements: Politics, Law and Economics*, 5(2), 175-189.

Potter, R., Binns, T., Elliott, J., & Smith, D. (1999). Geographies of Development. Harlow, *England: Pearson Education Limited;* Prentice Hall.

princeton.edu. (2008). *WordNet: "Grass roots".*   *Princeton, NJ: Cognitive Science Laboratory,   Princeton   University.* Retrieved 19 December, 2008, from http://wordnetweb.princeton.edu/perl/webwn?s=grass%20roots

Raskin, P., Banuri, T., Gallopín, G., Gutman, P., Hammond, A., Kates, R., et al. (2002). *Great Transition: The Promise and Lure of the Times Ahead.* Boston, MA: Stockholm Environment Institute - Boston and the Tellus Institute.

Rodriguez-Garavito, C. A. (2005). Global Governance and Labor Rights: Codes of Conduct and Anti-Sweatshop Struggles in Global Apparel Factories in Mexico and Guatemala. *Politics & Society,* 33(2), 203-233.

Rumchev, K., Spickett, J. T., Brown, H. L., & Mkhweli, B. (2007). Indoor air pollution from biomass combustion and respiratory symptoms of women and children in a Zimbabwean village. *Indoor Air,* 17(6), 468-474.

Russell, D. (1989). Environmental racism: minority communities and their battle against toxics. *The Amicus Journal,* 11(2).

Shaver, S. (1996). *Liberalism, Gender, and Social Policy: Social Policy Research Centre.*

Smith, Z. A. (2004). *The Environmental Policy Paradox* (4th ed.). Upper Saddle River, NJ: Prentice Hall.

Speth, J. G. (2002). The Global Environmental Agenda: Origins and Prospects. In D. C. Esty & M. H. Ivanova (Eds.), *Global Environmental Governance: Yale Center for Environmental Law & Policy.*

St. Clair, A. L. (2006). The World Bank as a Transnational Expertised Institution. *Global Governance,* 12, 77-95.

Stokke, O. S. (1997). Regimes as Governance System. *In O. Young (Ed.), Global Governance: Drawing Insight from the Environmental experience* (pp. 27-63). Cambridge, MA: MIT Press.

uoguelph.ca. (2008*). U of G First to Offer Environmental Governance Degree.* Guelph, Ontario, Canada: University of Guelph. Retrieved 1 December, 2008, from http://www. uoguelph.ca/news/2008/02/u_of_g_first_to.html

Vogler, J. (2003). Taking Institutions Seriously: How Regime Analysis can be Relevant to Multilevel Environmental Governance. *Global Environmental Politics,* 3(2), 25-39.

Vogler, J., & Stephan, H. R. (2007). The European Union in global environmental governance: *Leadership in the making? International Environmental Agreements,* 7, 389-413.

Wallström, M. (2003). *Commissioner Wallström's speech on "Environmental Governance and Civil Society".* Brussels, Belgium: europa-eu-un.org. Retrieved 1 December, 2008, from http://www.europa-eu-un.org/articles/en/article_1958_en.htm

Weiss, G. T. (2000). Governance, good governance and global governance: conceptual and actual challenges. *Third World Quarterly,* 21(5), 795-814.

worldbank.org. (2008). Global Environmental Governance in 2050. The World Bank, News & Broadcast. Retrieved 1 December, 2008, from http://go.worldbank.org/ FX39REP010.

In: Handbook of Environmental Policy
Editors: Johannes Meijer and Arjan der Berg

ISBN 978-1-60741-635-7
© 2010 Nova Science Publishers, Inc.

Chapter 5

ENVIRONMENTAL GOVERNMENTALITY AS A POLICY APPARATUS: THE CASE OF SHRIMP AQUACULTURE IN BANGLADESH

Md Saidul Islam[*]

Division of Sociology, Nanyang Technological University
Singapore

ABSTRACT

Shrimp industry is contested as it is identified with negative social and environmental legacies. Bangladesh, being one of the major shrimp producing countries of the world, has been facing resistance and criticism from local and international environmental NGOs. In response, Bangladesh government along with its donor agencies has come up with a series of environmental agendas and programs to ensure "environmentally sound shrimp aquaculture". This process of institutionalizing environmental domain pertaining shrimp industry have some positive impacts in terms of creating awareness among the people regarding environment, but at the same time it marginalizes others. It benefits a fortunate few, but the fate of people affected by the industry remains almost the same. The study demonstrates a trend and development, which is quite common to all environmental issues today, that is, the trend of moving towards a domain of managerialism, bureaucratization, and governmentality.

INTRODUCTION

Bangladesh enjoys an advantageous natural setting for shrimp culture. Its vast floodplain stretches about 25,000 square kilometers along the coastline and is nourished by spring tides. The low-lying areas of Sundarban, home to the world's largest mangrove forest, are important hub for shrimp-fry (Deb 1998, p. 65). The beginning of the present shrimp culture dates back to the late sixties when a number of fish freezing plants were set up in Chittagong and

---

[*] Email: msaidul@ntu.edu.sg

Khulna. Shrimp export and cultivation in Bangladesh has undergone rapid expansion over the last two decades. Since the 1980s there has been a dramatic increase in shrimp farming, especially in the coastal areas where this has been termed as the "blue revolution" (Deb 1998). Between 1980 and 1995, the area under shrimp cultivation increased from 20,000 hectors to 140,000 hectors (Metcalfe 2003, p. 433). Between 1983 and 2003 the volume of shrimp cultivated in inland aquaculture has increased more than 14 times. Over the same period, the area of ponds dedicated to shrimp production has more than trebled. In 2003, the Department of Fisheries estimates that there are approximately 203,071 hectares of coastal shrimp farms producing an average of 75,167 metric tons of shrimp annually and an average of 370 kg/ha/year.[*] Currently, shrimp cultivation in Bangladesh is concentrated in the coastal areas of Khulna and Chittagong regions, with 80 percent of the total shrimp lands in Khulna region especially in the districts of Khulna, Satkhira and Bagerhat, with the remaining 20% in Chittagong region (USAID Bangladesh, 2006).

During the mid-seventies, production and export of shrimp witnessed an exponential growth, owing to a combination of several factors, such as increased demand from abroad and a large supply of low-wage local labors. Demand for shrimp products stemmed from Europe, United States, Japan and the Middle East; while the European Union continues to be the largest importer of Bangladeshi shrimp (Pokrant and Reeves 2003). Shrimp, being the second largest industry next to garments, brings a foreign exchange of US$360 million annually that accounts for 4.9 percent of exports in 2004 (Gammage et al 2006, p. 1). Besides earning foreign exchange, the sector also employs significant numbers of rural workers and provides a livelihood for households throughout Bangladesh. The Bangladesh Shrimp and Fish Foundation estimate that there are over 600,000 people employed directly in shrimp aquaculture who support approximately 3.5 million dependents. The estimates of my study indicate that there may be as many as 1 million individuals engaged directly in production and exchange throughout the shrimp value chain and a further 5 million household members whose wellbeing is linked to the sector.

Approximately 35,000 tonnes of shrimp are exported to the world annually. Experts estimate that the volume can be raised up to 300,000 tonnes just through proper utilization of shrimp fry. The country's shrimp sector has now set an ambitious target to raise export earnings by 2008 to Taka[*] 10,000 crore or about US$1.7 billion, (i.e., five times up from the present annual exports) (NewAge, April 3, 2004).

Several research in Asia and elsewhere have shown that the emergence and growth of shrimp culture is identified with various social and environmental problems. In many areas the land now under shrimp farming was used for growing paddy for generations together. Paddy growers, who were not able to shift towards shrimp farming, had to lease or sell their lands. This resulted in the displacement of many local residents. In addition to social repercussions, shrimp farming has resulted in environmental devastation. For example, research has shown how vast areas of mangrove forest and its biodiversity are being destroyed due to shrimp industry (Shiva 1995, Vandergeest et al 1999, Flaherty et al 1999).

---

* These data are for 2003-2004, quoted in Gammage et al (2006).
* Bangladesh's local currency.

Export of Bangladeshi Shrimp (Percentage)

Country	Percentage of earning in terms of value	
		Subtotals
USA	34.78	34.78
Japan	11.26	11.26
EU Countries		48.74
Belgium	16.22	
U.K.	11.26	
Netherlands	8.59	
Germany	6.86	
Denmark	2.07	
France	1.15	
Norway	0.92	
Switzerland	0.76	
Italy	0.91	
South East Asia		2.21
Thailand	0.12	
Singapore	0.29	
Malaysia	0.08	
Taiwan	0.05	
Hong Kong	1.67	
Australia	0.03	0.03
Middle East		0.05
Saudi Arabia	0.03	
UAE	0.02	
Other countries including India	2.93	2.93
Total	100	100

Source: Bangladesh Export Promotion Bureau, 2004

The shrimp aquaculture in Bangladesh is not an exception from this pattern. The expansion of shrimp culture in Bangladesh "has been unregulated, uncontrolled and uncoordinated" (Metcalfe 2003, p. 433). The general impact of this unregulated expansion has been fourfold: first, over fishing (as the industry is mainly dependent on wild sources of fry); second, the loss of shellfish and fin fish (due to the catching of wild tiger shrimp fry); third, the destruction of fragile mangrove ecosystems (which provide many other benefits to society); and finally, soil acidity and salt incursion with flow-on effects on other coastal agricultural activities that include paddy farming. In addition, intensive shrimp farming also requires large amounts of fresh water (approximately 25-30 million litres to raise one tonne of shrimp), and shrimp pond effluent (ammonia, nitrate, phosphate, bacteria) makes significant contribution to both inorganic and organic coastal pollution. The socio-economic effects of shrimp cultivation include the traditional resource-use rights of coastal people being eroded and labour utility disequilibrium (Deb 1998; Metcalfe 2003).

Local and international environmental activists, especially environmental NGOs have succeeded in bringing the issue of environmental and social damage caused by the shrimp industry as well as the issue of food-safety to the world attention. Bangladeshi authorities, in turn, responded by presenting a series of environmental agendas as well as inspection and quality-control institutions for quality shrimp that demonstrate their adherence to the principle of "environmentally-sound shrimp aquaculture". Not only that, in response to buyer and consumer requirements for quality shrimp, the leading processing plants are continuously shifting towards new regulations (Pokrant and Reeves 2003). This has led to the emergence of a regime of environmental governance, which has profound sociological implications on the people involving in the industry.

Prevalent literature on the topic are either from the activist perspective, which contests the industry for its social and environmental costs or from the perspective of the state/government that continuously defends the industry by presenting different policies and procedures. Through a combination of qualitative methodologies, this study thus analyses discourses that have emerged from the various stakeholders in Bangladesh's shrimp industry. By using "environmental governmentality" as conceptual metaphor, the paper addresses the following research questions: (a) What are the environmental and social damages caused by the shrimp farming in Bangladesh that local and international NGOs claim are serious? (b) What are the politics behind the rhetorical shift of Bangladeshi government towards the principle of "environmentally sound shrimp aquaculture"? In other words, how does institutionalized environmental managerialism contribute to privilege a fortunate few, and marginalize others? And (c) how is the regime of governmentality established and amplified, to the possible exclusion of other forms of engagement in the politics of nature?

ENVIRONMNETAL GOVERNMENTALITY

The term 'governmentality' ('*gouvernementalite*') is a neologism Foucault presented and explored at the end of the 1970s (Foucault 1979; 1991 and 1984) that implies the establishment of complex social techniques and institutions to intensify and expand the mechanism of control and power over the population in the name of what became known as the 'reason of state'. Governmentality, for Foucault, referred famously to the "conduct of conduct" (2000, p. 211), a more or less calculated and rational set of ways of shaping conduct and securing rule through a multiplicity of authorities and agencies in and outside of the state and at a variety of special levels, which he calls "art of government" (1979, p. 5), albeit negatively.

There are two aspects to governmentality in the Foucault's writings. First, it is a concept based on the European historical context. Secondly, it implies a novel definition of power, which has profound implications for our understanding of contemporary political power and particular public policy. For Foucault, governmentality is the unique combination of three components: institutional centralization, intensification of the effects of power, and power/knowledge (Foucault 1979; Pignatelli 1993), that denotes "governmental rationality" (Gordon 1991). In speaking of governmentality, Foucault was referring not only to the domain of civil/political government as it is conventionally understood but to a broader domain of discourses and practices that create and administer subjects through the presence of a variety of knowledge-making apparatuses. Most significantly, the focus of a Foucaultian study of policy is on the broader impact of state policy or more specifically on the power effects across the entire social spectrum (macro level) down to individual's daily life (micro level). Governmentality for Foucault refers not to sociologies of rule, but to quote Rose (1999, p. 21), to the:

> studies of stratums of knowing and acting. Of the emergence of particular regimes of truth concerning the conduct, ways of speaking truth, persons authorized to speak truth... of the invention and assemblage of particular apparatuses for exercising power... they are concerned with the conditions of possibility and intelligibility for ways of seeking to act upon conduct of others.

For Foucault, governmentality is a fundamental feature of the modern state. Most significantly, he sees state authorities and policies as mobilizing governmentality which tries to incorporate the economy and the population into the political practices of the state in order to be able to govern effectively in a rational and conscious manner (Foucault 1991; Luke 1995). Governmentality, then, applies techniques of instrumental rationality to the arts of everyday management exercised over the economy, the society, and, in the context of this study, the environment.

Recently there have been attempts to extend the concept of governmentality into the realm of environment (see Luke 1999; Brosius 1999; Escobar 1995; Agrawal 2003; Darier 1999).   Éric Darier, for example, deploys Foucault's analytic tools to deconstruct contemporary environmental discourses, specifically the relations and technologies of power/knowledge that underpin them and the effects they have on individual conduct in private, daily life (cf. 1996a, 1996b; 1999). He applies Foucaultian frame to the deployment of citizenship in Canadian environmental discourse to theorize what he calls "environmental governmentality"[*]:

> Environmental governmentality requires the use of social engineering techniques to get the attention of the population to focus on specific environmental issues and to instill, in a non-openly coercive manner, new environmental conducts... [T]he challenge for the state is to find ways to make the population adopt new forms of environment conduct. If coercion is not the principal policy instrument, the only real alternative is to make the population adopt a set of new environmental values, which would be the foundation of new widespread environmental ways of behaving. These new environmental values will be promoted by the establishment of an "environmental citizenship" (Darier, 1996b, p. 595).

One of the basic ways of extending the realm of governmentality is institutionalization, which is, to Foucault, a process of centralizing power around the government (army, education, governmental ministries, justice etc.), and thereafter intensifying the effects of power at the levels of entire population, the economy and the individual. The process also

---

[*] The term "Environmental/ Green Governmentality" or "Environmentalty" has first been used by Luke (1995, 1997) who views it as an attempt by transnational environmental organizations to control and dominate environmental policy and activities around the world, but especially in developing countries. See also the collection of essays in Darier (1999). Agrawal's (2003) use of the term is indebted to Luke for the coinage, but is different both in intent and meaning. Agrawal attempts to examine more insistently the shifts in subjectivities that accompany new forms of regulation rather than see regulation as an attempt mainly to control or dominate.

needs new forms of (scientific) knowledge, which eventually creates 'institutionalized subjectivity'. Brosius (1999) explicates the remarkable scale of the process of institutionalization in the realm of environment. It takes many forms: the Montreal Protocol, the Convention of Biological Diversity, and the Agenda 21, to name just a few. Acronyms proliferate: UNEP (United Nations Environment Programme), UNCEP (United Nations Conference on Environment and Development), CSD (Commission on Sustainable Development), WCMC (World Conservation Monitoring Centre), TFAP (Tropical Forest Action Plan), IPCC (Intergovernmental Panel on Climate Change), ICDPs (Integrated Conservation and Development Programmes), NEAPs (National Environmental Action Policies) and so on ad infinitum. Each of these represents or supports regimes for the institutionalization of environmental surveillance and governance.

To exemplify the indicators of the scale of contemporary environmental institutionalization, Brosius turns our attention to "the accelerating pace of professionalization" (1999, p. 38), specially "the remarkable growth of the field of environmental management and... the proliferation of environmental studies programme at universities" (p. 38). With respect to the proliferation of environmental studies programme at universities, Brosius argues that it represents efforts to train a transnational cadre of planners to design and execute various forms of environmental intervention. This process of environmental institutionalization can be viewed as a positive development especially in terms of raising environmental concerns to the level of legitimacy that was previously lacking. However, some concerns remain, as Brosius (1999) puts it:

> There are reasons to be concerned about this process of institutionalization. Such institutions whatever they may do, inscribe and naturalize certain discourses. While they create certain possibilities of ameliorating environmental degradation, they simultaneously preclude others. They privilege certain actors and marginalize others. Apparently designed to advance an environmental agenda, such institutions in fact often obstruct meaningful change through endless negotiation, legalistic invasion, compromise among "stakeholders", and the creation of unwidely projects aimed at top-down environmental management. More importantly, however, they insinuate and naturalize a discourse that excludes moral or political imperatives in favour of indifferent bureaucratic and/or technoscientific forms of institutionally created and validated intervention. [p. 39]

This study examines the institutional development as part of environmental governmentality in Bangladesh with reference to the global campaign of the different environmental NGOs against the severe social and environmental damage caused by shrimp farming. In other word, the study examines the rhetorical shift of the Bangladeshi authorities towards the domain of institutionalized environmental managerialism. This, in fact, provides a number of important insights into how regime of environmental governmentality is established and amplified, to the possible marginalization of other forms of engagement, like Non-Governmental Organizations (NGOs), and Community-led institutions, in the politics of nature and environment.

Methods and Procedures

A triangulation of methods- secondary sources substantiated by ethnography and qualitative interviews- was employed in this study. A combination of methods is deemed necessary in order to gain a deeper understanding of shrimp culture. Secondary sources were obtained from journals, books, national newspapers, internet search, government reports, and publications by NGOs in Bangladesh. The archival library at the Daily Sangram office in Dhaka was particularly helpful as it contains a wealth of up-to-date data on shrimp. While taking any data from newspapers, a serious scrutiny was made as reports in the newspapers are purely journalistic than academic. Ethnography and qualitative interviews in Bangladesh were conducted in two different phases.

Phase I: (May- December 2005)

Ethnography and interviews were conducted in three districts of greater Khulna region- Satkhira, Bagerhat, and Khulna. These sites are selected because of two main reasons: first, these are areas within the highly productive "rice bowl" region of Bangladesh currently shifting towards shrimp farming. Secondly, these areas contain about 80 percent of total shrimp ponds in Bangladesh (Khatun 2004; Fleming 2004). In order to gain an understanding of the size, success, regulations, dynamics of power relations, and problems and prospects associated with the introduction of the shrimp culture into rice growing areas, semi-structured qualitative interviews were conducted with five target groups: villagers, shrimp farmers, shrimp traders and processors, NGO workers, and government officials. The sample size was 9 shrimp farmers, 9 villagers, 6 government officials, 6 processors, and 5 NGO workers, a total of 35 respondents, in three regions of Bangladesh namely Satkhira, Khulna and Bagherhat. It is not practical to obtain a formal random sample, but I endeavored to get a representative and diverse sample from all three districts.

Snowball sampling was employed in getting access to the respondents. However, before getting any subject for interview, a discussion was done with District Fisheries Officers (DFOs) in all three areas with prior appointments as well as with local NGOs. Not all areas in those three regions contain high concentration of shrimp farming, and therefore a suggestion was necessary from the DFOs and local NGOs as to what particular areas in those districts should be selected for ethnography and interview. Three sub-districts (*upajilla*) were randomly selected among areas recommended by DFOs and NGOs, namely Satkhira Sadar in Satkhira district, Ovoinagar in Khulna district, and Mollarhat in Bagherhat districts.

Farmers were selected via snowball sampling, but certain criteria were employed while interviewing farmers. Three farmers were selected from three representative groups: (a) Small scale farmers having up to 5 acres of land under shrimp farming, (b) Medium scale farmers who have 5-10 acres of land, and (c) Large scale farmers with more than 10 acres of land under shrimp farming.* Three villagers from three different randomly selected villages (key informant about their areas) in three local districts were accessed via snowball sampling and were interviewed.

---

* This categorization may seem arbitrary, but it was done to ensure representativeness.

Processing centres are mostly located in Khulna city. Six processing workers (four males and two females) were selected via snowball sampling and interview was conducted with them. Among the six government officials interviewed, two were from the Ministry of Fisheries and Livestock, two DFOs of Satkhira and Bagherhat, and two UFOs (Upajilla Fisheries officers) of Kaliganj and Molarhat respectively. Five NGO workers were chosen via snowball sampling from each of five different NGOs, namely, Agragati Sangstha, Coastal Development Partnership, Bangladesh Rural Advancement Committee (BRAC), Grameen Bank, and Nijera Kori - all of them are among the major NGOs working in shrimp sector albeit differently.

As explicated, the study involves both stratified (in terms of areas) and snowball sampling. Apart from this semi-structured qualitative interview with selected respondents, I also talked to the many other people directly or indirectly related to shrimp in a 'group interview' manner to generate some important data. The search for additional respondents ended when the author found that little or no new information was being obtained.

Phase II (April –August 2006)

The second phase of research was grounded on the research done in the first phase, and focused mainly on gender, equity, and poverty issues in shrimp aquaculture in Bangladesh. A semi-structured interview with the processing workers, with an emphasis on gender and equity issues, was conducted in the greater Khulna regions.

There are 124 processing factories in Bangladesh with a capacity for processing 825 ton per day. Of these 35 are presently operational in Khulna region, (65-70% of the total production capacity) 38 are operational in Chittagong district and 51 plants have either closed down or are waiting for approval for quality inspection license (Halim 2004). Four factories, namely Atlas Sea Food, Rupali Sea Food, Asian Sea Food, and Fresh Food Ltd. were randomly selected out of 35 factories operating in Khulna region. Semi-structured interview was conducted with the owners/ managers as well as male and female workers working in those factories. One person (owner/manager) from each factory management (a total of four) was interviewed with prior appointment.

Having no access to the workers while they were in factories*, I talked to local people and knew that processing workers live in mainly two areas of Khulna city, namely Char Village (a slum), and Lobon Chora Village. First, I went to Char village and found a slum of approximately 6000 people, many of them work in the processing factories. I first contacted with the slum leader to get access to the workers. I interviewed 2 male workers, 2 female permanent workers and 4 causal female workers. In the similar manner, I interviewed 2 male workers, 3 female permanent workers and 5 causal female workers in the Lobon Chora vicinity. Separate semi-structured questionnaires for factory owners/ managers (inside factory) and workers (both male and female outside of factory) were used to elicit information about processing, storage, wages, prices, profits, productivity, cost-structures, problems and prospects along the commodity chain.

---

* In June 2006, Bangladesh Frozen Food Exporters Association called a meeting and passed a resolution that no factory would disclose any information of its workers to anyone. There is a four-member council in every

Among the farming areas, three areas (sub-districts) were randomly selected, namely Shyam Nagar (Satkhira), Mongla (Khulna) and Faqirhat (Bagherhat), and a series of focus group discussions was done with representatives from key nodes of shrimp commodity chain as well as conscious people living in the shrimp firming vicinity. The following table illustrates my focus group interview:

	Type	Number	Place
1.	Fry catcher	12	Mongla
2.	Fry traders	9	Mongla
3.	Fry hatchery owners	7	Shyam Nagar
4.	Shrimp traders and Middlemen	8	Faqirhat
5.	Women workers in shrimp ponds	12	Shyam Nagar
6.	Upajilla Fisheries Officers	5	Faqirhat
7.	Conscious people (hotel managers, local businessmen, teachers etc.)	10	Faqirhat
8.	Conscious people (College professors, and Human rights activists)	8	Shyam Nagar
	Total	71	

The data collected do not constitute a statistically representative sample of all shrimp producers and processors in Bangladesh. They are, however, illustrative of the types of the production and marketing that takes place in the lower end of shrimp commodity chain in the areas sampled in Bangladesh. In addition to the above, I had informal and occasional meetings with other government officials, NGO workers, national intellectuals, other resource persons with similar interests, shrimp seal of quality organization, Bangladesh Shrimp Farmers Association (BSFA), Bangladesh Shrimp and Fish Foundation (BSF), and Bangladesh Frozen Food Exporters Association.

In both phases, semi-structured interviews were conducted in conversational mode. This allowed for rapport and trust to develop between the interviewer and respondents, as well as allowed maneuverability in exploring emergent themes. Moreover, respondents in the obtained sample have been identified as 'key' individuals to interview. The respondents therefore speak with relative authority on the subject matter. In addition to primary data, collected through key informants and focus group interviews and ethnography (my direct observation in the field), I also rely on secondary sources for information about environmental and social costs and benefits of shrimp production and data verification. Where possible, I attempted to triangulate to verify my findings by comparing data and information from these different sources.

RESULT

1. Environmental and Social Damage Generated by Shrimp Industry

Historically, farmers have confined shrimp culture to ocean coastlines because shrimp require large volumes of saltwater to reproduce and mature. Recent developments in

factory. For any information/ data, permission is needed from the council. Because of this, no worker in the factory was allowed for interview.

Bangladesh, suggest, however, that this once purely coastal activity could soon be a thing of the past. Many Bangladeshi rice farmers are adopting low-salinity culture systems that rely upon sea water or salt farm effluent that is trucked inland. This innovation, combined with low farm gate prices for rice and high prices for shrimp, has led increasing numbers of rice farmers in Bangladesh to convert paddy fields to shrimp ponds. As a result, the amount of land under shrimp production has skyrocketed in recent years. According to Kabir (2003), a total of 375,000 acres (150,000 hectors) of land have been brought under saline water shrimp cultivation, most of which has been land reclaimed from rice paddies. As the shrimp ponds take the place of rice paddies, locals have been increasingly isolated from both water and food sources. Due to constant pressure of the shrimp-firm owners, many paddy farmers have to sell or lease their lands for the shrimp farms.

According to the respondents, shrimp farming in Bangladesh is sustained through a complex system of political patronage, a group of corrupt officials, local politicians and a handful of drug-dealers. They are lured by the short-term enormous profit at the expense of the local people. There are various instances of the shrimp producers evicting small and marginal rice farmers with the help of hired musclemen.

Sundarban, located at the southern region of Bangladesh, is the biggest mangrove forest of the world. However, this forest has been endangered due to the proliferation of shrimp industry. Thousands of people go to Sundarban for fry collection and thereby damage the bio-diversity in the forest area. Mangrove destruction has a major negative impact on marine fisheries. Parkins (1999) estimates, each acre of mangrove forest destroyed leads to an estimated 676 pounds loss in marine harvest. According to Pasha (2000):

> In the Khulna region one shrimp-fry collector for each fry destroys 25 different species of shrimps, and 16 fish species. The collectors take only their desired shrimp fry, and dump rest of the species on the bank of the rivers. Shrimps farms in Bangladesh require at least 5000 million shrimp fry, of which only 1,500 million are supplied by the hatcheries (p. 14).

Shrimp farms have displaced many local people. According to Mike Hagler of Greenpeace, 'Shrimp farming in Bangladesh's Shatkhira region displaced 40% of the areas 300,000 inhabitants into the country's overcrowded cities' (1997, p. 2). These numbers might be exaggerated; however, there is no doubt that there is some displacement. In an interview, Mukul, an NGO worker in Satkhira, reported, "Previously there was nice greenery; flora and fauna in these areas, now the only things we can see are shrimp ponds. Many people who used to work in the paddy fields and earn their livelihood, moved to the towns and the place for most of them in the towns is the urban slums."

My village informants believe that shrimp farms play a significant role in causing flood in the regions of Khulna and Satkhira. During the rainy season, water cannot move due to hundreds of shrimp ponds. The mud brought by flood is covering the rivers, and consequently the navigation in the rivers is severely affected. A village elder said, "There were floods, but not like there are now- the waves were usually supported by the forests. After the proliferation of shrimp farming, the destruction increased and so did the intensity of flood."

In summing up the social impacts of export-oriented shrimp farming, it can be said that its benefit accrue substantially to a minority directly involved in exploitation of coastal resource systems, while a series of direct costs are paid by the majority who reside in these areas, and who make their daily living from the resources. Neither social nor the ecological costs of shrimp culture development are paid by the investors, who pocket high profits during the growth phase of the industry, but socialize the costs as society at large is left with the bill for the considerable environmental and social damage.

2. Rhetorical Shift towards 'Environmentally Sound Shrimp Aquaculture'

The explosive growth of shrimp industry has generated mounting criticisms over its negative social, economic, and environmental consequences. The escalating conflicts between critics and supporters of industrial shrimp farming have transcended local and national arenas. These tensions have catalyzed the formation of environmental and peasant-based non-governmental organisations (NGOs) opposed to shrimp farming and industry groups seeking to counter the claims and campaigns of the resistance coalition.

Alarmed at the rapid growth of destructive types of aquaculture, like shrimp farming, NGOs from shrimp producing and consuming countries around the world are organizing to halt the spread of destructive shrimp farms, since their governments are failing to act (Princen 1994)[*]. The NGOs are also demanding that governments enforce prohibitions on the wholesale conversion of agricultural or cultivable land to aquaculture use, the use of toxic and bio-accumulative compounds in aquaculture operations, pollution of surrounding areas, the development and use of genetically modified organisms in aquaculture and the use of exotic/alien species. The NGOs also call on governments to prohibit aquaculture practices that cause the salinization or depletion of fresh water supplies, and ban the use of feeds in aquaculture consisting of fish that is or could be used as food for people (Low 1994). There is also a need to ensure that the collection of wild larvae to stock shrimp ponds does not adversely affect species biodiversity in the areas where collection takes place. Another key demand of the NGOs is put to the multi-lateral development banks, bilateral aid agencies, the UN Food and Agriculture Organization (FAO) and other relevant national and international development assistance organisations that they stop funding or otherwise promoting aquaculture development that is inconsistent with the above criteria.

Apart from this global awareness created by different Global NGOs, several other local and international NGOs in Bangladesh are working locally to create a voice for those who are severely affected and whose basic rights are compromised due to the growth of shrimp farming. Some international NGOs like Nijera Kori, ADAB, CARE, Uttaran etc. are working for creating awareness among the people with regard to the negative consequences of shrimp industry. Nijera Kori is playing a vociferous role in this regard. National NGOs like Coastal

---

* In May 1995, Greenpeace and 24 other NGOs, some representing people living in the communities that are directly feeling the impacts of the shrimp farming boom, submitted an unprecedented "NGO Statement on Unsustainable Aquaculture" to the Commission on Sustainable Development (CSD) meeting in New York at the United Nations. The group urged their governments to move quickly to ensure the development of aquaculture that is compatible with the social, cultural and economic interests of coastal communities, and ensure that in future such developments are sustainable, socially equitable and ecologically sound. The NGO statement to the United Nations was followed up by an NGO Forum on Shrimp Aquaculture held in Choluteca, Honduras (Oct. 13-16, 1996), organized by Greenpeace together with CODDEFFAGOLF, a Honduran grassroots group. Twenty-one NGOs from Latin America, India, the US and Sweden concluded the meeting with the adoption of the Choluteca Declaration. The Choluteca Declaration reaffirms the demands contained in the NGO Statement on Unsustainable Aquaculture submitted to the United Nations earlier in the year (See, Hagler 1997).

Development Network (CDN) have various agendas on environmental and social protection. It includes 25 local NGOs of which 23 are working in Satkhira region.

Faced with severe criticisms from local and international environmental NGOs, the government of Bangladesh responded with a series of environmental agendas for shrimp aquaculture. In order to counter different international and local environmental NGOs, the international development agencies as well as the state government have promoted several institutions to endorse environmentally friendly shrimp aquaculture[+]. With a view to displacing the politics from shrimp industry in Bangladesh, the government officials always vociferously reiterate that the government is committed to ensure an environmentally sound shrimp aquaculture. These statements have been published in different pro-government national dailies. There are several government-funded environmental magazines, such as, the "Bangladesh Environmental News Letter", "Agribusiness Bulletin" and so forth. These take the lead to discipline the people according to what the government thinks. An attempt to extend power over the realm of environment is evident in the initiation of powerful institutions like the "Ministry of Environment" and the "Ministry of Fisheries and Livestock". The government has its own experts employed in different institutions for knowledge production. The purpose is to amplify the government's power in the realm of fisheries and environment. Some notable research and educational institutions are: Department of Fisheries: Government of Bangladesh, Bangladesh Fisheries Research Institute in Mymensingh, Bangladesh Agriculture Research Council, and so on, and each institution has specialized experts. Apart from all these, there are few hundreds experts in the Bangladesh Agriculture University (BAU), and other universities of Bangladesh who are also involved in various research programmes initiated and financed by the government.

New quality, social and environmental regulations and programmes are slowly being extended to suppliers, shrimp farmers, hatcheries, and wild fry collectors. The government has various programmes in the realm of environment. The Sustainable Environment Management Programme (SEMP) is one of the initiatives of the Ministry of Environment (MoE) to minimise the environmental degradation facing the country. The Ministry of Environment prepared the National Environment Management Action Plan (NEMAP). SEMP consists of five major institutional components: Policy and Institution, Participatory Ecosystem Management, Community based Environmental Sanitation, Awareness and Advocacy, and Training and Education (Jilani 1999; Karim 2000). In short, the following institutional bodies are involved in aquaculture and fisheries in Bangladesh:

- Department of Fisheries (DoF) under the Ministry of Fisheries and Livestock (MoFL) is the sole authority with administrative control over aquaculture in Bangladesh. The DoF is managed by a Director General and has two main sub-departments namely, inland and marine. The main responsibilities held by the DoF include planning, development, extension and training, DoF has six divisional offices, 64 district offices and 497 upazilla (sub-districts) offices and in addition it has 118 hatcheries and four training centers (Mazid, 2002).

[+] On July 21, 1999, the World Bank announced the approval of a US$28 million equivalent credit and a US$5 million Global Environmental Facility (GEF) grant for the Bangladesh Fourth Fisheries Project to increase "environmentally-friendly and sustainable" fish and shrimp production. The credit is provided by the International Development Association (IDA), the World Bank's concessional lending affiliate (The World Bank Group, News Release No. 2000/009/SAS, January 20, 1999).

- Bangladesh Fisheries Research Institute (BFRI) conducts and coordinates research and to some extent training.
- Bangladesh Rural Development Board is responsible for the fisheries component of integrated rural development.
- Land Administration and Land Reform Division is responsible for the leasing of public water bodies.
- Export Promotion Bureau is responsible for export of fisheries products, along with the Bangladesh Frozen Foods Exporters Association which is also involved in the export of frozen shrimp, fish and fish products.
- The country's universities are responsible for higher level fisheries education.
- External Resource Division under the Ministry of Finance is responsible for external aid for aquaculture development.
- Bangladesh Krishi (Agriculture) Bank, Bangladesh Samabay (Co-operative) Bank and some other commercial banks are responsible for issuing credit to the aquaculture sector.
- Many of the national and international NGO's provides credits to the fish farmers and as well as takes up projects for aquaculture extension and development.
- International organizations (DFID, Danida, NORAD, JICA, World Bank, IMF, ADB etc.) provide grants and credits for aquaculture development.
- Youth Development Training Centers, under the Ministry of Youth, deals with extension and the training of unemployed young people and fish farmers.

For governing regulations, the government enacted laws and enforced them. The basic act regulating inland fisheries is the Protection and Conservation of Fish Act (1950), as amended by the Protection and Conservation (Amendment) Ordinance (1982) and implemented by the Protection and Conservation of Fish Rules (1985). The Marine Fisheries Ordinance (1983), as implemented by the Marine Fisheries Rules (1983), is the basic act regulating marine fisheries. Although the basic fisheries legislation does not have separate sections on aquaculture, some of its provisions are relevant to the subject. The Protection and Conservation of Fish Rules, for instance, specifically deal with the protection of certain carp species, prohibit certain activities to facilitate their augmentation and production and stipulate that licenses for their catch shall only be issued for purposes of aquaculture. In Bangladesh, seeding is traditionally by wild post larval and juvenile shrimps, or fish fry, which are trapped in ponds during tidal exchanges or which are gathered from the estuaries in the vicinity and used to stock the ponds. In recognition of the fact that fry collection from nature may result in long term ecological destruction, in 2000 the government reportedly prohibited the collection of fry or post larvae of fish, shrimp and prawns of any kind, in any form and in any way in estuary and coastal waters.

Other legislation that is relevant to aquaculture includes the Tanks Improvement Act (1939), which provides for the improvement of tanks for irrigation and aquaculture purposes. The Shrimp Culture Users Tax Ordinance (1992) stipulates that shrimp cultivation areas developed by the government by construction of embankments, excavation of canals or other water management structures shall be liable to payment of tax. In addition to these laws, aquacultures, and the conditions of its development, are affected by a variety of other laws, such as land laws, water laws and environmental regulations.

The Ministry of Fisheries and Livestock (MoFL), through its Department of Fisheries (DoF), has overall responsibility for fisheries and aquaculture development, management and conservation. Its functions, which are both regulatory and development oriented, are defined in Schedule 1 of the Rules of Business (1975) and include, inter alia, the preparation of schemes and the coordination of national policy in respect of fisheries, the prevention of fish disease, the conservation, management and development of fisheries resources, the management of fish farms and training and collection of information. The activities of DoF are supported by the Bangladesh Fisheries Research Institute (BFRI), which is responsible for fisheries research and its coordination. In addition, the Bangladesh Fisheries Development Corporation (BFDC), established under the Bangladesh Fisheries Development Corporation Act (1973), supports DoF in developing the fishing industry. Functions of BFDC include, inter alia, the establishment of units for fishing and for the preservation, processing, distribution and marketing of fish and fishery products.

In 1998, a National Fisheries Policy was adopted to develop and increase fish production through optimum utilization of resources, to meet the demand for animal protein, to promote economic growth and earn foreign currency through export of fish and fishery products, to alleviate poverty by creating opportunities for self-employment and by improving socio-economic conditions of fisher folk, and to preserve environmental balance, biodiversity and improve public health. The Policy extends to all government organizations involved in fisheries and to all water bodies used for fisheries. It includes separate policies for inland closed water fish culture and for coastal shrimp and fish culture. The Policy touches on many contentious issues. For instance, it addresses conflicts over shrimp cultivation and underscores the need for formulation of suitable guidelines. To help conservation efforts, it prescribes a moratorium on further cutting of mangrove for shrimp cultivation. It also supports an integrated culture of fish, shrimp and paddy in paddy fields. In addition, the Policy deals with many other relevant issues such as quality control, industrial pollution and the use of land.

The strategies and action plans adopted by the government of Bangladesh seems to be positive towards environment as they create awareness among people concerning environment, but they have an attempt to displace the politics and campaign of the environmental and human rights NGOs. The strategies were directed to discipline people as well. While asked about the severe social and environment damages caused by the shrimp industry in Bangladesh, one government official working at the Ministry of Fisheries and Livestock, said, "The social and environment effect of shrimp industry that we constantly hear is the construction of different NGOs. Basically they have no ideas; they do not go the farms areas as they lead a luxurious live in Dhaka city. If the NGOs are to run, they have to talk, and raise issues." Regarding the saline water that affects the paddy farmers, another officer said in a defensive way, "In Bangladesh shrimp industry is based on a very natural and traditional way. The shrimp farmers use the saline water that automatically comes from rivers. It doesn't affect the paddy farmers. The paddy farms, which are affected by the saline water, would be affected even though there were no shrimp ponds." The third government officer added, "It is the NGOs and some newspapers that are playing politics with the industry. Once there was a hue and cry made by some newspapers that different species are being lost due to the untrained shrimp-fry collectors who frequently throw away non-shrimp fry on the beaches and riverbanks. The government then issued a ban on the collection of shrimp fry from the rivers and seas. Alternatively, the government provided huge loan for the development of hatcheries to meet the demands of shrimp fry. Now the same newspapers wrote again that numerous fry-collectors in the coastal areas have no livelihood due to the government policy." Responding vigorously, the government officials accused NGOs of politicizing the issue. What is significant is that both sides of the debate were framed in the most resolute moral and political terms. As the campaign continues, however, a series of striking rhetorical shifts and changes in regulation began to occur.

The global environmental NGOs have not only put pressure on the shrimp producing countries, but also turned their efforts towards reducing shrimp consumption in the consumer countries. They raised the issue of food-safety, and created awareness among shrimp consumers. In 1997, the EU imposed a ban on shrimp exports from Bangladesh because of its sub-standard product. At the same time, the USA and Japan created pressures for using HACCP (Hazard Analysis Critical Control Point) to maintain shrimp freshness and quality (Chowdhury and Islam 2000; Karim 2000). Thus this vulnerable export industry faced more challenges.

To face them other institutions emerged, and new regulations were imposed. Many shrimp factory owners renovated their facilities and converted them into modern plants. According to Karim (2000), the shrimp programme coordinator of the Agro-based Industries and Technology Development Project (ATDAP)[*], the EU imposed ban on the shrimp imports from Bangladesh because of the following reasons:

- Exported shrimp does not retain the desired level of freshness. Salmonella, E-coli and other harmful bacteria and germs are found at an alarming rate in the shrimp. These germs attack the shrimp through animal waste and polluted water.
- Flies; mosquitoes or bodies of other insects; hairs of dogs, cats, cattle, goats or mice; feathers of chickens and ducks; bamboo sticks; leaves; jute fibres and sand are found in shrimp.
- Pieces of iron and glass, sticks of coconut and other unacceptable things are found in shrimp bodies.
- Bodies of shrimp become soft, spongy or bruised; colour of the shell changes or becomes black; shell is broken or become soft or meat hangs from the body.
- Shrimp of the grade lower than what is referenced on cartoons are sent. Also weights are lower than what is written on the cartoons.
- Besides, if the shrimp is found to contain any trace of insecticides or antibiotics, the product will be treated as poisonous (p. 5).

The ban was later on lifted when Bangladesh showed its commitment to adhere to the principles of HACCP. According to a report by the New Nation, a national Daily of Bangladesh (January 15, 2005), the credibility of Bangladesh's frozen shrimp export has again been questioned by a delegation of the European Commission (EC). The EC delegation members alleged that in frozen shrimps exported from Bangladesh they found among others Nitrofuran, a toxic material, which causes cancer. The Daily added, the US importers recently

---

[*] ATDAP is a project of the Ministry of Agriculture, the Peoples Republic of Bangladesh. The ATDAP publishes a monthly bulletin named "Agribusiness Bulletin" which is distributed to over 7,000 readers in Bangladesh and other countries. The bulletin is free and published in both Bengali and English. The internet version of this bulletin can be found in www.agrobengal.org

have said that they might drop import of shrimp, prawn and lobster from Bangladesh because they have been finding bacteria in Bangladesh shrimps and in other varieties. This vulnerable export industry thus faced further challenge.

Bangladesh government claims that it is very careful to enforce quality standards and has implemented HACCP and quality assurance measures in the fish processing industry. Due to the pressure from the consumer countries, Bangladesh government established an institution known as Fish Inspection and Quality Control (FIQC) under the Department of Fisheries, Ministry of Fisheries and Livestock. FIQC has three stations and all are equipped with modern library facilities and technical personnel. The government has undertaken the following programs to uplift the quality and safety of the fishery products to meet the HACCP system:

- Fish and Fish Products (Inspection and Quality Control) Rules of 1989 were amended in 1997 based on the HACCP system required by importing countries
- More than 24,000 field level people were trained on post harvest, handling, transportation, hygiene and sanitation
- Raw material suppliers of the processing plants have been brought under compulsory registration
- Follow-up training programs were arranged for the personnel of fish processing plants on HACCP
- Quality of water and ice of fish processing plants has been standardized
- Infrastructural facilities of fish processing plants have been renovated and modified in accordance with the HACCP system
- Competent authority has been strengthened with proper laboratory facilities and other logistical support
- Government has formed a supervisory audit team to monitor the work of the competent authority (Chowdhury and Islam, 2000, p. 3-4)

Fifty-eight licensed factories developed their HACCP-based, quality assurance protection (QAP) manuals, which have been provisionally approved by the competent government authority. The factory personnel are very keen to implement the HACCP system in their respective plants by following sanitation standard operating procedures (SSOP) and good manufacturing practice (GMP) (Chowdhury and Islam, 2000; Huq and Mallick, 2000). Apart from these, there are extensive plant inspections from time to time and pre-shipment inspections.

Two more recent developments pertaining to regulation are remarkable. First is the proposal by the Bangladesh government in its recent formulated "Shrimp Action Plan" and in conjunction with an international quality-control agency to introduce a "Seal of Quality", which is to guarantee that shrimp are free from antibiotics, chemicals and growth hormones and are produced in work environments that respects the human rights of workers. Second, in May 2003, a "Bangladesh Shrimp Foundation" was established to work with the shrimp industry to improve its economic, social and environmental performance. The mission statement of the foundation says that it is an independent, non-profit making body directed by major stakeholders involved or affected by the Bangladesh shrimp industry. The Shrimp Foundation addresses the key threats facing the sustainable and equitable development of shrimp industry through research and development, environmental and social advancement projects in shrimp producing areas, education and dialogue. Government and international agencies have shown more interest in changing the quality and regulatory environments of shrimp production than they have in regulating conditions of work (Pokrant and Reeves, 2003). All these show how Bangladesh government is moving towards a regime of new regulation and managerialism as the demands of the consumers grow.

3. A Regime of Environmental Governmentality

From the discussion above we see that Bangladesh government came up with a series of programmes, and institutions to ensue an 'environmentally sound shrimp aquaculture' in order to respond to the local and global campaign of different environmental NGOs and activists against destructive shrimp industry. The institutions developed by the government have some positive outcomes; however, the intention to establish those seems different. It appears clear that governments came up with a series of environmental agendas and institutions in order to dislocate and displace politics surrounding the industry. The process, policies and discursive practices of the government as discussed above shows the infiltration of state apparatus throughout the everyday life of the people involving shrimp industry. The effects are far-reaching, to discipline and make them environmental subjectivities as echoed in Anwar (2000):

> During 1980s shrimp industry was at the pick in Bangladesh. Lured by enormous profit, many big shrimp pond owners occupied the lands of the small farmers to expand their farms. This resulted in the violence, fighting and even killing, apart from other environmental disasters. At that time several NGOs campaigned against shrimp industry surrounding the social and environmental damages the industry caused. However, *now the resistance is no longer strong. It is as if the people have accepted the industry. The government has not only managed to normalize the issue, but also constantly said that extensive development of shrimp industry is of great need in order to boost the national economy* (p. 13, emphasis added).

The government of Bangladesh is trying its best to amplify this profitable, yet vulnerable, industry. However, the real barrier is some NGOs which campaign against negative social and environmental impacts of the industry. In many cases, NGOs, which are talking against destructive consequences of shrimp industry, are under attack. NFB (News From Bangladesh, September 19, 2001) reports:

> Bangladesh shrimp sector has long been facing a conspiracy hatched by different vested quarters, particularly some non-government organisations (NGOs), it was alleged. Sources in the sector said the NGOs, to create a new field of their work for hauling donors' money, started propaganda against shrimp culture, saying it was destroying bio-diversity and causing environmental hazard. They presented imaginary pictures of adverse effect of shrimp culture in different international forums (p. 1).

Due to the state regulations and initiation of different programmes and institutions with regard to fisheries and environment, some NGOs have shifted their activities from direct resistance. Now anti-shrimp groups are not calling for a boycott– this would hurt the few people shrimp farming does employ. Campaigners are working instead to develop common strategies of resistance among villagers to keep resources accessible to, and managed by, local people. Some have been successful and can declare their communities 'shrimp-free zones'. Two NGOs are remarkable here: Nijera Kori (Dhaka), and Uttaran (Satkhira).

Some other NGOs have shifted their activities from resistance to co-operation playing an ambivalent and ambiguous role. They have adopted an integrated approach of both confronting and engaging. Ideologically these NGOs are confrontational to shrimp industry due to its destructive environmental and social outcomes, but they do not position themselves as diametrically opposed to the new institutionalized regime of the government. They believe in changing the system and its effects by working with it. Using the tools of the system such as management processes and public relations, they aim to generate good outcomes for the people.

The new institutions and programmes of the government with regard to environment and shrimp industry have 'alienated' some NGOs from the politics of environment. An NGO worker working in 'Agragati Sangstha', a local NGO in Satkhira region, reported,

> Satkhira is one of the biggest regions of Bangladesh for shrimp industry, and we know how much social and environmental damages are caused by the industry. But we have no way to involve in the environmental issues, and hence our NGO does not have any environmental agenda.

Apart from displacing politics from the environment, the government regulations and institutions have led to a high degree of managerialism, and a complex web of networks among the bureaucrats and shrimp-firm owners. Even within this network, the influential few are mostly benefited. One shrimp farmer reported, "even though we are producing shrimps, we are getting less than ten percent of the actual price." Secondly these institutions preclude other forms of engagements. These institutions are, according to Peter Brosius (1999),

> both enabling and limiting. Defining themselves as filling particular spaces of discourse and praxis, they in effect define (or redefine) the space of action; they privilege some forms of action and limit others; they create spaces for some actors and dissolve spaces for others (p. 50).

It is remarkable to see how thoroughly the rhetoric of environmentally sound shrimp aquaculture has today insinuated itself into the official environmental discourse in Bangladesh. What we observe in the Bangladeshi response to the environmental NGOs campaign against shrimp farming is a rhetorical shift towards an environmentally sound shrimp aquaculture, different institutional developments that privilege certain groups and preclude others. We observe a broader domain of environmental discourses and practices that create and administer subjects through the presence of a variety of knowledge-making apparatuses. It is an extension of environmental/ green governmentality, as government formed the 'bio-power' that disciplines the people and many NGOs involving shrimp farming, and turn them to subjected individuals, which Eric Darier (1996b) calls "environmental citizenship".

CONCLUSION

From the discussion above, we see that Bangladesh's effort towards instituting the appearance of regime of "environmentally sound shrimp aquaculture" became an extension of an elaborate public relation apparatus focussed on dissipating concern about environmental and social damage caused by shrimp industry among the consumers and diverting attention from the political to the technical and institutional. This institutionalization, however, creates some possibilities especially in terms of growing awareness among the people regarding environment. It has also some negative consequences. As we found the "rhetoric of environmentally sound shrimp aquaculture" has become a legitimate guise and protection for the shrimp producers under which they continue and extend their production despite of the industry's various negative social and environmental legacies. We also find that the environmental NGOs began participating in the process of environmentally sound shrimp aquaculture, and as the institutional foundations of certification were established, the role of such images and individuals could play was diminished in most cases. The new institutionalization and managerialism privileges a fortunate few. This also amplifies the state's power, as the state captures the domain of environment by different discursive practices, creating new knowledge, and disciplining people involving in the industry. However, the fundamental question remains: What direction should the NGOs go if not participating in the new regime? Brosius (1999) reminds us here:

> Much the same could be said of the dilemma that faces us in the encounter between grassroots environmentalism and institutionalization. The question is not whether we should make a choice between one or the other, or whether one is an intrinsically better alternative than other. Rather, it is a call for us to weigh carefully the terms under which institutionalization occurs, and to make an effort to discern what is gained and what is lost, who is heard and who is silenced, as the process continues (p. 51).

Environmental governmentality is, as we have seen, one of the conspicuous trends in the study of environmental movements today. Despite the state's effort to normalize the environmental discourse and displace politics from shrimp industry of Bangladesh, the negative consequences are obvious. Therefore, the type of institutionalization and the environmental managerialism, which we have seen above, is subject to contestation. In the era of democracy, the comprehensive environmental policies cannot survive without large public legitimacy and trust, and these require open and just policy procedures, communicative actions, and possibilities for consensus building.

REFERENCES

Agrawal, A. 2003. "Environmentality: Technologies of Government and the Making of

Subjects". Working chapters as presented as part of YCAR's *Asian Environments Series* lectures, September 29, 2003.

Anwar, Khairul. 2000. Shrimp Cultivation: Problem and Prospects. *Daily Ittefaq*, (National Daily), Dhaka. Nov. 11 issue.

Bailey, C. 1988. The Social Consequences of Tropical Shrimp Aquaculture Development. *Ocean and Shoreline Management*. 2: 31 44.

Bangladesh Export Promotion Bureau. 2004. Available from http://www.epb.gov.bd/ index.html (Accessed December 12, 2005).

Brosius, Peter J. 1999. Green Dots, Pink Hearts: Displacing Politics from the Malaysian Rain Forest. In *American Anthropologist* 101 (1):36-57

Cato, James C. and Carlos A. Lima dos Santos. 1998. "Costs to Upgrade Bangladesh Frozen Shrimp Processing Sector to Adequate Technical and Sanitary Standards and to Maintain HACC Programme", presented in poster session at *"Economics of HACC" Conference*. Washington DC, USA. June 15-16.

Chowdhury, SN and Md Rafiqul Islam. 2000. "HACCP Implementation and Quality Control in the Fish Processing Industry of Bangladesh", *Agribusiness Bulletin*, Vol. 45, 25 January.

Darier, Eric 1996. Environmental Governmentality: The Case of Canada's Green Plan. Environmental Politics 5(1996):4, Winter, 585-606.

Darier, Éric. 1996a.. The Politics and Power Effects of Garbage Recycling in Halifax, Canada. Local Environment, 1(1), 63-86.

Darier, Éric. 1996b. Environmental Governmentality: The Case of Canada's Green Plan. Environmental Politics 5(4), 585-606.

Darier, Éric. 1999. Foucault and the Environment: An Introduction. In É. Darier (Ed.), Discourses of the Environment. Oxford: Blackwell.

Deb, A.K. 1998. "Fake Blue Revolution: Environmental and Socio-Economic Impacts of Shrimp Culture in the Coastal Areas of Bangladesh" in Ocean and Coastal Management. Vol. 4, pp. 63-88.

Duplisea, Bradford. 1998. *"What's Behind Shrimp Farming"*www.perc.flora.org/PEN/1998-10/duplisea.html

Escobar, Arturo. 1995. *Encountering Development: the Making and Unmaking of the Third World*. Princeton, NJ: Princeton University Press.

Flaherty, Mark, Peter Vandergeest and Paul Miller. 1999. Rice Paddy or Shrimp Pond: Tough Decision in Rural Thailand. *World Development* 27(12):2045-2060.

Foucault, M. 2000. 'Governmentality' in J. Faubian (ed.). Focault/Power. New York: New Press

Foucault, Michel. 1979. "On Governmentality", *Ideology and Consciousness* 6: 5-21; Foucault, "The Subject and Power", 208-26.

Foucault, Michel. 1984. Space, Knowledge, and Power. In the book: The Foucault Reader. Edited by Paul Rabinow. Pantheon Books, New York. Pp. 239-256.

Foucault, Michel. 1991. Governmentality. In the book: The Foucault Effect. Stidues in Governmentality. Edited by Graham Burchell, Colin Gordon and Peter Miller. Harvester Wheatsheaf, London. Pp. 87-104.

Gammage, S., Swanburg, K., Khandkar, M., Islam, M. Z., Zobair, M., & Muzareba, A. M. 2006. *A Gendered Analysis of the Shrimp Sector in Bangladesh*. Dhaka: Greater Access to Trade and Expansion, USAID.

Gordon, Colin. 1991. *Governmental Rationality: An Introduction. In the Foucault Effect: Studies in Governmentality*. Graham Burchell, Colin Gordon and Peter Miller, eds. Pp. 1-51. Chicago: University of Chicago Press.

Guin, Philip. 1995. Bangladesh: Attack on the Shrimps. *Third world Resurgence* 59:18-19. Global Agricultural Alliance (GAA).

Halim, S. 2004. *Shrimp Processing in Bangladesh: Socio-Economic Overview*. Dhaka: International Labour Organization Regional Office.

Halger, Mike (et al). 1997. "Shrimp: The Devastating Delicacy", submitted to *Greenpeace Delicacy International*. www.greenpeace.org

Hossain, Kamal. 2000. Illegal Shrimp Cultivation: Sonadia is now at Stake. *Daily Sangram*, (National Daily), Dhaka. September 2 issue.

Huq and K Mainuddin. 2000. "Sustainable Trade and Environmental Concerns in Bangladesh". In *Bangladesh Environmental News Letter*. January 2001. www.bcas.net

Huq, Saleemul & Dwijen Mallick. 2000. "Sustainable Trade in Bangladesh" in *Daily Star*, (National Daily), Dhaka, March 1.

Jilani, Golam. 1999. "Sustainable Environment Management Programme (SEMP)", *Bangladesh Environmental News Letter*. Volume 10 No. 1 July.

Kabir, A.K.M. Enayet (2002). "Shrimp Farming: Ecological Impact Has to be Assessed" in *The Independent*, Dhaka, May Issue.

Karim, Dr. Mahmudul (Shrimp Program Coordinator of ATDP/IFDC). 2000. "Preserving Shrimp Quality", *Agribussiness Bulletin*, Vol. 61, 25 February 2000

Khor, M. 1994. *The Aquaculture Disaster: Third World Communities Fight the Blue Revolution*. India: (Publisher unknown)

Kunstadter, Peter., Eric C. F. Bird, and Sanya Sabhasri (eds.). 1986. *Man in the Mangroves: The Socio-Economic Situation of Human Settlement in Mangrove Forests*. Tokyo: United Nations University.

Low, J. A., Arshad and K.H. Lim. 1994. "ASEAN Mangroves as Important Centres of Biodiversity and Habitats for Endangered Species". In: Wilkinson, C.R. (ed) *Living Coastal Resources of Southeast Asia: Status and Management*. Report of the Consultative Forum of the Third ASEAN-Australia Symposium on Living Coastal Resources. Bangkok, May.

Luke, Timothy W. 1999. Environmentality as Green Governmentality. *Discourses of the Environment*. Eric Darier (eds.) Malden, Penn: Blackwell.

Mainuddin, K. 2000. "Environmental Concern in Bangladesh" in *Bangladesh Environmental Newsletter*. January.

Maxwell, J.A. 1996. *Qualitative Research Design: An Interpretive Approach*. London: Sage Publication

Mazid, M.A. 2002. Development of Fisheries in Bangladesh. Plans and strategies for income generation and poverty alleviation. Dhaka, Bangladesh. 176. pp.

Metcalfe, Ian. 2003. "Environmental Concerns for Bangladesh" in *South Asia: Journal of South Asian Studies*. Vol. XXVI, No. 3, December 2003, pp. 423-438.

NewAge. April 3, 2004. http://www.newagebd.com/apr1st04/030404/index.html

News From Bangladesh (NFB). 2001. *"NGOs accused of plotting against shrimp sector"* September 19. *http://bangladesh-web.com/news/sep/19/ev4n692.htm#A1*

Parkins, Keith. 1999. *Tropical Shrimp Farms*. In www.heureka.clara.net/gaia/shrimps.htm visited on November 18, 2000.

Pasha, Niaz. 2000. The Destruction of 56 Species while Collecting Shrimp Fry. In *Daily Inqilab*, Dhaka. Nov. 19 Issue.

Pignatelli, Frank. 1993. "Dangers, Possibilities: Ethico-Political Choices in the Work of Michel Foucault", *Philosophy of Education.* www.ed.uiuc.edu

Pokrant, Bob and Peter Reeves. 2003. "Work and labour in Bangladesh Brackish-Water Shrimp Export Sector" in *South Asia: Journal of South Asian Studies*. Vol. XXVI, No. 3, December 2003, pp. 359-389.

Princen, Thomas. 1994. NGOs: Creating a Niche in Environmental Diplomacy. In *Environmental NGOs in World Politics: Linking the Local and the Global*. Thomas Princen and Matthias Finger, eds. Pp. 29-47. London: Routledge.

Rose, N. 1999. Powers of Freedom: Reframing Political Thought. London: Cambridge University Press

Rubin, H.J. and I. S. Rubin. 1995. Qualitative Interviewing: The Art of Hearing Data. Thousands Oaks, London: Sage.

Scott, David. 2000. "The Environmental and Social Impacts of Commercial Shrimp Farming" in *Reports*. November 8.

Shiva, Vandana. 1995. The Damaging Social and Environmental Effects of Aquaculture. *Third World Resurgence* 59:22-24

Thamina, Q. 1995. Profit by Destruction. *International Workshop on Ecology, Politics and Violence of Shrimp Cultivation for Export*. Dhaka, Bangladesh.

The World Bank Group. 1999. *World Bank to Help Bangladesh Boost Environmentally Friendly and Sustainable Fish and Shrimp Production*. http://www.worldbank.org/html/extdr/extme/009.htm

Vandergeest, Peter, Mark Flaherty, and Paul Miller. 1999. A Political Ecology of Shrimp Aquaculture in Thailand. *Rural Sociology* 64:573-596

In: Handbook of Environmental Policy
Editors: Johannes Meijer and Arjan der Berg

ISBN 978-1-60741-635-7
© 2010 Nova Science Publishers, Inc.

Chapter 6

AN INTERNATIONAL COMPARISON OF PUBLIC PARTICIPATION IN FOREST POLICY AND MANAGEMENT

Kati Berninger

Centre d'Étude de la Forêt (CEF), succursale Centre-ville, Montréal, Canada

ABSTRACT

This review compares the public participation regulations and practices in forest policy and forest management in Canada, the USA and two countries in Northern Europe, Finland and Sweden. The countries studied all have extensive forest cover and forestry is important to the economy and local people. They were selected to represent different forest ownership structures. The comparison revealed that the countries with a strong private forest ownership regulate public participation less than the countries where most of the forest is publicly owned. Public participation in relation to private land needs to be based on voluntary processes like certification or it needs to occur at a more political level concerning a relatively large area. Most countries studied had established public participation practices and the participants had a moderate possibility to change the management plan. Only in Sweden is participation in forestry almost nonexistent.

INTRODUCTION

Public participation in forestry is considered an important tool for achieving sustainable forest management (SFM, Hunt and Haider 2001; Chambers and Beckley 2003). It is seen as a means of achieving better management decisions through improved information on local conditions or different values related to forests (Lawrence et al. 1997). According to Harshaw and Tindall (2005) foresters have a relatively limited diversity of forest values and thus incorporating non-foresters into forest planning processes is essential if different values are to be considered. Participation may also promote communication between the forestry professionals and other groups (Leskinen 2004). Improved communication may help to prevent or resolve conflicts (Thompson et al. 2005).

Participation is also considered a means of achieving democracy in a more direct form than voting in elections. According to Fiorino (1990) the democratic criteria for a participation process are: direct participation of lay people, existing structure for face-to-face discussion, the participants have a share in decision making and all participants are treated equally. The last criterion has to do with the fairness of the process that may enhance the acceptance of outcomes not always ideal for all participants (Smith and McDonough 2001). The authors argue that fairness in decision-making could even be the purpose of participation.

Most of the research concerning public participation in natural resources management and forestry is done on public lands (for example Lawrence et al. 1997; Germain et al. 2001; Lecomte et al. 2005). In these cases participation is justified by saying that citizens should have a say on publicly owned forests because they, themselves, are the owners of the forests (Tanz and Howard 1991; Sinclair and Smith 1999). Less emphasis has been given to studying public participation in forestry on privately owned land. This study will explore some possibilities of participation in questions concerning private forests.

Public participation in forest policy and management goes back to the US National Environmental Policy Act (NEPA) of 1969 (Germain et al. 2001). In many other countries and internationally, a wider use of public participation in forest policy and management started in the 1990's with the emergence of sustainable forest management and forest certification (Duinker 1998; Hytönen 2000; Tittler et al. 2001; Beckley et al. 2005). Since then, practices have evolved and an extensive literature has developed describing different approaches and trying to evaluate participation processes from different angles (Sinclair and Smith 1999; Germain et al. 2001; Côté and Bouthillier 2002; Wellstead et al. 2003; Nadeau et al. 2004). There is little existing literature based on international comparison and those that exist are collections of case studies rather than a comparison of legal frameworks and general practices (Nordic Council of Ministers 2002; Selfa and Endter-Wada 2008) or mainly deal with other questions than participation (Tittler et al. 2001, Messier and Leduc 2004).

In this paper I compare public participation in regional and local forest policy as well as in forest management planning in Canada, the USA, Finland and Sweden. These countries were selected because they all have extensive forest cover at least partly in the boreal forest, forestry is important to the economy and local people, and these countries represent different forest ownership structures.

I review existing legislation, government and private sector publications, forest certification standards and scientific papers in order to describe the requirements and practices of public participation in forestry in each of the countries and after understanding each system compare them. In some cases, documented sources were complemented with personal communication with experts. Essentially, only general outlines can be given on the differences and similarities between the countries in question. Participation in the processes related to revision of forest legislation, elaboration of national forest policy, definition of criteria and indicators of SFM or definition of certification criteria are not dealt with here.

The main research question considered here is the following:

- What are the similarities and differences of public participation practices related to a) public land, and b) private land in Canada, the USA, Finland and Sweden?

The additional research question is the following:

-   Who has power in the processes in the countries studied?

DEFINING PARTICIPATION AND THE PUBLIC

What is meant by Public Participation?

Several terms are used interchangeably on the role of the public in natural resources management or environmental issues. Public involvement has often been used as a generic term. According to the European Aarhus Convention (1998), public involvement includes access to information, public participation in decision-making and access to justice. The World Bank makes a further division between public consultation that is seeking the advice of the public and public participation that means people's involvement in decision-making (AEETEC 2002). In this work, I use the term public participation in its wider meaning that includes different intensities from consultation to full decision-making partnership.

Participation can be defined from a wider citizen's perspective as "activities that affect formulation, adoption and implementation of public policies and/or that affect the formation of political communities in relation to issues or institutions of public interest" (Andersen et al. 1993, p.32; cited by Nordic Council of Ministers 2002, p.27). Defined in this way, participation includes a wide range of activities from membership in associations, to writing readers' letters, to lobbying or even to civil disobedience.

In forestry, participation is often seen from an administrator's perspective and it refers to a situation when people are invited to take part in some kind of planning or policy formulation process (Nordic Council of Ministers 2002, p.27), for example according to the CIF (1998) "Public participation may be defined as any situation where people other than resource-management professionals and tenure holders in forest decision-making are invited to give opinions on any matter in the decision process".

In this paper I use the definition of ILO (2000, p.6): "Public participation is a voluntary process whereby people, individually or through organized groups, can exchange information, express opinions and articulate interests, and have the potential to influence decisions or the outcome of the matter at hand".

Who is the Public?

Defining the public, who should be involved in a participation process, is controversial. Concerning the public forests in North America, some feel that local stakeholders[*] should be at the centre of the participation process (Sinclair and Smith 1999), while others argue that most of the owners of the public forests live in urban areas and that they should be included (McClosky 1999). Another perspective is from Northern Europe, where there is a lot of private land but everybody is allowed an access to the forest: Primmer and Kyllönen (2006) suggest that the participation of stakeholders would be defined through the effects that the decision has and everybody potentially affected by the decision should be included. In this way also people from further away who are either using the forest or think it has important non-use values, for example conservation values, could be involved in the participation process concerning a specific forest area.

The local, regional and national publics may have conflicting objectives for the same area (Côté and Bouthillier 1999), especially when the local people are economically connected to the forest industry and regional or national publics are more concerned about the recreational or ecological values of the forest, as has been shown by empirical studies (for example McFarlane and Boxall 2000). On the other hand, surveys conducted in both Canada and Finland show that people generally agree with the statement that the concerns of communities close to the forest should be given higher priority than more distant communities (McFarlane and Boxall 2000; Berninger 2007).

The definition of the public also depends on the institutional level of the planning process and the size of the planning area. A regional forest policy formulation process would be more likely to profit from the input of a national public than a specific forest management plan on a small area.

There are two possible approaches of organizing public participation that can both be used in the same process. In the first approach, all interested members of the public are invited to participate ("the general public") and in the second approach selected members of the public are engaged as members of advisory committees (Duinker 1998). Balance is needed between the amount of deliberation and the broadness of participation. Advisory committees cannot include everyone (Duinker 1998) and persons included in them are often representatives of interest groups only (Konisky and Beierle 2001).

Even if people are invited to participate, they don't always have sufficient time, interest or knowledge to participate (McDonough et al. 2003). Typical participants in meetings are middle-aged white men with above-average education and income levels (Ovendevest 2000; McFarlane and Boxall 2000; Thompson et al. 2005). Typically excluded groups, especially in the meetings held in the evening, are women and young parents that would need child care (Buchy and Hoverman 2000). The inclusion of excluded groups in participation processes has a potential for personal and community empowerment (McDonough et al. 2003).

PARTICIPATION BASED ON LEGAL REQUIREMENTS IN THE FOUR COUNTRIES STUDIED

In this section I compare the public participation regulations in some provinces of Canada, in the USA, in Finland and in Sweden. The results of the comparison are summarized in Table 1. It is not always easy to distinguish between activities directly required by law and the ones based on government policy or guidelines. In this section I concentrate strictly on the content of actual legislation. The participation strategies based on government policy without direct legal requirements are described in the section on voluntary measures of participation.

---

* According to Côté and Bouthillier (1999) "A stakeholder is any individual and group with objectives and legitimate interests in the goods or services of a specific forest area".

Examples of Provincial Regulations in Canada

In Canada, forestry is regulated at a provincial level. I provide here examples of provincial legislation from British Columbia, Quebec, Ontario, New Brunswick as well as Newfoundland and Labrador concerning Crown lands. Forestry on private lands is generally poorly regulated. These five provinces cover 65 % of the forest land in Canada and were selected to represent different approaches to public participation.

In British Columbia (BC), objectives for the use and management of Crown resources or Crown land may be established by order under the Land Act (RSBC 1996, c. 245). These objectives are subject to public review and comment. The BC Forest and Range Practices Act (SBC 2002, c. 69) requires the licensees to make Forest Stewardship Plans publicly available for review and comment before submitting them for approval. Forest Stewardship Plans determine which areas will be used for road building, harvesting or silvicultural activities, but do not necessarily show exact locations.

The Forest Act of Quebec (Loi sur les forêts 1986) requires the government to elaborate a consultation policy on forestry. The consultation policy includes all individuals or organisms interested in the protection, management or use of the public forests and concerns general forest policy issues in both public and private forests, though its main emphasis is on public forests (Gouvernement du Quebec 2003). The forest industry is required by law to invite certain forest users like municipalities, aboriginal communities and hunting organizations, to participate in the elaboration of a management plan (MNRF 2004). Also other people may be invited to participate. A public hearing is organized before the plan is accepted, but the general public does not necessarily have an opportunity to have a say in the early stages of the plan (Lecomte et al. 2005). There is a possibility for mediation if conflicts occur during the public hearing (Loi sur les forêts 1986). Municipalities may regulate the management of private woodlands (for example AMFM 2006).

Forest management planning on Crown lands in Ontario is regulated by the Environmental Assessment Act (1990) and the Crown Forest Sustainability Act (1994). Under the Environmental Assessment Act, approval is given to the Ontario Ministry of Natural Resources (OMNR) to undertake forest management planning and implement it. The Forest Management Planning Manual for Crown Forests (OMNR 2004), a regulated requirement of the Crown Forest Sustainability Act, describes the process of forestry planning in Ontario. According to the manual, forest management plans are prepared in an open and consultative fashion by OMNR or by an organization authorized by OMNR. The planning process has several stages of participation that start in the phase of identification of objectives and management alternatives (Saunders 2003, OMNR 2004). A legally required advisory committee called local citizens committee assists the plan author and the planning team during the production of the forest management plan. A representative of the local citizens committee may be a member of the planning team (OMNR 2004). Aboriginal communities in or adjacent to the management unit are provided an opportunity to participate in the planning team (OMNR 2004).

In Ontario, private land stewardship is encouraged through information and incentives, but there are no provincial-level regulations concerning management of private forests. Municipalities may regulate the management of private woodlands (OMNR 2003).

The New Brunswick Crown Lands and Forests Act of 1980 requires establishment of an Advisory Board that gives advice on matters related to the management of Crown lands and dispute resolution related to this management. The Advisory Board has representatives of

First Nations, environmental organisations, forest industry, private woodlot owners and scientists.

The Forestry Act of 1990 requires that the Forest Service of Newfoundland and Labrador (FSNL) consult with different government agencies regarding the planning, development and use of the forest resources of the province. FSNL should also consult with residents of the province who may be directly affected by the preparation of a sustainable forest management (SFM) plan, the issuing of a Crown timber licence or a timber-sale agreement or the preparation of a sustainable forest management strategy for the province. When preparing a SFM plan, members of the public, the governments of the province and Canada and other agencies having an interest in the management of the area are given an opportunity to meet for consultation and to record and respond to the concerns of the local community (The Forestry Act 1990). The Environmental Protection Act (2002) requires that all five-year operating plans should be registered for environmental assessment. District Planning Teams are required to have a central role in providing assistance to the district manager in preparing the SFM plan (FSNL 2003). The planning team includes local citizens and representatives of local businesses, non-governmental organizations and federal, provincial, and municipal governments (Nazir and Moores 2001). The team uses a consensus-based decision-making process (Moores and Duinker 1998; Nazir and Moores 2001). By 2001, Planning Teams have been established for most management districts (FSNL 2003).

The USA

In the USA, the National Forest Management Act (1976) regulates the management of national public forests. The USDA Forest Service that manages the public forests is required to establish advisory boards that give advice on the planning and management of the national forests. The Act requires that advisory boards are representative of different interest groups.

The USDA Forest Service planning rules provide more detailed regulations on public participation. Land management planning must be collaborative and participatory, and appropriate combinations of government agencies, consultants, contractors, federally recognized Indian Tribes, or other interested or affected communities, groups, or persons should be included (USDA 2005). The public should be given opportunities to collaborate and participate openly and meaningfully in the planning process. Specifically, as part of plan development, plan amendment, and plan revision, the public should be involved in developing and updating the comprehensive components of the plan, and designing the monitoring program (USDA 2005). The National Environmental Policy Act (NEPA) of 1969 requires federal agencies to integrate and consider the environmental impacts of their proposed actions. Since 1982, the public has participated in forest planning through a NEPA process. According to the new planning rules (USDA 2005), a separate NEPA evaluation is no longer needed in forest management planning, but it is integrated into the management planning process.

In the USA the management of private forests may be regulated at the state level and according to Ellefson et al. (2006) 64 % of the states do so at least to some extent.

Northern Europe

In the Finnish forest legislation, participation is only mentioned in connection with the Regional Forest Programs. According to the Forest Act (1093/1996), Regional Forestry Centres should "cooperate with the parties representing forestry in the area and other relevant parties". The Forest Decree (1200/2000) specifies that the Forestry Centre "must cooperate with the key parties representing the forest sector in the area, with the nature conservation authorities and other parties relevant for compiling the programme." The Finnish Forest and Park Service must take part in drawing up the program as regards the lands it governs. The Forest Act also regulates the management of private forests.

The Swedish Forestry Act (429/1979) does not include a direct mention of public participation in forestry planning or policy formulation. The only mention of participation is for the areas with reindeer husbandry, where the Sámi village that is affected should be given the opportunity to participate in joint consultations. The Forestry Act also regulates also the management of private forests.

Voluntary Measures of Participation

The less public participation is regulated by legislation, the more its implementation depends on diverse voluntary measures. This section describes some voluntary approaches used in the countries studied. The results are summarized in Table 1.

In Canada, differences in provincial legislation lead also to differences in voluntary measures. In the British Columbia, the Ministry of Forests and Range has a policy to provide two opportunities for public review during the timber supply review process which is used to determine the Annual Allowable Cut (AAC) (Government of British Columbia 2008). In Ontario, stewardship councils are volunteer groups of landowners and other interested parties that work with an Ontario Ministry of Natural Resources (OMNR) resource person. Councils get funding from OMNR for their community forests (OMNR 2003). The stewardship councils work with a wide range of environmental issues, including forestry.

In the USA, there are several locally–based natural resource collaboration and cooperation projects to address cross-boundary questions (Selin et al. 2000). One example of this is the Oregon watershed councils that promote ecosystem management and have a broad representation of local actors including NIPF (Rickenbach and Reed 2002). The participation of NIPF is voluntary, but essential for the achievement of management goals (Rickenbach and Reed 2002).

In Finland the state organization Finnish Forest and Park Service has a hierarchical planning system that includes a large-scale strategic-level natural resources management plan which defines the land use objectives and the target levels of forestry activities and a landscape ecological plan that defines the ecological constraints to forest management. Both plans are elaborated using participatory planning methods that include a wide range of stakeholders (Karvonen et al. 2001).

Table 1. Comparison of public participation in the forest policy and forest management planning in Canada, the USA, Finland and Sweden. CSA=Canadian Standards Association, FCS=Forest Stewardship Council, SFI=Sustainable Forestry Initiative, FFCS=Finnish Forest Certification System.

	Canada	USA	Finland	Sweden
Legal basis of participation	Strong	Strong	Weak	Non-existent
Participation on public land	Mandatory or voluntary advisory groups Possibility to review and comment management plans for the general public	Mandatory advisory boards. Possibility to review and comment management plans for the general public	Mandatory in regional planning Voluntary processes of Finnish Forest and Park Service	Discussion groups without mandate No public participation in the forestry planning processes
Regulation of forest management on private land	Weak	Weak	Strong	Strong
Support for non-industrial forest owners	Weak	Medium	Strong	Strong
Participation on private land	Voluntary certification-related processes by forest companies	Voluntary collaboration in local ecosystem management	Mandatory in regional planning, voluntary processes in local forestry planning	Non-existent
Certification	Public participation strong in the CSA and FSC criteria, which are primarily made for public forests	The FSC criteria include public participation requirements for both public and large and medium-scale private forests. The SFI standards do not require more than the legislation.	Public participation not included in the current FFCS criteria, the draft FSC criteria include the minimum international requirements	Public participation not included in the current FSC criteria, the draft FSC criteria include the minimum international requirements
Does the public have power to change the plans?[1]	From 2.2. to 3.7 depending on province	From 2.5 to 3.2 depending on state	3.2	1[2]

[1] Estimate of a local expert in each region (Messier and Leduc 2004). 1=lacking, 5=excellent
[2] Estimate of the author of this paper. There is no public participation in forestry planning.

Sveaskog, the organization managing Sweden's state forests, organizes so-called forum Sveaskog that are local meetings for those interested in forests, for example politicians, environmentalists, government authorities and local forest users. From five to seven meetings are held every year in different parts of the country since 2003 (Sveaskog 2006). However, these meetings do not give an opportunity to comment on or to any specific plan or policy (Lindahl 2006).

In Finland, forest management planning at the local level concerning several forest holdings is carried out by forestry professionals. The participation of the forest owners is voluntary. The forest planner hears the views of forest owners and different stakeholder groups separately; they never discuss at the same negotiation table and the stakeholders do not have the opportunity to see or comment on the draft plans (Leskinen 2004). In Finland, collaboration networks of forest owners are formed for forest biodiversity conservation and the forest owners may voluntarily invite other actors to participate (Kurttila et al. 2005).

Certification

In Canada, both the Forest Stewardship Council (FSC) National Boreal standard (FSC 2004) and the Canadian Standards Association (CSA) standard on Sustainable Forest Management (CSA 2002) include extensive criteria on public participation. They include detailed requirements on both the participation process and the contents of participation. There are separate sections on the rights of indigenous peoples (CSA 2002; FSC 2004).

In the USA, the different FSC regional standards (FSC 2005b; FSC 2005c; FSC 2005d) require medium or large forest owners (the limit varies between 400 and 20 000 ha) to provide opportunities for people and groups affected by management operations to provide input into management planning. People and groups affected by management operations are apprised of proposed forestry activities and associated environmental and aesthetic effects in order to solicit their comments or concerns (FSC 2005b). The Lake States Regional Standard (FSC 2005d) requires clearly defined and accessible methods for public participation in both the strategic and tactical planning of the public forests. About 61 million ha of forest in the USA are certified under the Sustainable Forestry Initiative (SFI) program of the American Forest and Paper Association, but the standards do not require more than the legislation in public participation (SFI 2005).

About 95% of the Finnish forest area is certified by the Finnish Forest Certification System (FFCS) that is approved by the Programme for the Endorsement of Forest Certification schemes (PEFC). The Finnish system includes the possibility of group certification either at the regional or local level (FFCS 2003a; 2003b). The only criterion related to participation is that the Finnish Forest and Park Service should safeguard the Sámi culture and the traditional livelihoods in Sámi homelands while managing the state forests (FFCS 2003a; 2003b; 2003c). The national draft FSC Standard (FSC 2005a) requires municipalities with at least 1 000 ha of forest land and forest owners with at least 10 000 ha of forest land to compose a landscape ecological plan using participatory planning principles established by the Forest and Park Service. Forest owners with less than 10 000 ha are required to negotiate with the people and groups directly affected by management operations. Forest management planning of areas designated for recreation or areas with high ecological values should use participatory planning methods at the municipal and state lands (FSC 2005a).

In Sweden, FSC certification is relatively common, especially among the large forest companies, but also among municipalities that own forest and small-scale land-owners. About 45% of the Swedish forests are certified by the FSC. The FSC certification standard in use does not generally mention public participation. The only exceptions are the Sámi villages whose involvement is already required by the Swedish Forestry Act (Svenska FSC –rådet

2000). The proposition for a new standard includes social impact assessment and meetings with individuals or groups that are directly affected by forestry activities, but it does not include much more than the international minimum requirements (Svenska FSC rådet 2005).

PUBLIC PARTICIPATION IN RELATION TO FOREST OWNERSHIP PATTERNS

Different land ownership patterns may lead to different systems of forest management planning and different forms of public participation. Forest land can be publicly owned by the state, the provinces or by municipalities. In Canada most of the forests are publicly owned, but the situation varies from province to province (Table 2). About one third of the forest in the United States is on federal land, the rest is owned by states, counties, tribal governments, corporations or non-industrial forest owners (Table 2, Ellefson 2000). For example in Minnesota 56 % of the forests are on public land and 44% on private land (5% corporate and 39% family forest, Kilgore et al. 2005). The national forests, however, are considered more in the national forest policy than the private or municipally owned forests, for example (Ellefson 2000). In Finland and Sweden about one fourth of the forest is state-owned and there is some municipal ownership, but the majority of the forests are privately owned (Table 2). In Canada, licenses are given to companies to exploit public forests, whereas in the USA and Europe public forests are managed by state (Tittler et al. 2001; Messier and Leduc 2004) or the municipality that owns the forest.

Private forest land may be owned by corporations like forestry companies. For example in Sweden forest companies own 26% of the forest (National Board of Forestry 2005). There are examples of corporations carrying out voluntary public participation processes as a part of a certification process (Côté and Bouthillier 1999; 2002) or as a means of avoiding conflicts (Chambers and Beckley 2003).

Other private lands are owned by families or individuals commonly called non-industrial private forest owners (NIPF). They have typically small forest holdings and differing goals for their forest ownership ranging from mainly economic interests to nature conservation (Karppinen 1998; Hugosson and Ingemarson 2004). Small forest holdings are common in Europe and in parts of the United States and Canada (Table 2). In Finland and Sweden, the NIPF are well organized and supported both by their own and by government organizations (Koskela and Ollikainen 1998). The situation is different in North America. In the United States, a variety of technical assistance, financial subsidies and outreach services intended to encourage sustainable forestry is offered to the NIPF (Rickenbach et al. 2005). The impacts of the efforts are, however, limited, since only 20% of the NIPF have sought professional assistance and fewer than 10% have a written management plan (Rickenbach et al. 2005). Similar assistance programs are provided in various provinces in Canada, but there is no exact information on the outcomes of the programs (Danserau and deMarsh 2003).

Tittler et al. (2001) compared forest planning processes in several provinces in Canada and some European countries and concluded that the planning hierarchy is more complex and regimented for public than for private lands. Planning on public lands included broader policy issues, multiple forest values and public input. Public involvement in private forest management decisions in North America has traditionally relied on policy tools, for example education, financial assistance, and regulations to ensure regeneration of forests and to minimize negative environmental effects (Boyle and Teisl 1999). For example the Minnesota Forest Resources Council formed under the Minnesota Sustainable Forest Resources Act (1995) and including representatives of various interest groups has prepared voluntary forest management guidelines in cooperation with various groups (MFRC 1999).

Table 2. Forest ownership structure in the countries studied (MacFarlane 1997, Ellefson 2000, Finnish Forest Research Institute 2005, Gouvernement du Québec 2005, NRCAN 2001, 2006). For Canada, examples of provinces with different ownership structures are provided. Municipally owned land is here included in the corporate and industrial category, since it is classified as private land in the statistics.

	Public	Corporate and industrial	Non-industrial private
Sweden	25 %	26 %	50 %
Finland	26 %	13 %	60 %
USA	43 %	8 %	49 %
Canada			
Newfoundland and Labrador	99 %	0 %	1 %
British Columbia	97 %	0 %	3 %
Ontario	92 %	1 %	7 %
Quebec	85 %	2 %	13 %
New Brunswick	50 %	20 %	30 %

Public participation cannot conflict with ownership rights, but a public participation process needs to be accepted by the forest owners (ILO 2000). That is why most public participation processes in relation to private land are on a voluntary basis and each land owner can decide whether to be involved in the process. The use of public participation processes on private land also depends on the scale of the planning: More extensive public participation can be used for example in regional forest policy questions than in local forest management planning (Leskinen 2004). The size of the forest holdings are closely related to the scale of planning. Big forest holdings or groups of small holdings will probably be more willing to allow a public participation processes on their forests than individual small holdings.

POWER AND CONTROL IN THE PROCESS

In all the countries studied with established public participation processes, the public has moderate power to change the forest management plans (Table 1). This goes well together with the statement of Beckley et al. (2005) that the most common levels of public influence on decisions are consultation and collaboration. For example, advisory groups are widely used in forest management planning in Canada, but they have limited possibilities of influencing decisions (McGurk et al. 2006). In the USA, the collaborative planning approach is widely used in the management planning of national forests (Selin et al. 1997; Carr et al. 1998; Selin et al. 2000). Participants have differing views on what collaboration should mean. Some participants would like to have more decision-making power while Forest Service employees think the participants have an advisory role (Carr et al. 1998).

The timing of public participation defines the level of decisions in which the public has a potential to influence. No information on timing of public participation was found for the USA. In Canada, only three provinces, BC, Alberta and Ontario, search for public input before determining the annual allowable cut (AAC, Saunders 2003). For example, McGurk et al. (2006) report from Manitoba that the three public participation processes they studied started after the important strategic decisions had already been made. The role of the participation processes was thus commenting on existing drafts and refining the plans. In Finland, in the processes in the state forests by the Finnish Forest and Park Service, the participation starts early and the stakeholders have the possibility to take part in the planning of the process, especially the methods and timing of the process, as well as in the definition of alternatives and their assessment (ILO 2000, p. 73; Wallenius 2006).

In most of the cases in all countries reviewed, the control over the process of forestry planning or forest policy formulation remained in the hands of forestry professionals (Carr et al. 1998; Leskinen 2004; McGurk et al. 2006), also in the cases of co-management (see previous section on empowerment). Even in the forests of private land owners, forestry professionals seem to control the process of forest management planning in Finland (Leskinen 2004). Forestry professionals take the land owners' management objectives into consideration, but "occasionally it seems that good forest management and experts' views are the most important criteria in planning" (Leskinen 2004).

CONCLUSION

This paper has reviewed public participation regulations and practices in four countries in North America and in Northern Europe. Each country reviewed has its own characteristics of public participation in forest management and forest policy that partially reflect the forest ownership structure and partially the history of forest policy in each country. Canada is characterized by provincial legislation that regulates public participation to different extents. The strongest focus on participation is in Ontario and BC (Saunders 2003). The focus on participation in Canada is almost entirely on public lands. There are many innovative processes and extensive research in the field of participation. The USA has a long tradition of about 40 years of public participation in forestry (Leach 2006). It is regulated at a federal level. There is a little more emphasis on private land than in Canada. Finland has little regulation on participation both on public and private lands. The exception is the mandatory participation processes in the preparation of regional forest programs. There are voluntary processes both on public and private land. Sweden has neither regulation nor practice on public participation in forestry.

According to the information obtained on the countries reviewed, it seems that public participation in forest policy and forest management is less regulated in countries with a strong private ownership of forests than in countries where most of the forest land is publicly owned. Public participation on private lands is possible, but it needs either to be based on voluntary processes or to occur at a relatively large scale at a more political level. Among the countries studied, public participation in relation to private forests seems to be most developed in Finland with a high percentage of non-industrial private forest owners and every man's right that allows everybody an access to the forest.

Access to forests is an important question when considering the land ownership questions in forestry. While every man's right ensures a free access to all the forests in Finland and Sweden (Nordic Council of Ministers 2002, p. 24), access to private forests is restricted in North America (Hunt 2002). This may mean that there is more need for public participation on private land in Northern Europe, where the forests have multiple users.

Sweden, however, has a similar distribution of forest ownership and similar general access to forests than Finland, and there is no effort to organize public participation in forest management on private or public lands. The lack of participation processes is surprising, since Sweden is known for its consensus politics (McBride 1988). It has to be remembered, however, that in Sweden there is public participation in forest policy at the national level on issues like revision of forest legislation that were out of the scope of this study (Nordic Council of Ministers 2002). The difference could be in the history of forest management planning and its regulation. The forest policy in Sweden changed in 1993, and it now focuses strongly on the maintenance of biodiversity and landscape ecological planning (Lämås and Fries 1995). It may be that the focus on biodiversity has been so strong that no other issues have fitted on the forest policy agenda at the same time. Another explanation could be the strong forest ownership of forest industry in Sweden in comparison with other countries reviewed (Table 2).

Forest certification with its potential market benefits may be a motivation to arrange public participation in relation to private forests or to require a more participatory process than required by the legislation. Industry-driven certification programs like FFCS and SFI do not include any requirements for public participation, whereas the FSC program has some minimum requirements for hearing people directly affected by forestry activities. National or regional FSC standards vary a lot in their requirements for public participation and generally they don't require more participation than is done anyway. They could even be used as an indicator of public participation, since they seem to reflect the general "climate" for participation in forestry questions.

The timing of public participation is an important determinant of the power the public has to influence the decisions. The earlier in the process participation begins, the more strategic decisions the public is able to discuss. For example in Canada, the participation does not always start early enough for influence over strategic decisions. In most Canadian provinces the definition of the annual allowable cut (AAC) is defined before the public participation processes start. The definition of the AAC, that is how much wood is going to be cut, is the most important decision when defining the forest management in a region. In contrast, the Finnish Forest and Park Service has established clear principles for the early involvement of the participants.

The final decision-making power in forest management planning or forest policy formulation is generally in the hands of forestry professionals, representatives of either government or forest industry. According to this review, the public has a moderate power to influence the outcome of the planning or policy formulation processes. Should the power of the public then be increased? The answer is not easy, since with decision-making power comes also the responsibility for the decisions made. The arrangement of public participation is a balancing act between power and responsibility, between technical and popular knowledge as well as between the amount of deliberation and the number of persons involved. At its best the public participation processes combine the knowledge and views of many parties and the outcome is better than with a restricted participation only (Beierle 2002). The result may be more than the sum of its parts. This is a great challenge to the forestry professionals who are required new skills to act as negotiators or in some cases as facilitators (Côté and Bouthillier 1999; Leskinen 2004).

The public participation in forestry in Canada, as well as in the USA, relies heavily on advisory groups, which is an intensive form of participation. According to Beierle (2002), the more intensive processes have the most potential for improving the quality of decisions. The participants of advisory groups have, however, not always been satisfied with the representation or the degree of influence on the plan or policy in question (Nadeau at al. 2004; McGurk et al. 2006; Parkins et al. 2006).

The present study has revealed that forest ownership structure may have an effect on the degree of regulation of public participation in forestry. This finding is in line with Tittler et al. (2001) who did an international comparison of forest planning hierarchy, but it should be confirmed with further research including more countries with different forest ownership structures. There should be more theoretical and applied research on participation processes in relation to private forests. Questions that need more theoretical work are for example the balance between democracy or access to forests and private property rights. More applied research is needed on the possible mechanisms for public participation on private lands and the acceptability of it among the land owners.

REFERENCES

Aarhus Convention. (1998). Convention on Access to Information, Public Participation in Decision-making and Access to Justice in Environmental Matters [Www -page]. http://www.unece.org/env/pp/treatytext.htm Viewed 25.10.2006.

L'Agence Régionale de Mise en Valeur des Forêts Privées Mauriciennes (AMFM). (2006). Résumé du règlement sur l'abattage d'arbres en forêt privée [Www –page]. http://www.agence-mauricie.qc.ca/reforestier.asp Viewed 18.8. 2006.

Andersen, J., Christensen, A-D., Siim, B., & Torpe, L. (1993). Medborgerskab. Demokrati og politisk deltagelse. Systime, Herning.

Asia-Europe Environmental Technology Centre (AEETEC). (2002). Public involvement in environmental issues in the ASEM – background and overview. Helsinki: AEETEC. 89 p.

Berninger, K. (2007). Neljän intressiryhmän näkemyksiä Kaakkois-Suomen metsien hoidosta. (Views of four interest groups on forest management in Southeastern Finland, in Finnish). Alue ja ympäristö 36(1), 45-50.

Beckley, T., Parkins, J., & Sheppard, S. (2005). Public Participation in Sustainable Forest Management: A Reference Guide. Edmonton, *Alberta: Sustainable Forest Management Network*. 55 p.

Beierle, T. (2002). The quality of stakeholder-based decisions. *Risk Analysis* 22 (4), 739-749.

Boyle, K.J., & Teisl, M.F. (1999). Public preferences on timber harvesting on private forest land purchased for public ownership in Maine. Orono: Maine Agricultural and Forest Experiment Station, Miscellaneous report 414. 18 p.

Buchy, M., & Hoverman, S. (2000). Understanding public participation in forest planning: a review. *Forest Policy and Economics* 1, 15-25.

Carr, D.S., Selin, S.W, & Schuett, M.A. (1998). Managing Public Forests: Understanding the Role of Collaborative Planning. *Environmental Management* 22(5), 767–776.

Chambers, F.H., & Beckley, T. (2003). Public involvement in sustainable boreal forest management. In P.J. Burton, C. Messier, D.W. Smith, D.W., & W.L. Adamowicz (Eds.), Towards sustainable management of the boreal forest (pp. 113-154). Ottawa: NRC Research Press.

Canadian Institute of Forestry (CIF). (1998). Public participation in decision-making about forests. Position paper of Canadian Institute of Forestry [Electronic document]. http://www.cif-ifc.org/pdfs/policypos/E-Pos-6-Public_Participation.pdf Viewed 25.10. 2006.

Côté, M.-A., & Bouthillier, L. (1999). Analysis of the relationship among stakeholders affected by sustainable forest management and forest certification. *Forestry Chronicle* 75(6), 961-965.

Côté, M.-A., & Bouthillier, L. (2002). Assessing the effect of public involvement processes in forest management in Quebec. *Forest Policy and Economics* 4, 213-225.

Canadian Standards Association (CSA). (2002). CSA Standard Z809-02 *Sustainable Forest Management: Requirements and Guidance*. Missisauga, Ontario. 58 p.

Dansereau, J.-P., & deMarsh, P. (2003). A portrait of Canadian woodlot owners. *Forestry Chronicle* 70(4), 774-778.

Duinker, P. (1998). Public participation's promising progress: advances in forest decision-making in Canada. *Commonwealth Forestry Review* 77 (2), 107-112.

Ellefson, P.V. (2000). Integrating science and policy development: case of the national research council and US national policy focused on non-federal forests. *Forest Policy and Economics* 1, 81-94.

Ellefson, P.V., Kilgore, M.A., & Granskog, J.E. (2006). Government regulation of forestry practices on private forest land in the United States: An assessment of state government responsibilities and program performance. *Forest Policy and Economics* 9, 620-632.

Finnish Forest Certification System (FFCS). (2003a). Criteria for group certification for the area of a Forestry Centre. *FFCS standard 1002*-1:2003. 17 p.

Finnish Forest Certification System (FFCS). (2003b). Criteria for groups certification for the area of a Forest Management Association. *FFCS standard* 1002-2:2003. 17 p.

Finnish Forest Certification System (FFCS). (2003c). Criteria for certification of holdings of individual forest owners. *FFCS standard* 1002-3:2003. 14 p.

Finnish Forest Research Institute. (2005). Finnish Statistical Yearbook of Forestry. Finnish Forest Research Institute, Helsinki, Finland. 366 p.

Fiorino, D.J. (1990). Citizen participation and environmental risk: A survey of institutional mechanisms. *Science, Technology & Human Values* 15 (2), 226-243.

Forest Stewardship Council (FSC). (2004). Canada working group. *National boreal standard.* 181 p.

Forest Stewardship Council (FSC). (2005a). The Draft FSC standard for Finland. *Finnish FSC Association.* 46 p.

Forest Stewardship Council (FSC). (2005b). Final Appalachia (USA) Regional Forest Stewardship Standard, Version 4.2. *Appalachia Working Group of the Forest Stewardship Council – US.* 67 p.

Forest Stewardship Council (FSC). (2005c). Revised Final Forest Stewardship Standard for the Northeast Region (USA). Northeast Region Working Group Forest Stewardship Council, *U.S. Initiative*. 59 p.

Forest Stewardship Council (FSC). (2005d). Revised Final Regional Forest Stewardship Standard for the Lake States-Central Hardwoods Region (USA). *Lake States Working Group of the Forest Stewardship Council – US. 57 p.*

Forest Service of Newfoundland and Labrador (FSNL). (2003). *Provincial sustainable forest management strategy.* 81 p.

Germain, R.H., Floyd, D.W., & Stehman, S.V. (2001). Public perceptions of the USDA Forest Service public participation process. *Forest Policy and Economics* 3, 113-124.

Gouvernement du Québec. (2003) Politique de consultation sur les orientations du Québec en matière de gestion et de mise en valeur du milieu forestier. Charlesbourg: Ministère des Ressources Naturelles. 24 p.

Gouvernement du Québec. (2005). Ressources et industries forestières. Portrait statistique, edition 2004. Résumé. Ministère des Ressources Naturelles et de la Faune, Québec, 83 p.

Government of British Columbia. (2008). Timber Supply Review of Kootenay Lake Timber Supply Area. Information Report. *British Columbia Ministry of Forests and Range.*

Harshaw, H.W., & Tindall, D.B. (2005). Social structure, identities and values: A network approach to understanding people's relationship to forests. *Journal of Leisure Research* 37(4), 426-449.

Hugosson, M., & Ingemarson, F. (2004). Objectives and motivations of small-scale forest owners; Theoretical modeling and qualitative assessment. *Silva Fennica* 38 (2), 217-231.

Hunt, L.M. (2002). Exploring the availability of Ontario's non-industrial private forest lands for recreation and forestry activities. *Forestry Chronicle* 78 (6), 850-857.

Hunt, L., & Haider, W. (2001). Fair and Effective Decision Making in Forest Management Planning. *Society Nat. Resources* 14, 873-887.

Hytönen, L.A. (2000). Osallistamismenetelmät metsätalouden päätöksenteossa. (In Finnish) Metsätieteen aikakauskirja 3/2000, 443–456.

International Labor Organization (ILO) (2000). Joint FAO/ECE/ILO committee on forest technology, management and training. Public participation in forestry in Europe and in North America. Report of the team of specialists on participation in forestry. Sectoral working paper 163. Geneva: ILO Sectoral Activities Department. 130 p.

Karppinen, H. (1998). Values and objectives of non-industrial private forest owners in Finland. *Silva Fennica* 32 (1), 43-59.

Karvonen, L., Eisto, K., Korhonen, K.-M., & Minkkinen, I. (2001). Alue-ekologinen suunnittelu Metsähallituksessa. Yhteenvetoraportti vuosilta 1996-200. Metsähallituksen julkaisuja 40, 2001. Vantaa, Finland: Metsähallitus. 127 p.

Kilgore, M., Lealy, J., Hibbard, C., Donnay, J., Flitch, K., Anderson, D, Thompson, J., Ellefson, P., & Ek, A. (2005). Developing a certification framework for Minnesota's family forests. Staff paper series no. 183. St. Paul, *Minnesota: Dept. Forest Resources,* Univ. Minnesota. 212 p.

Konisky, D.M., & Beierle, T.C. (2001). Innovations in public participation and environmental decision making: Examples from the Great Lakes Region. *Soc. Nat. Resources* 14, 815-826.

Koskela, E., & Ollikainen, M. (1998). A game-theoretic model of timber prices with capital stock: an empirical application to the Finnish pulp and paper industry. *Can. J. For. Res.* 28, 1481–1493.

Kurttila, M., Pykäläinen, J., & Leskinen, L.A. (2005). Metsäluonnon monimuotoisuuden yhteistoimintaverkostot ja yksityismetsien aluetason metsäsuunnittelu. (In Finnish) Metsätieteen aikakauskirja 1/2005, 33-49.

Lawrence, R.L., Daniels, S.E., & Stankey, G.H. (1997). Procedural justice and public involvement in natural resource decision making. *Society Nat. Resources* 10, 577-589.

Leach, W.D. (2006). Public involvement in USDA Forest Service policy making: A literature survey. *Journal of Forestry* 1, 43-49.

Lecomte, N., Martineau-Delisle, C., & Nadeau, S. (2005). Participatory requirements in forest management planning in Eastern Canada: A temporal and interprovincial perspective. *Forestry Chronicle* 81 (3), 398-402.

Leskinen, L. (2004). Purposes and challenges of public participation in regional and local forestry in Finland. *Forest Policy and Economics* 6: 605-618.

Lindahl, K. B. (2006). Jokkmokk: Department of Urban and Rural Development, Swedish Agricultural University. *Personal communication* 6.6. 2006.

Lämås, T., 6 Fries, C. (1995). Emergence of a biodiversity concept in Swedish Forest Policy. Water, *Air and Soil Pollution* 82 (57-66), 189-196.

MacFarlane, D. (1997). The forest resource in Atlantic Canada and an overview of issues that impact wood production. Canadian Forest Service, Atlantic Forestry Centre, Information Report M-X-202E, Socio-economic Research Network. Fredericton. 14 p.

McBride, S. (1988). The comparative politics of unemployment. Swedish and British Responses to Economic Crisis. *Comparative Politics* 20 (3), 303-323.

McClosky, M. (1999). Local communities and the management of public forests. *Ecology Law Quarterly* 25, 624-629.

McDonough, M., Russell, K., Burban, L., & Nancarrow, L. (2003). Dialogue on diversity: Broadening the voices of urban and community forestry. St. Paul, MN: *USDA Forest Service, Northeastern Area*. NA-IN-04-03. 66 p.

McFarlane, B.L., & Boxall, P.C. (2000). Forest values and attitudes of the public, environmentalists, professional foresters, and members of public advisory groups in Alberta. Northern Forestry Centre Information Report NOR-X-374. Edmonton: Natural Resources Canada, *Canadian Forest Service*. 17 p.

McGurk, B., Sinclair, A.J., & Diduck, A. (2006). An assessment of stakeholder advisory committees in forest management: case studies from Manitoba, Canada. *Society Nat. Resources* 19, 809-826.

Messier, C., & Leduc, A. (2004). Éléments de comparaison des politiques et norms d'aménagement forestier entre le Québec, différentes provinces canadiennes, quelques États américains et la Finlande. Document présenté à la Commission d'étude sur la gestion de la forêt publique québécoise. 43 p.

Minnesota Forest Resources Council (MFRC). (1999). Sustaining Minnesota forest resources: voluntary site-level forest management guidelines for landowners, loggers and resource managers. St. Paul, Minnesota. [Www page] http://www.state.mn.us/ebranch/frc/FMgdline/Guidebook.html Viewed 1.11. 2006.

Ministère des Ressources Naturelles et de la Faune (MRNF). (2004). Des forêts en partage. La contribution des utilisateurs a la planification forestière. [Electronic document]http://www.mrnfp.gouv.qc.ca/publications/forets/comprendre/foret-partage.pdf Viewed 18.10. 2006.

Moores, L., & Duinker, P. (1998). Forest planning in Newfoundland: Recent progress with public participation. *Forestry Chronicle* 74 (6), 871-873.

Nadeau, S., Martineau-Delisle, C., & Fortier, J.-F. (2004). La participation publique à la gestion forestière par l'entremise des comités : portrait de la situation dans quelques régions du Québec Rapport préparé pour la Commission d'étude sur la gestion de la forêt publique québécoise. 74 p.

National Board of Forestry. 2005. *Swedish statistical yearbook of forestry* 2005. Jönköping, Sweden. 282 p.

Nazir, M., & Moores, L. (2001). Forest policy in Newfoundland and Labrador. *Forestry Chronicle* 77 (1), 61-63.

Nordic Council of Ministers. (2002). Have a "good participation": Recommendations on public participation on forestry based on literature review and Nordic experiences. *TemaNord* 2002:515. 88 p.

Natural Resources Canada (NRCAN). (2001). Canada's Forest Inventory 2001. [Www -page] http://cfs.nrcan.gc.ca/subsite/canfi/home Viewed 13.3.2009.

Natural Resources Canada (NRCAN). (2006). The State of Canada's Forests 2004-2005. Natural Resources Canada, *Canadian Forest Service*. 97 p.

Ontario Ministry of Natural Resources (OMNR). (2003). *Ontario's forests: Sustainability for today & tomorrow.* 18 p.

Ontario Ministry of Natural Resources (OMNR). (2004). Forest Management Planning Manual for Ontario's Crown Forests. Toronto: Queen's Printer for Ontario. 442 p.

Ovendevest, C. (2000). Participatory democracy, representative democracy, and the nature of diffuse and concentrated interests: A case study of public involvement on a National Forest District. *Society Nat. Resources* 13, 685-696.

Parkins, J.R., Nadeau, S., Hunt, L., Sinclair, J., Reed, M., & Wallace, S. (2006). Public participation in forest management: results from a national survey of advisory committees. Nat. Resour. Can., Can. For. Serv., North, For. Cent., Edmonton, AB. Inf. Rep. NOR-X-409.

Primmer, E., & Kyllönen, S. (2006). Goals for public participation implied by sustainable development, and the preparatory process of the Finnish National Forest Programme. *Forest Policy and Economics* 8 (8), 838-853.

Rickenbach, M., Zeuli, K., & Sturgess-Cleek, E. (2005). Despite failure: The emergence of ''new'' forest owners in private forest policy in Wisconsin, USA. *Scandinavian Journal of Forest Research* 20, 503-513.

Saunders, K. (2003). Forest management planning in Canada. Corner Brook, NL: *Western Newfoundland Model Forest Network.* 34 p.

Selfa, Y., & Endter-Wada, J. (2008). The politics of community-based conservation in natural resource management: A focus for international comparative analysis. *Environmental and Planning* A 40 (4), 948-965.

Selin, S.W., Schuett, M.A., & Carr, D.S. (1997). Has collaborative planning taken root in the National Forests? *Journal of Forestry* 95 (5), 25-28.

Selin, S.W., Schuett, M.A., & Carr, D. (2000). Modeling Stakeholder Perceptions of Collaborative Initiative Effectiveness. *Society Nat. Resources* 13, 735–745.

Sustainable Forestry Initiative (SFI). (2005). *Sustainable Forestry Initiative Standard* 2005-2009. 16 p.

Sinclair, A, J., & Smith, D.L. (1999). The Model Forest Program in Canada: Building consensus on sustainable forest management? *Society Nat. Resources* 12, 121-138.

Smith, P.D., McDonough, M.H. (2001). Beyond public participation: Fairness in natural resource decision making. *Society Nat. Resources* 14, 239-249.

Sveaskog. (2006). Hållbarhetsredovisning 2005. Available at: http://www.sveaskog.se/upload/PDF/Sveaskog%20Hallbarhets%2005.pdf Viewed 2.10. 2006.

Svenska FSC –rådet. (2000). Svensk FSC –standard för certifiering av skogsbruk. Andra upplagan. 37 p.

Svenska FSC –rådet. (2005). Svensk FSC –standard för certifiering av skogsbruk. Standardutkast 050905. 43 p.

Tanz, J.S., & Howard, A.F. (1991). Meaningful public participation in the planning and management of publicly owned forests. *Forestry Chronicle* 67 (2), 125-130.

Thompson, J.R., Elmendorf, W.F., McDonough, M.H., & Burban, L.L. (2005). Participation and conflict: Lessons learned from community forestry. *Journal of Forestry* 103 (4), 174-178.

Tittler, R., Messier, C., & Burton, P.J. (2001). Hierarchical forest management planning and sustainable forest management in the boreal forest. *Forestry Chronicle* 77 (6), 998-1005.

United States Department of Agriculture (USDA). (2005). National Forest System Land Management Planning. *Forest service rule. Federal Register* 70 (3), 1023-1061.

Wallenius, P. (2006). Finnish Forest and Park Service. *Personal communication* 16.11. 2006.

Wellstead, A.M., Stedman, R.C., & Parkins, J.R. (2003). Understanding the concept of representation within the context of local forest management decision making. *Forest Policy and Economics* 5, 1-11.

Reviewed by John Parkins, University of Alberta, Edmonton, Canada

In: Handbook of Environmental Policy
Editors: Johannes Meijer and Arjan der Berg

ISBN 978-1-60741-635-7
© 2010 Nova Science Publishers, Inc.

Chapter 7

QUANTIFYING ECO-EFFICIENCY
WITH MULTI-CRITERIA ANALYSIS

Prof. Dr. Jutta Geldermann[1]*, *Dr. Martin Treitz*[2]+

[1]Chair for Production and Logistics, University of Göttingen, Göttingen, Germany
[2]Institute for Industrial Production (IIP), University of Karlsruhe (TH),
Karlsruhe, Germany

ABSTRACT

Based on the efficiency definition by (Koopmans, 1951) a case study is presented in this paper comparing the results of a multi-criteria method and an eco-efficiency analysis for emerging technologies for surface coating. Multi-criteria analysis aims at resolving incomparabilities by incorporating preferential information in the relative measurement of efficiency during the course of an ex-ante decision support process. The outranking approach PROMETHEE is employed in this paper for the case study of refinish primer application with data from an eco-efficiency analysis presented by (Wall et al., 2004; Richards and Wall, 2005). Comprehensive sensitivity and uncertainty analyses (including the first implementation of the PROMETHEE VI sensitivity tool) elucidate the variability in the underlying data and the value judgements of the decision makers. These advanced analyses are considered as the distinct advantage of MCA in comparison to the eco-efficiency analysis (Saling et al., 2002), which just comprises various types of normalisation of different criteria.

INTRODUCTION

The term eco-efficiency was coined by the World Business Council for Sustainable Development (WBCSD), comprising almost 200 international companies in a shared commitment to sustainable development through economic growth, ecological balance, and social progress (WBCSD, 2006). The concept of eco-efficiency has emerged as one of the

* Tel.: +49-551 39-7257, Fax: +49-551 39-9343
+ Tel.: +49 721 608-4406, Fax: +49 721 758909

crucial themes linking the economy and environment and presenting opportunities for joint improvement in economic and environmental performance. However, methods for quantified eco-efficiency analyses for the comparison of the sustainability of different alternatives are in their early stages of development although the need for comprehensive evaluations of different technological options is well acknowledged. Such analyses require the simultaneous consideration of different mass and energy flows and economic performance, leading to a multi-criteria problem cause by various units of measurements and goal conflicts. This became most obvious in the European Union when the Directive on the Integrated Pollution Prevention and Control (IPPC 96/61/EG) was adopted, aiming "*to achieve a high level of protection for the environment taken as a whole*" (art. 1). The assessment and comparison of effects of industrial installations call for suitable approaches to gauging the effectiveness of these measures. A special technical challenge is to avoid the shift of environmental problems from one medium to another. Thus, an information exchange on Best Available Techniques (BAT) was organised by the European Commission for all industrial activities with a significant contribution to environmental pollution as listed in Annex I of the Directive. BAT covers all aspects of the technology used in production and in the way that installations are designed, built, maintained, and decommissioned. BAT means using the most effective economically and technically viable means to achieve a high level of protection for the environment and for human health and safety. In this way, BAT delivers a comprehensive description of aspects relevant for eco-efficiency.

The first round of 31 "BAT reference documents" (BREFs for short) has been completed by end of 2006, with the last three BREFs related to ceramic manufacturing, large volume inorganic chemicals and surface treatment using organic solvents. Altogether, around 55.000 installations are covered by the IPPC Directive, encompassing an immense economic dimension. It is a significant challenge both in terms of environmental protection and competitiveness to regulate and operate all these industrial installations in a successful way. In spite of some criticism, (Hitchens et al., 2001) come to the conclusion that the IPPC Directive did not hamper the competitiveness of the European Industry but rather promoted innovation and deployment of environmentally friendly technologies. Especially linkages between environment and energy savings are important aspects given the current developments on energy markets.

While the definition of BAT is focused on "available" techniques[*], the information exchange also includes sections on so-called "*Emerging Technologies*", being developed either by companies or institutes and having the potential to become available in the near future. As companies keep their innovations confidential for competition reasons, institutes provide an open policy in publication but might lack practical experience in scale-ups. Thus, if companies want to introduce innovative technologies in order to improve the eco-efficiency of their production processes, it is important to have credible and reliable information for prospective analyses. Since innovative or emerging technologies have to compete with existing technologies on economic, technical, ecological, and social aspects, the effectiveness in all these dimensions needs consideration.

---

[*] Article 2, para. 11 gives the following definition: "'available' techniques shall mean those developed on a scale which allows implementation in the relevant industrial sector under economically and technically viable conditions, taking into consideration the costs and advantages, whether or not the techniques are used or produced inside the Member State in question, as long as they are reasonably accessible to the operator".

This paper describes and discusses the quantification of eco-efficiency by Multi-Criteria Analysis (MCA), especially for emerging technologies. Section 2 gives an introduction to the problem of defining eco-efficiency. The application of MCA is illustrated in Section 4 with a case study about emerging technologies for surface treatment. Six types of primers and their application techniques in vehicle refinishing are being compared on the basis of data delivered by (Wall et al., 2004; Richards and Wall, 2005) with an eco-efficiency analysis. Preferential information is modelled by weighting factors and preference functions based on paired comparisons within the Outranking approach PROMETHEE. Special emphasis is put on comprehensive sensitivity analyses. Finally, Section 5 summarises the findings.

The Problem of Defining Eco-Efficiency

The optimal allocation of resources to maximise the desired output for the given input is the core question in business economics (Koopmans, 1951; Koopmans, 1975). In the context of thermodynamic processes (which underlie many environmentally relevant production processes) the input and output parameters are often limited to energy quantities, such as the transferred or converted energy compared to the employed energy (for example in a power station, wind turbine, etc.). However, this definition considers only the heat quantity and not the quality, e.g. temperature, which is relevant for defining its convertibility in, for example, refrigeration systems and thus for its economic value (for a discussion see (Grassmann, 1950)). If it is not possible to define a single common denominator, such as the heat content, the definition of the degree of efficiency is quite difficult and can only be based on relative comparisons. Consequently, the allocation of resources is efficient if no improvement (i.e., an addition to the output of one or more goods at no cost to the others) is possible. This relative efficiency definition is called 'Pareto efficiency' and a possible improvement is referred to as a 'Pareto improvement' or 'Pareto optimisation' (cf. e.g., (Moffat, 1976)). Mathematically, every Pareto efficient point in the commodity space is equally acceptable. Trade-offs and compromises are to be made when moving from one efficient point to another.

The definition of eco-efficiency in the context of technique assessment is complex since ecological, economic, technical and social parameters must be considered and representative ones selected. As the discussion for public goods and external costs shows, no competitive markets exist which could guide resources to their maximised utility (cf. e.g. (Rabl and Eyre, 1998; Schleisner, 2000)). Hence, no common denominator for eco-efficiency exists, and only relative comparisons can lead to value judgements. Relative efficiency measurements are the starting point for the Data Envelopment Analysis (DEA) and Multi-Criteria Analyses (MCA), two different approaches to resolving incomparabilities in a technique assessment, which are briefly introduced in the following. After that, approaches for Life Cycle Assessment (LCA) and particularly the so-called eco-efficiency methodology are being compared to the more formal MCA approach.

Data Envelopment Analysis (DEA)

The Data Envelopment Analysis (DEA) is an approach to comparing the relative efficiency of so-called 'decision making units' (DMUs) in general. The decision making units are characterized by their vector of external inputs and outputs. By using scalarizing functions, the inputs and outputs are aggregated to an efficiency measure for each unit (Charnes et al., 1978; Belton and Stewart, 1999; Kleine, 2001; Cooper et al., 2004; Kleine, 2004). DEA has been developed for the evaluation of non-profit organisations, whose inputs and outputs can hardly be monetarily valued with market prices and are therefore more difficult to compare.

In general, DEA assumes that inputs and outputs are 'goods', but from an ecological perspective also pollutants with negative properties have to be considered. Thus, „ecologically extended DEA models" have been derived by incorporating a multi-dimensional value function (Dyckhoff and Allen, 2001). The fact that no explicit weights are needed to aggregate the indicators is seen as an advantage. Nevertheless, it is possible to integrate preferential information into DEA (Korhonen et al., 2002; Mavrotas and Trifillis, 2006).

Recently, there are more and more applications of DEA for technique assessment or technology selection and environmental performance measuring (Sarkis and Weinrach, 2001; Keh and Chu, 2002; Zaim, 2004; Zhou et al., 2006; Kuosmanen and Kortelainen, 2007). Especially in the context of the regulation of the energy market and particularly the electricity distribution, DEA benchmarking has been tested for various large samples (Korhonen and Luptacik, 2004; Estellita Lins et al., 2007).

It can be concluded that the application of DEA is useful for the ex-post evaluation of many similar organisations but is less suitable for the comparison of few emerging technologies. A crucial point of the DEA is the determination of the efficiency frontier, and thus the *virtual* efficient production process to which the real existing organisations are compared. Such a virtual efficient technology cannot be constructed by any combination of existing technologies and will never exist in reality.

Multi Criteria Analysis (MCA)

The efficiency definition is also the starting point for Multi Criteria Analysis (MCA). In a decision problem, all non-dominated alternatives are called efficient.[*] Through special focus on the dominance relation multi-criteria methods seek to reduce incomparabilities by explicitly incorporating preferential information of the decision maker (Brans and Mareschal, 2005). The research field of Multi Criteria Analysis comprises methods for *Multi Attribute Decision Making* (MADM), covering the assessment of a finite set of alternatives (discrete solution space), and *Multi Objective Decision Making* (MODM) focussing on alternatives restricted by constraints (continuous solution space). The comparison of emerging technologies calls for MADM, for which two main streams exist (Belton and Stewart, 2002; Figueira et al., 2005):

- the *'classical'* approaches, which are based on the assumption that clear judgements exist about utility values of the attributes and their weightings, which can be

---

[*] Alternatives are dominated if there is another alternative that is not worse in any attribute and better in at least one.

formalised within the multi-criteria technique. Examples are the Multi Attribute Value/Utility Theory (MAVT/MAUT) or the Analytical Hierarchy Process (AHP).

- the *Outranking* approaches, which suppose that the preferences are not apparent to the decision maker, and therefore the decision support aims at giving insights into the consequences of different weightings. The main difference to the 'classical' MCA methods lies in the consideration of weak preferences and incomparable criteria. The most prominent Outranking models are *ELECTRE* and *PROMETHEE*.

Both 'classical' and Outranking approaches structure the decision making process and thus support the understanding of preferences. During the last decades, behavioural aspects of decision making became more important (French et al., 1998; Pöyhönen et al., 2001; Hodgkin et al., 2005), while comparisons of different algorithms are no longer in the focus of the scientific debate (Lootsma, 1996; Simpson, 1996). Thus, MCA can be considered as mature, which explains its wide use in environmental contexts (Miettinen and Hämäläinen, 1997; Seppälä et al., 2002), in technique assessment (Geldermann et al., 2000; Geldermann and Rentz, 2001), and in technology foresighting (Gustafsson et al., 2003). It can provide support for the decision maker in his/her quest for better understanding of the interdependencies in the weighting of environmental criteria. However, this discussion is highly controversial and it is important to note that some authors favour a more technical approach, whilst others stress the importance of detailed stakeholder involvement because of context sensitivity and the significant influence on the overall results (Joubert et al., 1997; Prato, 1999; Brouwer and van Ek, 2004; Munda, 2004b).

In contrast to DEA, which aims at an *ex-post* evaluation of many similar units for the purpose of monitoring and control, the objective of MCA is the *ex-ante* assessment of a few individual options by explicitly considering the subjective preferences of a decision maker for the purpose of decision support, planning and choice, (Belton and Stewart, 1999). The selection of one final solution out of the mathematically equivalent set of Pareto optimal solutions is not a purely mathematical question which can be addressed objectively. Thus, the explicit acknowledgement of the remaining subjectivity within decision processes is the goal of multi-criteria methods (see also (Munda et al., 1995; Martinez-Alier et al., 1998; Munda, 2004a).

Concerning the efficiency definition it can be said that MCA provides a partial ordering of the alternatives. Multi-criteria methods try to reduce these incomparabilities by incorporating preference models. A wide range of literature about elicitation of value judgements, preference modelling, and the ability of the decision maker to provide this information exists (see (Weber et al., 1988; Weber and Borcherding, 1993; Belton and Stewart, 1999; Mustajoki and Hämäläinen, 2000) for more information). It is also acknowledged that the weighting within MCA is context specific since no objective values exist. Apart from the classical weighting techniques (e.g. direct ratio, SWING, SMART (Edwards, 1977; Winterfeld and Edwards, 1986), SMARTER (Edwards, 1977; Barron and Barret, 1996), eigenvector method (Saaty, 1980) etc.), specialised methods are discussed in the case of environmental impacts. The normalisation of attribute weights in these various approaches has been identified as a procedural source of biases (Salo and Hämäläinen, 1997; Pöyhönen and Hämäläinen, 2001).

Another classical problem in MCA is the phenomenon of rank reversals, in particular, when a new decision alternative, which is added to an otherwise unchanged decision problem, causes a reordering of the previous rank order (Keyser and Peeters, 1996; Simpson, 1996). But (Leskinen and Kangas, 2005) show that rank reversals caused by inconsistency are natural and acceptable and that geometric-mean aggregation (instead of the traditional arithmetic-mean aggregation rule) does not cause undesired rank reversals.

Life Cycle Assessment (LCA)

Over the last 35 years, life cycle assessment (LCA) has been developed in varying forms to evaluate the environmental impacts of products, services, or processes 'from cradle-to-grave'. The International Standards Organisation (ISO) has established guidelines and principles for LCA in order to provide information for decisions regarding product development and eco-design, production system improvements, and product choice at the consumer level (cf. (Wrisberg et al., 2002; Ness et al., 2007). LCA may provide *qualitative* or *quantitative* results. The latter makes it easier to identify problematic parts of the life-cycle and to specify what gains can be made with alternative ways of fulfilling the function. Choosing between ecological profiles involves balancing different types of impact and is thus typical of a multi-criteria decision problem when explicit or implicit trade-offs and pair-wise comparisons are needed to construct an overall judgment. Consequently, the combination of LCA with a subsequent Multi Criteria Analysis has been proposed by various authors (Maystre et al., 1994; Geldermann and Rentz, 2001; Belton and Stewart, 2002; Seppälä et al., 2002; Keefer et al., 2004; Hämäläinen, 2004).

Table 7. Modules of an integrated technique assessment.

Phases of Life Cycle Assessment according to ISO 14044	*Modules of an integrated technique assessment*	*Tasks*
Goal and scope definition	Goal definition	Selection of scenarios Setting of system boundaries Definition of functional units
Inventory analysis	Mass and Energy Balance	Data collection Modelling of relevant mass and energy flows
Impact assessment	Calculation of Characteristic Figures	Calculation of technical and economic characteristic figures Calculation of impact potentials and critical volumes for ecological evaluation
Interpretation	Decision Support	Data editing and Normalisation Weighting of the criteria Multi Criteria Analysis Sensitivity Analysis

Within LCA four basic steps are identified in ISO 14044:2006 - Environmental management - Life cycle assessment - Requirements and guidelines. It resembles the ubiquitous decision analytic cycle with the main phases: problem formulation, evaluation of options, and review of the decision models (Geldermann and Rentz, 2004b; French and Geldermann, 2005). Table 7 summarises the modules for an integrated technique assessment, comprising ecological and techno-economic aspects.

Decision support is particularly necessary for the simultaneous consideration of ecological, economic, technical, and social criteria when quantifying the eco-efficiency of emerging technologies. In the majority of practical decision problems no alternative exists that is the best in all criteria. In fact, each alternative offers both strengths and weaknesses, which must be counterbalanced. Therefore, the use of multi-criteria analysis (MCA) is suggested in order to support structuring the problem, formalising the trade-offs between the alternatives and fostering the transparency of the decision.

The Eco-Efficiency Analysis

LCA has been further developed into the "eco-efficiency" analysis by BASF and has been applied in more than 220 analyses of products ranging from vitamins to basic chemicals (Saling et al., 2002; Wall et al., 2004; Saling et al., 2005). Based on the ISO14044 standards, some additional enhancements like an economic assessment allow for expedient review and decision-making at all business levels. A modular design aims at focusing on decision-relevant aspects by concentrating on specific events in a life cycle where the alternatives under consideration differ. The authors state that *"the representation of a multiplicity of individual results from the actual life cycle assessment is frequently opaque, difficult to interpret and thus not very meaningful."* (Saling et al., 2002), p.11) In order to overcome these difficulties, schemes of weighting factors have been developed, such as those based on relevance factors indicating how important the individual environmental compartment is for a particular eco-efficiency analysis. Societal views of the individual ecological impact categories were reflected through surveys, public opinion polling and expert interviews. Much emphasis is put on catchy presentations of the results as done by the Boston Consulting Matrix or Spider Diagrams.

Normalisation as a Common Difficulty in Quantifying Eco-Efficiency

Well known approaches addressing the issue of valuation of potential environmental impacts are the Cumulative Energy Demand, Eco Indicator method, environmental pressure indicators, MIPS (Material Input per Unit Service), the calculation of shadow prices based on abatement costs, determination of weights based on eco-taxes, panel methods, or the total relevance using verbal predicates based on ecological relevance and specific contribution (Berkel and Lafleur, 1997; Soest et al., 1998; Wrisberg et al., 2002; Ness et al., 2007). All these approaches have the general quest for normalisation in common. During the normalisation procedure the measure of the performance of alternatives is modified to be comparable, thus ensuring the applicability of preference or utility aggregation while still considering all decision-relevant criteria.

In LCA software several data conversion techniques are applied to transform the information supplied by the various data sources like the "person equivalent", which expresses the environmental impact or resource consumption that an average person contributes per year worldwide, in Europe, or in a specific country (van Oers and Huppes, 2001; Huijbregts et al., 2003). Similarly, the specific contribution is defined as the share of the collected results of the average impact potential in relation to the total impact potential for the EU in the respective impact category for a year. Data availability, however, limits the applicability of such normalisation approaches in many case studies.

In MCA normalisation is achieved by scale transformations through which measures of preference, value, or utility are derived. MCA can be applied to interpret the impact category weights within an LCA and can offer rules for aggregation and normalisation (Seppälä and Hämäläinen, 2001). Though weighting of criteria is unavoidable (since even 'no weighting' would imply an equal weighting of all considered criteria), it causes problems as case specific normalisation may not be compatible with generally established weighting factors.

In the context of MCA, effects like rank reversals (Keyser and Peeters, 1996; Simpson, 1996; Leskinen and Kangas, 2005), independence of irrelevant alternatives (Sen, 1970; Neumann, 2007), and splitting effects of weights (Weber and Borcherding, 1993) have been investigated for a long time. Some of these effects can be explained by the fact that the consideration of further alternatives in the current decision problem enhances the knowledge base, leading to a different view of the decision maker on the decision problem (Keyser and Peeters, 1996). Therefore, it is always emphasized that MCA can only support decision making by delivering a structured image of the often contradictory and inconsistent reality, but can never replace the decision maker.

Case Study

Though quantifying eco-efficiency still lacks a consistent theory and ranked preferences may be prone to undesired rank reversals,[*] decisions about emerging technologies must be made. The aim of this case study is the critical reflection of the eco-efficiency analysis as developed and applied by BASF. Although a company can base its decisions on whatever grounds it likes, so long as its shareholders and stakeholders do not contradict, this special approach is much being promoted (Saling 2002)(Saling et al., 2005). Thus, a genuine case study performed by (Wall et al., 2004; Richards and Wall, 2005) is re-analysed with MCA. The starting point for the recalculation of the case study was an initiative by an informal working group of various European coating producers on environmental assessment. They were uncertain if all coating producers should apply the BASF eco-efficiency analysis or if other approaches would be beneficial or even more meaningful. Though these practitioners had been aware of MCA in general, they were sceptical about too scientific and sophistic approaches.

Many manufactured items receive surface coatings for decoration and/or protection against damage. In a number of places along the production line emissions of VOC (Volatile Organic Compounds) can occur. Because VOC are a major contributor to photochemical smog, control of VOC emissions is a major concern for the industry's commitment to the environment. That is why a special BREF is dedicated to surface treatment using solvents.

---

[*] Various Nobel Price winners like K. Arrow, G. Debreu or A. Sen pointed out these difficulties in their famous works, and no satisfying solutions are ready at hand up until now. Thus, mankind will still have to live with some imperfections.

Not only large installations, such as serial coating of automobiles, are under consideration but also small and medium sized companies like vehicle refinishing, which refers to the commercial application of paint coating materials to automobiles, trucks, motorcycles, and other like equipment, often as a repair after accidents (Geldermann and Rentz, 2004a). Thus, coating manufacturers, such as BASF (Wall et al., 2004), have to supply appropriate coating materials and introduce innovative technologies. One example are UV-cured undercoats, which promise enhanced performance, higher productivity and an improved eco-efficiency compared to traditional thermally cured coatings. Ultraviolet (UV) curing is a process in which UV energy produced by a mercury discharge lamp is absorbed by a sensitizer, causing a reaction in the monomer which makes it hard and dry (Koleske, 2002). These combinations of new coating products and matched application technologies are currently considered as emerging technologies in the surface coating sector (Geldermann et al., 2007).

Table 8. General Data of the Case Study (Wall et al., 2004; Richards and Wall, 2005).

Criterion	Unit	Weight	1K UV	1K UV-A	2K U-T	2K U-IR	E-T	E-IR
Total Costs	[$/job]	50 %	14.58	22.00	27.65	19.05	24.94	18.37
Energy	[MJ/CB]	12.5%	30	25	800	35	690	35
GWP	[g CO$_2$-Eq. /CB]	5%	1 700	1 700	52 000	2 000	45 500	1 700
POCP	[g Ethene-Eq./CB]	5%	5.5	20	32	26	26	24
Health Effects	norm. scores	10%	0.94	0.96	0.94	0.92	0.87	0.85
Risk	norm. scores	5%	0.41	0.71	0.91	0.99	0.85	0.93
Resources	scores	12.5%	100	105	375	120	330	105

GWP = Global Warming Potential, POCP = Photochemical Oxidant Creation Potential, CB = Coated body

Although other coating layers (e.g. base-coat, clear-coat) also have to be applied during the refinishing process, only the primer application is considered in this case study. Primer means any coating that is designed for application to bare metal or existing finishes to provide corrosion protection prior to the application of a surfacer, which is applied later for the purpose of corrosion resistance, to ensure adhesion of the topcoat, and to promote the formation of a uniform surface finish by filling in minor surface imperfections. It is assumed that these subsequent process steps and products to be used are independent from the type of the primer applied. After its application the primer must be dried and cured, changing its state from liquid to a continuous solid film. Therefore, heat is applied either directly (by thermal drying) or indirectly (by UV or infra red drying), which results in the evaporation of the solvents and the film building of the primer. If the product contains no solvents, the coating layer is formed by chemical reactions between the components. In addition to the application step itself, related activities like, for example, cleaning of the surface and the equipment are taken into account. Furthermore, the extra fuel consumption of the automobile over its lifetime resulting from the weight of the dried primer is considered. Table 2 summarises the used data for the six alternatives (Wall et al., 2004; Richards and Wall, 2005),:

   (i).   One-Component UV Curing Primer (1K UV),
   (ii). One-Component UV Curing Primer, using an aerosol can (1K UV-A).
   (iii). Conventional Two-Component Urethane Primer with thermal drying (2K U-T),
   (iv). Two-Component Urethane with infra red drying (2K U-IR),
   (v). Epoxy with thermal drying (E-T),
   (vi). Epoxy with infra red drying (E-IR)

The weights of the environmental criteria have been taken form the underlying case study (Wall et al., 2004; Richards and Wall, 2005), while the weighting of costs has been assumed to be 50%.

Application of the Eco-Efficiency Analysis

As outlined in Section 0, the eco-efficiency analysis mainly follows the ISO 14044 phases (Wall et al., 2004; Richards and Wall, 2005). Following a company specific normalisation and weighting, the corresponding arithmetic values are summarized in a special plot, called "environmental fingerprint", which illustrates the relative ecological pros and cons of the analysed alternatives. The least favourable alternative is assigned a value of 1, and the further inward an alternative is located, the better it is. Thus, Figure 1 presents alternative 2K U-T and E-T as the two worst alternatives, since their profiles cover the largest areas, while 1K UV comes out best with the smallest area. The alternatives are, however, hardly distinguishable in such a graphical representation. The environmental fingerprint highlights the environmental impact drivers and points out improvement potentials. Finally, an Eco-Efficiency portfolio (in analogy to the growth-share matrix developed by Boston Consulting in 1970) consolidates all of the individual environmental and economic results into one representation(Walletal.,2004).

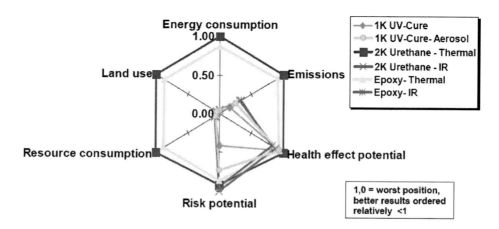

Figure 1. Ecological Fingerprint of six automotive refinish primers (Wall et al., 2004).

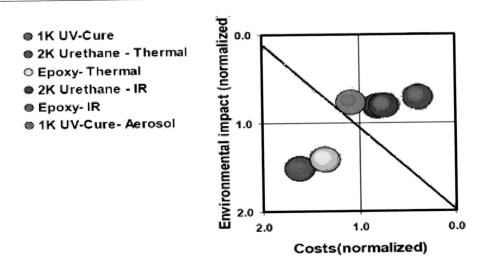

Figure 2. Ecological fingerprint of six automotive refinish primers (Wall et al., 2004).

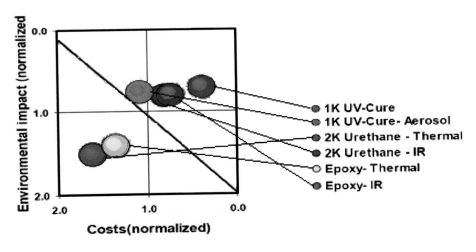

Figure 3. Results of the Eco-Efficiency Analysis (Wall et al., 2004).

Section 0 already commented on the difficulties inherent in normalisation in general. Since this eco-efficiency analysis employs BASF-specific normalisation and weighting factors, the transfer of this approach to quantifying the eco-efficiency of emerging technologies of all European coating producers is problematic. Another methodological question touches upon the interpretability of the graphical representations: Does the size of the "ecological fingerprint" really relate to the area marked by the connected points in Figure 1? This is certainly not the case because neither the angle between the axes nor their length reflects the weighting of the depicted alternatives.

Application of an Multi Criteria Analysis (MCA)

From the numerous approaches to MCA, the Outranking approach "PROMETHEE" has been chosen because of its flexibility and understandability. Outranking can be defined in the following way: alternative a_t outranks $a_{t'}$, if there is a *"sufficiently strong argument in favour of the assertion that a_t is at least as good as $a_{t'}$ from the decision maker's point of view"* (Brans et al., 1986). Accordingly, the outranking relation is the result of pair-wise comparisons between the alternatives with regard to each criterion (Stewart, 1992; Roy and Bouyssou, 1993). The outranking method PROMETHEE (Preference Ranking Organization METHod for Enrichment Evaluation) (Brans et al., 1986; Brans and Mareschal, 2005) derives the preference values by generalised criteria, which can be defined by the decision maker specifically for each considered criterion. Box 1 summarizes the PROMETHEE algorithm, and Figure 4 illustrates the six proposed types of generalised preference functions, as they have been implemented in Matlab (Schrader, 2005; Treitz, 2006). Although "Decision Lab 2000" is a commercial software package for the implementation of PROMETHEE, the academic implementation in Matlab allows for further sensitivity and uncertainty analyses, as will be shown later.

First of all, the basic PROMETHEE algorithm (see Box 1) has been applied to the decision table (as shown in Table 8). The preference function Type VI (Gaussian Criterion) has been chosen for all criteria, using half of the difference between the maximal and the minimal attribute value of each criterion as the inflection point s of the respective Gaussian function (cf. Figure 4)

On the basis of these parameters the One-Component UV Primer (1K UV) has the highest Net flow compared to the other alternatives and is ranked first. Figure 5 shows the aggregated outranking flows of Phi Plus (the relative strength of an alternative), Phi Minus (the relative weakness of an alternative) and the Phi Net (overall rating). The small Phi Minus flow for the alternative 1K UV indicates that it has a strong performance in most criteria, whereas the small Phi Plus flow of alternative 2K U-T is a sign that this alternative is weak in most attribute values.

This result can also be displayed as a spider diagram of the net flows of each criterion (cf. Figure 6). With the advent of more powerful computer aid, notably in the 1990s, various visualisation techniques for MCA have been investigated (Vetschera, 1994). Quite a few decision makers prefer spider diagrams. However, their readability decreases with the number of considered alternatives and criteria. Different from the ecological fingerprint in Figure 1, the outermost alternative is the better one while the dimension of the axes depicts the preference values of the alternatives (and not only a normalisation to the highest parameter value). The spider diagram shows that the 1K UV is the dominating alternative with respect to costs, risk, and POCP, that it shares this position with other alternatives for the criteria Energy, GWP and Resources (*Res*), and that it is outranked only with respect to the health criterion. Hence, also a change of the weight of the different criteria will show the 1K UV primer as the outstanding alternative.

Box 1. Outline of the PROMETHEE algorithm

As with most MADM methods, PROMETHEE is built on the basic notation, with a set A of T alternatives that must be ranked, and K criteria that must be optimised:

$A := \{a_1,...,a_T\}$: Set of discrete alternatives, scenarios or techniques a_t ($t = 1...T$)

$F := \{f_1,...,f_K\}$: Set of criteria relevant for the decision f_k ($k = 1...K$)

The resulting multiple criteria decision problem can then be concisely expressed in a matrix format. The goal achievement matrix or decision matrix $D := (x_{tk})_{\substack{t=1,...,T \\ k=1,...,K}}$ is a ($T \times K$) matrix whose elements $x_{tk} = f_k(a_t)$ indicate the evaluation or value of alternative a_t, with respect to criterion f_k :

$$D = \begin{bmatrix} x_{11} & \cdots & x_{1K} \\ \vdots & x_{tk} & \vdots \\ x_{T1} & \cdots & x_{TK} \end{bmatrix} := \begin{bmatrix} f_1(a_1) & \cdots & f_K(a_1) \\ \vdots & f_k(a_t) & \vdots \\ f_1(a_t) & \cdots & f_K(a_T) \end{bmatrix} \tag{1}$$

Thus, the algorithm for PROMETHEE can be summarized as follows (Brans et al., 1986): Specify for each criterion f_k a generalised preference function $p_k(d)$ (see also Figure 4).

(2) Define a vector containing the weights, which are a measure for the relative importance of each criterion, $w^T = [w_1, ..., w_k]$.

(3) Define the Outranking-Relation π for all the alternatives $a_t, a_{t'} \in A$:

$$\pi : \begin{cases} A \times A \to [0,1] \\ \pi(a_t, a_{t'}) = \displaystyle\sum_{k=1}^{K} w_k \cdot p_k(f_k(a_t) - f_k(a_{t'})) \end{cases} \tag{2}$$

(4) As a measure of the strength of the alternatives $a_t \in A$, the *leaving flow* (Phi plus) is calculated: $\Phi^+(a_t) = \dfrac{1}{T} \cdot \displaystyle\sum_{\substack{t'=1 \\ t' \neq t}}^{T} \pi(a_t, a_{t'})$ \hfill (3)

(5) As a measure of the weakness of the alternatives $a_t \in A$, the *entering flow* (Phi minus) is calculated: $\Phi^-(a_t) = \dfrac{1}{T} \cdot \displaystyle\sum_{\substack{t'=1 \\ t' \neq t}}^{T} \pi(a_{t'}, a_t)$ \hfill (4)

(6) Graphical evaluation of the outranking relation. Based on Phi net (=Phi plus – Phi minus), a total preorder can be derived.

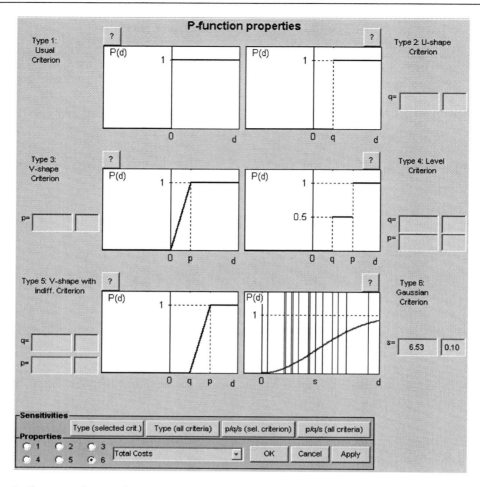

Figure 4. Six types of generalised preference functions within PROMETHEE as a MatLab Screenshot.

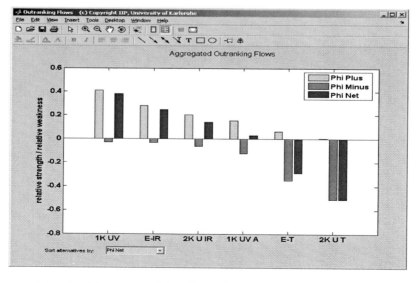

Figure 5. Aggregated Outranking Flows of the Alternatives.

A serious drawback of spider diagrams is their potential misinterpretation if the surface marked by the connecting lines is considered as a value judgement about the "size of the ecological damage". The screenshot in Figure 6 should certainly not evoke such an association since the axes show the PROMETHEE net flows, with the largest value as the best one (whereas the eco-efficiency analysis depicts the potential damage: the smaller, the better).

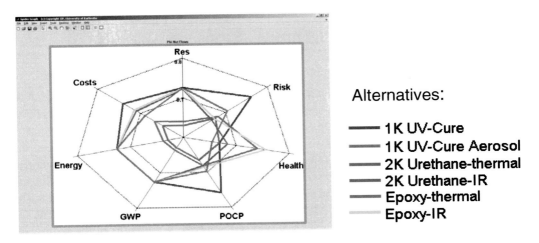

Figure 6. Spider Diagram of the Phi Net Flows.

Uncertainty Analysis

The majority of environmental decision problems involves uncertainty and risk. By their very nature, estimates and long-term forecasts, as required in LCA and for the evaluation of emerging technologies, are obviously uncertain; and a technology considered optimal on the basis of particular assumptions made today is highly unlikely to turn out optimal in the actual situation in a decade. For reviews discussing different types of uncertainty, variability, and risk see, for example, (French, 2003; Huijbregts et al., 2003; Bertsch, 2008).

Considering the above mentioned difficulties in modelling value judgements and normalisation, it is important to carry out sensitivity analyses as proposed in the ISO 14044. Sensitivity analyses investigate how the results of a model change with variations in the input variables and are therefore essential in any multi-criteria decision analysis problem. Using PROMETHEE several local and global (Saltelli et al., 2000) sensitivity analyses can be carried out (Treitz, 2006).

Especially the use of *Monte Carlo Techniques* and the *Principal Component Analysis* (PCA) (Timm, 2002) allows a simultaneous consideration of the uncertainty of the process data and the value judgements of the decision maker. Besides investigating the robustness of the decision, the strengths and weaknesses of each alternative can be addressed and so, the *distinguishability* of all alternatives can be evaluated (Basson, 2004). Consequently, sensitivity analyses are important to iteratively re-model the decision problem and to facilitate learning about the given problem.

The Principal Component Analysis (PCA) is based on the matrix M of the single criterion net flows in which the strengths and weaknesses of an alternative can be analysed for each criterion individually. By calculating the eigenvectors of the covariance matrix M'M and building a matrix of all eigenvectors sorted by the magnitude of the eigenvalues, the axes can be transformed (Timm, 2002). The transformation represents a rigid rotation of the old axes into the new principal axes based on the matrix of the sorted eigenvectors. The component scores in the new coordinate system are uncorrelated and have maximum variance. By projecting the cloud of alternatives from the R^n onto the plane of the two eigenvectors with the largest eigenvalues (henceforth called 1st and 2nd principal component), the so-called GAIA plane (Brans et al., 1986), as much information as possible is preserved. Apart from alternatives and criteria axes, the weighting vector (the PROMETHEE 'decision stick' or weighting vector π) can be projected on the GAIA plane. By defining upper and lower bounds for each weight, the convex hull of all valid weighting combinations can be projected on the GAIA plane. (Brans and Mareschal, 2005), p. 181) call the projection of the set of all extreme points of the unit vectors associated to all allowable weights the "Human Brain" or the PROMETHEE VI sensitivity tool. While this idea has been roughly sketched by (Brans and Mareschal, 2005), the implementation in Matlab is its first software realisation (Treitz, 2006).

The GAIA plane in Figure 7 illustrates the projection of all alternatives, criteria and the weighting vector. It can easily be seen that the criteria are not mutually independent (as claimed by (Saling et al., 2002) and (Wall et al., 2004)) since the projections of the criteria do not span the decision space equally, but rather point into one direction.

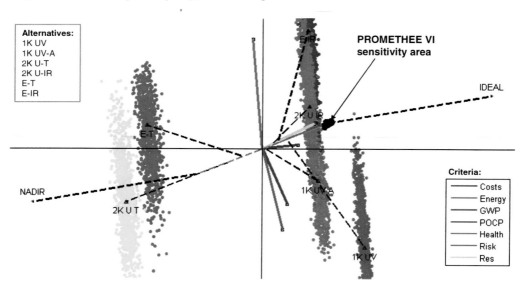

Figure 7. Illustration of Alternatives using PCA and Monte Carlo Simulation.

Figure 7 also shows the variation in value judgements with the grey area around the weighting vector π (as the PROMETHEE VI sensitivity tool) and the uncertainty in the attribute data with the scatter plot of 1,000 points created by changing all attribute values in a Monte Carlo Simulation using a standard deviation of 5 % and a normal distribution. This plot makes clear that a differentiation between the 1K UV-A alternative and the two IR alternatives (E-IR and 2K U-IR) might be difficult if the uncertainty in the underlying data is taken into account. This also becomes obvious when the small differences of their projections on the weighting vector π are compared. But even under these uncertain conditions, the 1K UV curing coat is always the best ranked alternative.

Furthermore, the selection of the preference function does not influence the ranking in its first position. This can be confirmed with a sensitivity analysis of all possible combinations of preference functions within PROMETHEE (see Figure 8).

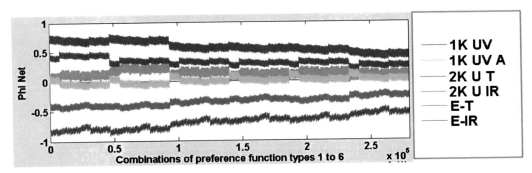

Figure 8. Influence of the selected type of preference function (279.936 combinations).

In summary, the analysis identifies the One-Component UV Primer (1K UV) as the best alternative even if uncertainties that are possibly caused by measurement errors or estimation methods and the difficulty to formalize value judgments are taken into account. It has the least total cost thanks to simpler coating preparation, less cleaning of equipment is required and shorter cure times are possible which results in labour and energy savings. The aerosol UV-primer has similar advantages but at higher material prices. With regard to their market penetration, additional criteria should be considered, such as extra training of the coaters or initial investment (which might be necessary to change to UV curable products since most of the body fillers used in refinishing at the moment are dried using infra red).

The PROMETHEE multi-criteria analysis helped to gain further insights into the given selection or ranking problem concerning different techniques, and enabled the decision maker to understand his own preferences and transparently model and communicate them. The sensitivity and uncertainty analyses in particular elucidate the variability in the underlying data and the subjective value judgments.

CONCLUSIONS

In the majority of practical decision problems – such as technique assessment - no alternative exists that is the best in all criteria, i.e., (almost) all alternatives are efficient. Multi-criteria analysis is one approach to resolving incomparability by incorporating preferential information in the relative measurement of efficiency during the course of an ex-ante decision support process. Various MCA approaches have been developed to support structuring the problem, formalising the trade-offs between the alternatives, and fostering the transparency of the decision since the final choice of one of the efficient alternatives is by its very nature subjective. The outranking approach PROMETHEE is employed for the case study of refinish primer application with data from an eco-efficiency analysis presented by

(Wall et al., 2004; Richards and Wall, 2005). Using a multi-criteria approach supports transparent and traceable decisions. With the help of sensitivity and uncertainty analyses the variability in the underlying data and the value judgements of the decision makers can be analysed and help re-model the decision problem. These advanced analyses are considered as the distinct advantage of MCA in comparison to the eco-efficiency analysis (Saling et al., 2002), which just comprises various types of normalisation of the different criteria. MCA offers mathematically sound approaches for the quantification of eco-efficiency and for the consideration of unavoidable subjective aspects in preference modelling. Extended distinguishability analyses can help to overcome the limitations of graphical illustrations, and suitable software tools (such as the proposed implementation in MatLab) allow for various modes of reporting.

REFERENCES

Barron, F.H., Barret, B.E., 1996. Decision Quality Used Ranked Attribute Weights *Management Science, 42,* 1515-1523.

Basson, L., 2004. Context, Compensation and Uncertainty in Environmental Decision Making; Ph. D Thesis, Department of Chemical Engineering, University of Sydney, Australia.

Belton, V., Stewart, T., 1999. DEA and MCDA: Competing or Complementary Approaches. Meskens, N, Roubens, M (eds.) *Mathematical Modelling: Theory and Applications Advances in Decision Analysis. (6),* 87-104; Kluwer Academic Publishers: Boston.

Belton, V., Stewart, T., 2002. Multiple Criteria Decision Analysis - An integrated approach; Kluwer Academic Press: Boston.

Berkel, R.v., Lafleur, M., 1997. Application of an industrial ecology toolbox for the introduction of industrial ecology in enterprises - II *Journal of Cleaner Production, 5 (1-2),* 27-37.

Bertsch, V., 2008. Uncertainty handling in multi-attribute decision support for industrial risk management; Universitätsverlag: Karlsruhe.

Brans, J.-P., Mareschal, B., 2005. PROMETHEE Methods. Figueira, J, Greco, S, Ehrgott, M (eds.) *Multiple Criteria Decision Analysis - State of the Art Surveys.,* 163-195; Springer: New York.

Brans, J.-P., Vincke, Ph., Mareschal, B., 1986. How to select and how to rank projects: The PROMETHEE method *European Journal of Operational Research, 24,* 228-238.

Brouwer, R., van Ek, R., 2004. Integrated ecological, economic and social impact assessment of alternative flood control policies in the Netherlands *Ecological Economics, 50 (1-2),* 1-21.

Charnes, A., Cooper, W.W., Rhodes, E., 1978. Measuring the efficiency of decision making units *European Journal of Operational Research, 2 (6),* 429-444.

Cooper, W.W., Seiford, L.M., Zhu, J., 2004. Data Envelopment Analysis: History, Models and Interpretations. Cooper, W W, Seiford, L M, Zhu, J (eds.) *International Series in Operations Research and Management Science Handbook on Data Envelopment Analysis. (1),* 1-40; Kluwer: Boston.

Dyckhoff, H., Allen, K., 2001. Measuring the ecological efficiency with data envelopment analyis (DEA) *European Journal of Operational Research, 132,* 312-325.

Edwards, W., 1977. How to use Multiattribute Utility Measurement for Social Decision Making *IEEE Transactions on Systems, Man, and Cybernetics, SMC-7,* 326-340.

Estellita Lins, M.P., Vervloet Sollero, M.K., Calôba, G.M., Moreira da Silva, A.C., 2007. Integrating the regulatory and utility firm perspectives, when measuring the efficiency of electricity distribution *European Journal of Operational Research, 181 (3)*, 1413-1424.

Figueira, J., Greco, S., Ehrgott, M., 2005. Multiple Criteria Decision Analysis - State of the Art Surveys; *Springer:* New York.

French, S., Simpson, L., Atherton, E., Belton, V., Dawes, R., Edwards, W., Hämäläinen, R.P., Larichev, O., Lootsma, F., Pearman, A., Vlek, C., 1998. *Problem formulation for multi-criteria decision analysis: Report of a workshop Journal of Multi-Criteria Decision Analysis, 7,* 242-262.

French, S., 2003. Modelling, making inferences and making decisions: the roles of sensitivity analysis; *TOP*, 11 (2), 229-252.

French, S., Geldermann, J., 2005. The varied contexts of environmental decision problems and their implications for decision support E*nvironmental Policy and Science, 8 (4),* 378-391.

Geldermann, J., Schollenberger, H., Rentz, O., Huppes, G., van Oers, L., France, C., Nebel, B., Clift, R., Lipkova, A., Saetta, S., Desideri, U., May, T., 2007. An integrated scenario analysis for the metal coating sector in Europe *Technological Forecasting and Social Change, 74 (8),* 1482-1507.

Geldermann, J., Rentz, O., 2001. Integrated technique assessment with imprecise information as a support for the identification of Best Available Techniques (BAT) *OR Spectrum*, 23, 137-157.

Geldermann, J., Rentz, O., 2004a. Decision support through mass and energy flow management in the sector of surface treatment *Journal of Industrial Ecology, 8 (4)*, 173-187.

Geldermann, J., Rentz, O., 2004b. Environmental Decisions and Electronic Democracy *Journal of Multi-criteria Analysis, 12 (2-3),* 77-92.

Geldermann, J., Spengler, T., Rentz, O., 2000. Fuzzy Outranking for Environmental Assessment, *Case Study: Iron and Steel Making Industry Fuzzy Sets and Systems, 115* (1), 45-65.

Grassmann, P., 1950. Zur allgemeinen Definition des Wirkungsgrades *Chemie Ingenieur Technik*, 22 (4), 77-96.

Gustafsson, T., Salo, A., Ramanathan, R., 2003. Multicriteria methods for technology foresight *Journal of Forecasting, 22 (2-3),* 235-256.

Hämäläinen, R.P., 2004. Reversing the perspective on the applications of decision analysis *Decision Analysis, 1 (1),* 26-31.

Hitchens, D., Farrell, F., Lindblom, J., Triebswetter, U., 2001. The Impact of Best Available Techniques (BAT) on the Competitiveness of European Industry, Report EUR 20133 EN; *Joint Research Centre (DG JRC) Institute for Prospective Technological Studies (IPTS):* Brussels.

Hodgkin, J., Belton, V., Koulouri, A., 2005. Supporting the intelligent MCDA user: A case study in multi-person multi-criteria decision support *European Journal of Operational Research, 160 (1),* 172-189.

Huijbregts, M.A.J., Breedveld, L., Huppes, G., de Koning, A., Suh, S., 2003. Normalisation figures for environmental life-cycle assessment - The Netherlands (1997/1998), Western Europe (1995) and the world (1990 and 1995) J*ournal of Cleaner Production,* 11 (7), 737-748.

Joubert, A.R., Leiman, A., de Klerk, H.M., Katua, S., Aggenbach, J.C., 1997. Fynbos (fine bush) vegetation and the supply of water: a comparison of multi-criteria decision analysis and cost-benefit analysis *Ecological Economics, 22 (2),* 123-140.

Keefer, D.L., Kirkwood, C.W., Corner, J.L., 2004. Perspective on Decision Analysis Applications, 1990 - 2001 (with discussion by R.P. Hamäläinen and S. Cantor) *Decision Analysis*, 1 (1), 5-24.

Keh, H.T., Chu, S., 2002. Retail productivity and scale economics at the firm level: a DEA approach *Omega*, 31, 75-82.

Keyser, W.d., Peeters, P., 1996. A note on the use of PROMETHEE multicriteria methods *European Journal of Operational Research, 89,* 457-461.

Kleine, A., 2001. Data Envelopment Analysis aus entscheidungstheoretischer Sicht *OR Spectrum*, 23 (2), 223-242.

Kleine, A., 2004. *A general model framework for DEA Omega, 32 (1),* 17-23.

Koleske, J.V., 2002. Radiation curing of coatings; ASTM manual series: West Conshohocken.

Koopmans, T.C., 1951. Analysis of Production as an Efficient Combination of Activities. Koopmans, T C (eds.) Activity Analysis of Production and Allocation, *33-97; John Wiley and Sons*: New York.

Koopmans, T.C., 1975. Concepts of optimality and their uses Nobel Memorial Lecture., 239-256; *Economic Sciences.*

Korhonen, P., Luptacik, M., 2004. Eco-efficiency analysis of power plants: An extension of Data Envelopment Analysis *European Journal of Operational Research, 154* (2), 437-446.

Korhonen, P., Siljamäki, A., Soismaa, M., 2002. Use of Value Efficiency Analysis in Practical Applications *Journal of Productivity Analysis, 17,* 49-64.

Kuosmanen, T., Kortelainen, M., 2007. Valuing environmental factors in cost-benefit analysis using data envelopment analysis *Ecological Economics, 62* (1), 56-65.

Leskinen, P., Kangas, J., 2005. Rank reversals in multi-criteria decision analysis with statistical modelling of ratio-scale pairwise comparisons *Journal of the Operational Research Society, 56,* 855-861.

Lootsma, F., 1996. Comments on Roy, B.; Vanderpooten, D.: The European School of MCDA: Emergence, Basic Features and Current Work. *Journal of Multi Criteria Decision Analysis, 5,* 37-38.

Martinez-Alier, J., Munda, G., O'Neill, J., 1998. Weak comparability of values as a foundation for ecological economics *Ecological Economics, 26 (3),* 277-286.

Mavrotas, G., Trifillis, P., 2006. Multicriteria decision analysis with minimum information: combining DEA with MAVT *Computers and Operations Research, 33,* 2083-2098.

Maystre, J., Pictet, J., Simos, J., 1994. Méthodes multicritères ELECTRE - Description, conseils pratiques et cas d'application à la gestion environnementale; *Presse Polytechniques et Universitaires Romandes*: Lausanne.

Miettinen, P., Hämäläinen, R.P., 1997. How to benefit from decision analysis in environmental life cycle assessment (LCA) *European Journal of Operational Research,* 102, 279-294.

Moffat, D.W., 1976. Economics Dictionary; Elsevier Scientific Publishing: New York.

Munda, G., Nijkamp, P., Rietveld, P., 1995. Qualitative multicriteria methods for fuzzy evaluation problems: An illustration of economic-ecological evaluation *European Journal of Operational Research, 82 (1),* 79-97.

Munda, G., 2004a. Social multi-criteria evaluation: Methodological foundations and operational consequences *European Journal of Operational Research, 158 (3),* 662-677.

Munda, G., 2004b. Social multi-criteria evaluation: Methodological foundations and operational consequences *European Journal of Operational Research, 158 (3),* 662-677.

Mustajoki, J., Hämäläinen, R.P., 2000. Web-HIPRE: Global Decision Support by Value Tree and AHP Analysis INFOR, 38 (3), 208-220.

Ness, B., Urbel-Piirsalu, E., Anderberg, S., Olsson, L., 2007. Categorising tools for sustainability assessment *Ecological Economics, 60,* 498-508.

Neumann, M., 2007. Choosing and Describing: Sen and the Irrelevance of Independence Alternatives *Theory and Decision,* 63, 79-94.

Pöyhönen, M., Vrolijk, H., Hämäläinen, R.P., 2001. Behavioral and procedural consequences of structural variation in value trees *European Journal of Operational Research, 134,* 216-227.

Pöyhönen, M., Hämäläinen, R., 2001. On the Convergence of Multiattribute Weighting Methods *European Journal of Operational Research, 129 (3),* 106-122.

Prato, T., 1999. Multiple attribute decision analysis for ecosystem management *Ecological Economics, 30 (2),* 207-222.

Rabl, A., Eyre, N., 1998. An estimate of regional and global O3 damage from precursor NOx and VOC emissions *Environment International,* 24 (8), 835-850.

Richards, B., Wall, C., 2005. Automotive Refinish Primers for Small Surface Damage Repair (http://corporate.basf.com/file/15376.file5).

Roy, B., Bouyssou, D., 1993. Aide multicritère à la décision; *Economica*: Paris.

Saaty, T.L., 1980. The Analytic Hierarchy Process; *McGraw Hill:* New York.

Saling, P., Maisch, R., Silvani, M., König, N., 2005. Assessing the Environmental-Hazard Potential for Life Cycle Assessment, Eco-Efficiency and SEE balance *International Journal of Life Cycle Assessment, 10 (5),* 364-371.

Saling, P., Kircherer, A., Dittrich-Krämer, B., Wittlinger, R., Zombik, W., Schmidt, I., Schrott, W., Schmidt, S., 2002. Eco-Efficiency Analysis by BASF: The Method *International Journal of Life Cycle Assessment, 7 (4),* 203-218.

Salo, A., Hämäläinen, R.P., 1997. On the measurement of preferences in the analytical hierarchly process (with discussion) *Journal of Multi Criteria Decision Analysis.*

Saltelli, A., Chan, K., Scott, E.M., 2000. Sensitivity Analysis; *John Wiley and Sons*: Chichester.

Sarkis, J., Weinrach, J., 2001. Using data envelopment analysis to evaluate environmentally consious waste treatment technology *Journal of Cleaner Production, 9,* 417-427.

Schleisner, L., 2000. Comparison of methodologies for externality assessment *Energy Policy, 28,* 1127-1136.

Schrader, B., 2005. Strategische Entscheidungsunterstützung: Implementierung und beispielhafte Anwendung eines Software Tools zur Evaluierung von Mehrzielentscheidungsproblemen.

Sen, A.K., 1970. Collective Choice And Social Welfare; *Holden-Day:* San Francisco.

Seppälä, J., Basson, L., Norris, G.A., 2002. Decision analysis frameworks for life-cycle impact assessment *Journal of Industrial Ecology, 5 (4),* 45-68.

Seppälä, J., Hämäläinen, R.P., 2001. On the meaning of the distance-to-target weighting method and normalisation in life cycle impact assessment International *Journal of Life Cycle Assessment, 6 (4),* 211-218.

Simpson, L., 1996. Do decision makers know what they prefer?: MAVT and ELECTRE *Journal of the Operational Research Society, 47,* 919-929.

Soest, J.P.V., Sas, H., De Witt, G., 1998. Apples, Oranges and the Environment: Prioritising Environmental Measures on the Basis of their Cost-effectiveness; *Centre for Energy Conservation and Environmental Technology:* Delft.

Stewart, T.J., 1992. A Critical Survey on the Status of Multiple Criteria Decision Making Theory and Praxis *OMEGA - International Journal of Management Science,* 20 (5/6), 569-586.

Timm, N.H., 2002. Applied multivariate analysis; *Springer*: New York.

Treitz, M., 2006. Production Process Design Using Multi-Criteria Analysis Dissertation, University of Karlsruhe (available online at: http://www. uvka.de/univerlag/volltexte/2006/178/); *Karlsruhe University Press:* Karlsruhe.

van Oers, L., Huppes, G., 2001. LCA Normalisation Factors for the Netherlands, Western Europe and the World *International Journal of Life Cycle Assessment, 6 (5),* 256.

Vetschera, R., 1994. Visualisierungstechniken in Entscheidungsproblemen bei mehrfacher Zielsetzung / Visualization techniques in multicriteria decision-making *OR Spektrum,* 16, 227-241.

Wall, C., Richards, B., Bradlee, C., 2004. The Ecological and Economic Benefits of UV Curing Technology *RadTech Report* 2004.

WBCSD, 2006. Eco-efficiency module; W*orld Business Council for Sustainable Development (WBCSD)*: Geneva.

Weber, M., Borcherding, K., 1993. Behavioural problems in weight judgements *European Journal of Operational Research, 67,* 1-12.

Weber, M., Eisenführ, F., Von Winterfeldt, D., 1988. The effects of splitting attributes on weights in multiattribute utility measurement *Management Science*, 34, 431-445.

Winterfeld, D.v., Edwards, D., 1986. *Decision Analysis and Behavorial Research*; Cambridge University Press: Cambridge.

Wrisberg, N., Udo de Haes, H.A., Triebswetter, U., Eder, P., Clift, R., 2002. Analytical Tools for Environmental Design and Management in a Systems Perspective; *Kluwer Academic Publishers*: Dordrecht.

Zaim, O., 2004. Measuring environmental performance of state manufacturing through changes in pollution intensities: a DEA framework *Ecological Economics, 48* (1), 37-47.

Zhou, P., Ang, B.W., Poh, K.L., 2006. Slacks-based efficiency measures for modeling environmental performance *Ecological Economics, 60* (1), 111-118.

In: Handbook of Environmental Policy
Editors: Johannes Meijer and Arjan der Berg

ISBN 978-1-60741-635-7
© 2010 Nova Science Publishers, Inc.

Chapter 8

ENVIRONMENTAL KUZNETS CURVES FOR CARBON EMISSIONS: A CRITICAL SURVEY

Nektarios Aslanidis[*]

Department of Economics, University Rovira Virgili, FCEE,
Avinguda Universitat 1, Catalonia, Spain

SUMMARY

The empirical finding of an inverse U-shaped relationship between per capita income and pollution, the so-called Environmental Kuznets Curve (EKC), suggests that as countries experience economic growth, environmental deterioration decelerates and thus becomes less of an issue. With more or less success, a large number of econometric studies have documented the existence of an EKC for pollutants such as sulfur dioxide, nitrogen oxide and suspended particulate matter. The baseline models estimated in the literature are linear polynomial models that include quadratic (and sometimes also cubic) terms of income as explanatory variables. Recently, these models have been criticized for being too restrictive, and alternative more flexible econometric techniques have been proposed. Focusing on the prime example of carbon emissions, the present chapter provides a critical review of these new econometric techniques. In particular, we discuss issues related to functional forms, heterogeneity of income effects across countries (regions), non-stationary ("spurious") regressions and spatial dependence in emissions. As for the functional form issue, some studies have addressed the nonlinearity of the income-emissions relationship by using a spline (piecewise linear) function, Weibull and smooth transition regression models, and more flexible parametric specifications, as alternatives to the polynomial model. The non-parametric models constitute one of the latest econometric tools used. Another important issue in panel data studies is the underlying assumption of homogeneity of income effects across countries. This assumption is too restrictive for large panels of heterogeneous countries. A further econometric criticism of the EKC concerns the issue of "spurious" regressions. As the model includes potentially non-stationary variables such as emissions and

---

[*] E-mail: nektarios.aslanidis@urv.cat.

GDP, one can only rely on regression results that exhibit the co-integration property. Finally, recent studies allow for spatial dependence in emissions across countries to account for the possibility that countries' emissions are affected by emissions in neighbouring countries. Despite these new approaches, there is still no clear-cut evidence supporting the existence of the EKC for carbon emissions.

JEL classifications: C20, Q32, Q50, O13

Key words: Environmental Kuznets Curve, carbon emissions, functional form; heterogeneity, "spurious" regressions, spatial dependence.

1. INTRODUCTION

The relationship between economic development and environmental quality has been extensively explored since the Grossman and Krueger (1991) finding of an inverse U-shaped relationship between per capita income and pollution, the so-called Environmental Kuznets Curve (EKC). The EKC suggests that as countries experience economic growth, environmental deterioration decelerates and thus becomes less of an issue. With more or less success, a large number of econometric studies have documented the existence of an EKC for pollutants such as sulfur dioxide (SO_2), nitrogen oxide (NO_x) and suspended particulate matter (SPM).[*] Apart from some exceptions, however, most of the EKC literature is statistically weak. The baseline models estimated in the literature are linear polynomial models that include quadratic (and sometimes also cubic) terms of income as explanatory variables. Recently, these models have been criticized for being too restrictive, and alternatives more flexible econometric techniques have been proposed.

Focusing on the prime example of carbon dioxide (CO_2) emissions, the present chapter provides a critical review of the new econometric techniques used. In particular, we discuss issues related to the functional form, the heterogeneity of income effects across countries (regions), "spurious" EKC regressions and spatial dependence in emissions across countries. To the best of my knowledge, no one has yet attempted to give an overview of the recent influential contributions and to determine whether and to what extent the EKC is robust regarding the new econometrics approaches employed.

On the functional form issue, some studies have addressed the non-linearity of the income-emissions relationship by using a spline (piecewise linear) function. The spline model has an advantage over the polynomial specification in that the approximation error is generally smaller. Others papers have considered Weibull distributions and smooth transition regression models and more flexible specifications as alternatives to the polynomial model. The non-parametric models, which do not require the specification of a functional form, constitute one of the latest econometric tools used. Yet, these new econometric approaches have not yielded conclusive results regarding the existence of the EKC for carbon emissions. Another important issue in panel data studies is the underlying assumption of homogeneity of

---

[*] Although it is essentially an empirical finding, some papers have also derived the EKC theoretically. See for example, Stokey (1998) and Jones and Manuelli (2001), among others. Levinson (2002) provides a review of the theoretical as well as the empirical literature.

income effects across countries (regions). As some studies show, not all countries display the same relationship between emissions and income. This is particularly true when developed and developing countries are compared, with the EKC holding for some developed countries only. A further econometric criticism of the EKC concerns the issue of "spurious" regressions. As the model includes potentially non-stationary variables such as emissions and GDP, one can only rely on EKC results that exhibit the co-integration property. The test for unit roots finds that carbon emissions and GDP per capita are integrated variables, although not always co-integrated, which casts doubt on the validity of the EKC. Finally, recent studies allow for spatial dependence in emissions across countries to account for the possibility that countries' emissions are affected by emissions in neighbouring countries. The results so far support the use of spatial econometric models over the polynomial EKC specification.

The main reason for studying carbon emissions is that they play a focal role in the current debate on environmental protection and sustainable development. CO_2 is a major determinant of the greenhouse gas implicated in global warming. While the physical effects of local pollutants such as sulphur dioxide or nitrogen oxide are conspicuous and can be accounted for by only domestic activity, the effects of carbon dioxide are far-reaching and cause an international externality. Thus the incentives to abate carbon emissions are clearly undermined by the free-rider problem, what makes the study of CO_2 emissions particularly interesting. Another reason is that CO_2 emissions are directly related to the use of energy, which is an essential factor in the world economy, both for production and consumption. Therefore, the relationship between carbon emissions and economic growth has important implications for environmental and economic policies.

The paper is organized as follows. Section 2 summarizes the basic idea of the emission-income relationship and surveys the first studies on the EKC. Section 3 discusses the standard polynomial specification and the reviews the studies using this methodology for carbon emissions. The new econometric techniques are presented in Section 4, while Section 5 discusses the policy implications emerging from the literature on the EKC. Finally, Section 6 concludes.

2. EKC: BACKGROUND IDEA

The basic idea of the EKC is that environmental degradation increases with income up to a threshold income level beyond which environmental quality improves as income continues to grow. This relationship is summarized by an inverted U-shaped curve (see Figure 1). It is known as the Environmental Kuznets Curve due to its resemblance to Kuznets's inverted U relationship between income inequality and economic growth (Kuznets, 1955). There are three main forces behind the EKC. First, growth exerts a scale effect on the environment: a larger scale of economic activity leads to increased environmental degradation as more energy is used. Second, income growth can have a positive impact on the environment through a composition effect: as a country grows and develops, the structure of its economy changes from a manufacturing based economy towards an information intensive and services based economy, and so increasing the share of cleaner activities in its GDP. Finally, as countries become richer, environmental awareness increases, and so does the demand for environmental regulations. This will generally lead to the substitution of obsolete and dirty technologies for cleaner ones, improving the quality of the environment. This is known as the induced technique effect of growth. The negative impact on the environment of the scale effect tends to prevail in the initial stages of countries' growth, but that it is eventually outweighed by the positive impact of the composition and induced technique effects that tend to lower emission levels.

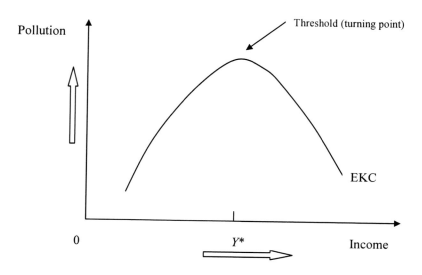

Figure 1. Environmental Kuznets Curve (inverse U-shaped relationship).

The EKC concept emerged during the early 1990s with three studies that appeared independently. Grossman and Krueger (1991) in an NBER working paper, published later in 1993 (Grossman and Krueger, 1993), tested the EKC hypothesis in the context of the much-debated North American Free Trade Agreement (NAFTA). At the time, many people feared that by opening the markets with Mexico companies would rush across the border to escape the stricter environmental standards of Canada and the United States. The authors already find an inverted-U relationship between pollutants such as sulphur dioxide or smoke and per capita income for the US previous to NAFTA. The emission-income relationship was also discussed by Shafik and Bandyopadhyay (1992) in the World Bank's inquiry into the growth and environment relation for the Bank's 1992 *World Development Report*. The authors argued that ''the view that greater economic activity inevitably hurts the environment is based on static assumptions about technology, tastes, and environmental investments'' and that ''as incomes rise, the demand for improvements in environmental quality will increase, as will the resources available for investment.'' The EKC was further popularized by Panayotou (1993) in a Development Discussion paper as part of a study for the International Labour Organisation. Panayotou was the first to name the relationship as the Environmental Kuznets Curve.

3. ECONOMETRIC METHODOLOGY

3.1. Baseline Model

The most prominent single-equation approach to the EKC is the estimation of linear polynomial models including quadratic (and sometimes also cubic) terms of income as explanatory variables. The standard quadratic polynomial model is given by[*]

$$p_{it} = \mu_i + \varphi_t + \beta_1 y_{it} + \beta_2 y_{it}^2 + u_{it} \quad i = 1,...,N \; ; \; t = 1,...,T \quad (1)$$

where $p_{it} = \ln(P_{it})$ is the logarithm of per capita emissions in region (country) i in year t, $y_{it} = \ln(Y_{it})$ is the logarithm of per capita GDP in region (country) i in year t, $\beta \equiv (\beta_1, \beta_2)'$ is the parameter vector and u_{it} is an error term.[+] If the coefficient on income, β_1, is positive and the coefficient on income squared, β_2, is negative, the relationship between income and emissions is not monotonic but displays an inverse-U shape. The term μ_i is a region-specific effect, which controls for unobserved factors that affect emissions at the regional level. The model accounts for heterogeneity in a limited way though. Although the level of emissions per capita may differ across regions, the income elasticity is assumed to be the same in all regions at a given income level. The time-specific (or year-specific) intercepts φ_t may reflect changes over time in relevant factors common across regions such as macroeconomic factors and stochastic shocks. In addition, φ_t may reflect common changes over time in the technology used as well as in the environmental policies and standards adopted. Some papers include a time trend, instead of year-fixed effects, in order to estimate a more parsimonious model. In this case, all years have an equal effect on emissions.

Some studies also control for other possible determinants of emissions such as trade openness and measures of international mobility of factors to account for the so-called "pollution haven hypothesis" (Grossman and Krueger, 1991, 1993, Jaffe et al., 1995, Janicke et al., 1997, Suri and Chapman, 1998, Cole and Elliott, 2003, Cole 2004). The "pollution heaven hypothesis" argues that heavy polluters move from high-income countries with strict environmental regulations to low-income countries with weaker environmental regulations. So, the shape of the EKC is a consequence of high-income countries "exporting" their pollution to low-income countries. Other studies have included measures of income inequality (Torras and Boyce, 1998; Magnani, 2000; Bousquet and Favard, 2005) and measures of corruption (Lopez and Mitra, 2000, Fredriksson et al., 2004; Cole, 2007). The reason for the inclusion of income inequality is that inequality may reduce a country's willingness to pay for environmental regulation and abatement, while corruption presumably reduces the stringency of environmental policy and, therefore, is likely to have a negative impact on the environment as well.

---

[*] The popular quadratic model appears to be due to Holtz-Eakin and Selden (1995), whereas Grossman and Kruger (1995) use a cubic polynomial model.
[+] The functional form takes typically either a log-linear or linear form, with a number of studies considering both. In general, the results are qualitatively the same.

The turning point or threshold level of income, where emissions are at a maximum is calculated by taking the derivative of $E(p_{it})$ in Eq. (1) with respect to y_{it}, setting it equal to zero and solving for y_{it} (or Y_{it})

$$Y^* = \exp(-\frac{\beta_1}{2\beta_2})$$

Estimation of the polynomial specification in Eq. (1) can be carried out by fixed effects (within-group estimator) or random effects (feasible generalised least squares). The fixed effects estimator treats the μ_i and φ_t terms as regression parameters, whereas the random effects estimator treats them as components of the error term u_{it}. The random effects estimator is more efficient than the fixed effects estimator. The important consideration here is whether μ_i and φ_t are correlated with per capita income. If they are, the random effects model yields inconsistent estimates and only the fixed effects estimator should be used. Many studies perform a Hausman test to choose between the fixed effects and random effects estimators.

3.2. Empirical Findings

Although evidence of an EKC has been found for several pollutants, these findings are not unanimously accepted in the literature. The case of CO_2 emissions is a good example. In this section we survey the early EKC literature using the polynomial model to study the carbon emissions-income relationship.[*] Table 1 summarizes the studies of carbon emissions, listed in chronological order. In early work, Shafik (1994) fits a country fixed effects model with a time trend for a panel of 149 countries over the period 1960–1990 and finds that carbon emissions do not improve with rising income, as the linear model has virtually all the explanatory power.[+] Holtz-Eakin and Selden (1995) estimate a quadratic polynomial model with country and year fixed effects for a panel of 130 countries during 1951–1986 and obtained some support for an EKC. However, their estimated turning point occurs at a very high level of per capita income ($35,428 in per capita 1986 dollars). An EKC model for CO_2 emissions is also estimated by Tucker (1995) on a cross-section of 131 countries for each year during the period 1971–1991. An inverted-U curve rises in statistical significance over time, and mainly during the 1980s. In particular, the coefficient of the linear income term is always positive and significant, while that of the quadratic income term is significant in 13 years out of 21, negative in 11 of those years, and becomes more negative and significant as time goes by.

---

[*] The list of references cited in this section is by no means exhaustive. For more general discussions, also on other pollutants, see the special issues of the Environmental and Development Economics (1997) and Ecological Economics (1998). See also the surveys of Stern (1998, 2004), Panayotou (2000), Dasgupta et al. (2002), Levinson (2002), Cole (2003), Copeland and Taylor (2004) and Dinda (2004).
[+] This paper was originally a background paper (Shafik and Bandyopadhyay, 1992) for the World Bank's inquiry into growth and environment relationships (see the 1992 *World Development Report*).

Table 1. CO₂ EKC studies using polynomial model.

Author (s)	Country/Time effects	Data sample	Time period	Shape of EKC
Shafik (1994)	Fixed country effects/Time trend	149 countries	1960–1990	Linear (positive) relationship
Holtz-Eakin and Selden (1995)	Fixed country/Time effects	130 countries	1951–1986	Inverse U-shaped (but turning point is too high)
Tucker (1995)	Cross-section regressions for each year	131 countries	1971–1991	Inverse U-shaped (stronger over time)
Cole et al. (1997)	Fixed country effects	7 world regions	1960–1991	Inverse U-shaped (but turning point is too high)
de Bruyn et al. (1998)	Time series regressions	4 OECD countries	1961–1990	Linear (positive) relationship
Hill and Magnani (2002)	Cross-section regressions for each year	156 countries	1970, 1980, 1990	Inverse U-shaped (but highly sensitive to dataset and turning point is too high)
Friedl and Getzner (2003)	Time series regressions	Austria	1960–1999	N-shaped
Lantz and Feng (2006)	Fixed region effects	5 Canadian regions	1970–2000	CO₂ is unrelated to income. Inverse U-shaped with population and U-shaped with technology

Cole et al. (1997) examine the EKC relationship for a wide range of environmental indicators using panel datasets. The study focuses on a quadratic polynomial model with country fixed effects estimated in both linear and log-linear versions. As in Holtz-Eakin and Selden (1995) they obtain an EKC relationship with significant income parameters but the turning points fall well outside the observed income range, and in the log-linear model the standard errors of the turning point are large. This implies that the estimates of the CO_2 turning point are quite unreliable, casting doubt on the possible downturn of CO_2 emissions. In general, their results suggest that a meaningful EKC exists only for local air pollutants.

In Hill and Magnani (2002) the EKC for carbon emissions is found to be highly sensitive to the dataset used. They use data for 156 countries and three separate years: 1970, 1980 and 1990. Cross-section estimation supports the EKC hypothesis for all three cross-sections, though the turning point is very high and near the upper end of the income distribution. However, when countries are split into low, middle and high income, carbon emissions seem to increase with income for all three groups of countries. The authors also test for omitted variables and find that openness, inequality and education are significant determinants of CO_2 emissions.

Other papers have focused on individual countries. de Bruyn et al. (1998) argue that the estimation of the EKC from panel data can not capture the dynamics of the relationship between income and emissions. By using a dynamic model and including energy prices to account for the intensity of use of raw materials, they consider an emission-income relationship separately for the Netherlands, the UK, the US and West Germany over the period 1961–1990. Their results show that economic growth has a positive direct effect on emissions and that emission reductions may be achieved as a result of structural and technological changes in the economy. In the context of a small open economy, Friedl and Getzner (2003) estimate an EKC for Austria over the period 1960–1999. They obtain the so-called N-shaped or cubic relationship, which exhibits the same pattern as the inverted-U curve initially, but beyond a certain income level the relationship between emissions and income is positive again (see Figure 2). The existence of an N-shaped curve suggests that at very high income levels, the scale effect of economic activity becomes so large that its negative impact on environment can not be counterbalanced by the positive impact of the composition and induced technique effects mentioned above. Lantz and Feng (2006) look at the EKC relationship for carbon emissions in Canada using a region-level panel dataset (five regions) with region fixed effects for the period 1970–2000. Their results show that carbon emissions are unrelated to GDP. Interestingly, they find an inverted U-shaped relationship between CO_2 emissions and population, and a U-shaped relationship between CO_2 emissions and technology.

On the whole, the variability of the empirical findings discussed leads to the conclusion that the standard polynomial model may not be the most adequate to capture the relationship between carbon emissions and income.

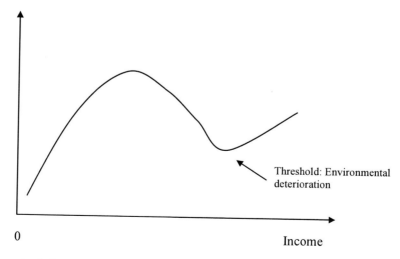

Figure 2. N-shaped relationship.

4. Econometric Issues Regarding The Estimation Of EKC

In this section we provide a critical review of the new econometric techniques recently used in the EKC literature. Table 2 summarizes the studies focusing on carbon emissions and listed in chronological order.

Table 2. CO₂ EKC studies using new econometric techniques.

Author (s)	Econometric issue addressed	Technique	Data sample	Time period	Shape of EKC
Schmalensee et al. (1998)	Functional form	Spline model	141 countries	1950–1990	Inverse U-shaped
Taskin and Zaim (2000)	Functional form	Non-parametric models	52 countries	1975–1990	Inverse U-shaped
Martinez-Zarzoso and Bengochea-Morancho (2004)	Heterogeneity	Pooled mean group estimator	22 OECD countries	1975–1998	N-shaped for majority of countries
Wagner and Müller-Fürstenberger (2004)	"Spurious" EKC relationship	Panel unit root & cointegration tests	107 countries	1986–1998	Results are mixed
Bertinelli and Strobl (2005)	Functional form	Non-parametric models	122 countries	1950–1990	Linear (positive) relationship
Dijkgraaf and Vollebergh (2005)	Heterogeneity	Polynomial & spline models	24 OECD countries	1960–1997	Inverse U-shaped in 11 out of 24 countries
Azomahou et al. (2006)	Functional form	Non-parametric models	100 countries	1960–1996	Linear (positive) relationship
Galeotti et al. (2006)	Functional form	Weibull model	125 countries	1960–1997 (OECD) 1971–1997 (non-OECD)	Inverse U-shaped for OECD Concave (but with no reasonable turning point) for non-OECD
Aslanidis and Iranzo (2009)	Functional form	Smooth transition regression models	77 non-OECD countries	1971–1997	Positive but at a slower rate after some income threshold
Auffhammer and Carson (2009)	Spatial dependence	Spline model augmented with spatial dependence	30 Chinese provinces	1985–2004	Linear (positive) relationship
Galeotti et al. (2009)	"Spurious" EKC relationship	Fractional panel unit root & cointegration tests	24 OECD countries	1960–2002	Inverse U-shaped in five out of 24 countries

4.1. New Functional Forms

Given the restrictiveness of the polynomial model in Eq. (1), alternative more flexible functional forms have been proposed. For instance, Schmalensee et al. (1998) use a spline (piecewise linear) function, which is a linear approximation to a non-linear function. The number of splines is based on a test, with the final model having 10-segment splines, each containing an equal number of observations. The spline model has the advantage over the polynomial specification in that the approximation error is generally smaller. Schmalensee et al. (1998) find evidence of an EKC for CO_2 emissions, with a within-sample turning point, for a dataset of 141 countries over the period 1950–1990.[*]

Galeotti et al. (2006) propose a Weibull functional form to estimate an EKC. The choice of the Weibull distribution is based on its easily interpretable parameters. The regression model is given by

$$\ln(P_{it}) = \mu_i + \varphi_t + (\alpha - 1)\ln\left(\frac{Y_{it} - \gamma}{\beta}\right) - \left(\frac{Y_{it} - \gamma}{\beta}\right)^{\alpha} + \delta\left(\frac{Y_{it} - \gamma}{\beta}\right)^{-\alpha} + u_{it}$$

where the shape parameter α governs the curvature of the function, while the scale parameter β is related to the height of the function, and therefore with the maximum level of emissions at the turning point, if the latter exists. Furthermore, the location parameter γ controls for the position of the function and, therefore, implies the turning point of income. As for δ, this parameter gives added flexibility to the model by allowing for different patterns in the shape of the function. The model is estimated by maximum likelihood (ML) on carbon emissions for 125 countries. The results are mixed. There is evidence of an EKC with reasonable turning point during 1960-1997 for OECD countries, while a concave pattern with no reasonable turning point is obtained for non-OECD countries over the period 1971–1997.

Aslanidis and Xepapadeas (2006) propose a 2-regime smooth transition regression (STR) model which is an even more flexible parametric specification, and as they show the quadratic polynomial model is just the linearized version of the STR. The STR model is given by

$$p_{it} = \mu_i + \varphi_t + (\beta_1 + \beta_2 F(y_{it})) y_{it} + u_{it}$$

$F(y_{it})$ is the transition function, in this case depending on income, which is assumed to be continuous and bounded between 0 and 1; y_{it} is the transition variable-income. An EKC exists if $\beta_1 > 0$ and $\beta_1 + \beta_2 < 0$. In words, emissions increase with income up to some threshold level of income after which they are reduced with further growth. To complete the model, consider the following logistic functional form for the transition function

$$F(y_{it}) = (1 + \exp(-\gamma(y_{it} - c)))^{-1}$$

---

[*] They use an extension of the Holtz-Eakin and Selden (1995) dataset.

where the parameter c is the threshold between the two regimes. The slope parameter γ gives flexibility to the model by determining the smoothness of the change in the value of the logistic function and thus the speed of the transition from one regime to the other. For instance, when $\gamma \rightarrow \infty$, $F(y_{it})$ becomes a step function and the transition between regimes is abrupt. Estimation of the STR is carried out by non-linear least squares (NLS). Aslanidis and Iranzo (2009) applied this methodology to CO_2 emissions from 77 non-OECD countries over the period 1971–1997. Although there is no evidence of EKC, they find two regimes; a low-income regime where emissions accelerate with economic growth and a middle-to high-income regime associated with a deceleration in environmental degradation.

The semi and non-parametric models constitute one of the latest econometric tools used to test for the EKC hypothesis. These models are appealing as they impose no parametric restrictions on the form of the relationship. For instance, the semi-parametric model considered by Millimet et al. (2003) is written as

$$p_{it}=\mu_i+\varphi_t+G(y_{it})+u_{it}$$

where $G(y_{it})$ is an unknown function of income, which *a priori* $G(.)$ can take any functional form. The estimation methods are based on standard kernel regressions. Taskin and Zaim (2000) estimate a non-parametric model for some measures of environmental efficiency. On the basis of cross-sectional data for carbon emissions, they compute environmental efficiency indices and show evidence of EKC for a panel of 52 countries over the period 1975–1990. However, other studies that use semi and non-parametric specifications obtain mixed results. For example, using a panel of 122 countries, Bertinelli and Strobl (2005) can not reject a linear (positive) relationship between per capita income and carbon emissions during 1950–1990. Azomahou et al. (2006) carry out an extensive analysis on a panel of 100 countries during 1960–1996 and find that the linear (positive) relationship between carbon emissions and GDP can not be rejected either. They formally test this hypothesis by performing a monotonicity test within their non-parametric framework. Moreover, they test and reject the polynomial functional form in favour of the non-parametric model. As shown from the previous studies, the use of a particular functional form does not yield conclusive results either.

4.2. Homogeneity across Countries

Besides the functional form, another important restriction of the polynomial model is the imposed homogeneous income effect across regions (or countries). List and Gallet (1999), Martinez-Zarzoso and Bengochea-Morancho (2004), and Dijkgraaf and Vollebergh (2005), among others, have relaxed such assumption.[*] The homogeneity assumption implies that except for the fixed (scale) effect all regions exhibit on average the same emission-income pattern. More precisely, all regions share the same turning point though the peak emission level may differ across regions via the individual specific effects (see Figure 3). This assumption is too restrictive for large panels of heterogeneous regions. Regions (or countries) vary in terms of resource endowments, infrastructure, public pressure, economic, social and political factors, etc., and thus so might vary their income-pollution relationship.

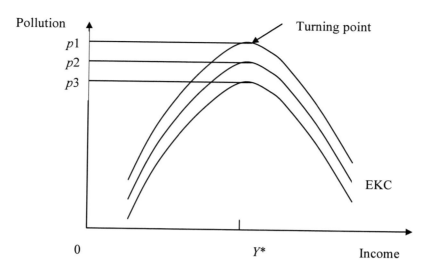

Figure 3. Environmental Kuznets Curve: Slope Homogeneity.

Using a panel of US state-level data on SO_2 and NO_x emissions List and Gallet (1999) address the homogeneity issue by allowing for different income slopes across states.[+] They use a polynomial seemingly unrelated regressions (SUR) model, which appears appropriate for long time series data (their sample period is 1929–1994). Their results reject the homogeneity assumption and provide some evidence of the EKC being robust across US states.

Martinez-Zarzoso and Bengochea-Morancho (2004) analyse carbon emissions for 22 OECD countries during 1975–1998. They employ a pooled mean group estimator that allows for slope heterogeneity across countries in the short run, while imposing restrictions in the long run. These long-run restrictions are tested and supported by the data. The results show a great deal of heterogeneity across countries, and in most cases an N-shaped relationship emerges.

In a similar spirit, Dijkgraaf and Vollebergh (2005) argue that even a cursory comparison of per capita CO_2 and GDP plots for Japan and France casts serious doubts on the homogeneity assumption. Using data for 24 OECD countries for the period 1960–1997, the authors fit polynomial and spline models to test the null hypothesis that income coefficients are the same for all countries. The homogeneity assumption is clearly rejected. When individual country time series models are estimated, only 11 out of 24 cases show a statistically significant turning point and confirm the EKC hypothesis.

---

* As mentioned before, de Bruyn et al. (1998) criticize the estimation of the EKC from panel data and argue for country-specific models. Effectively they are also challenging the homogeneity assumption.
+ This is the same data used by Millimet et al (2003) and Aslanidis and Xepapadeas (2006).

The firm rejection of the homogeneity assumption raises doubts not only on the homogeneous polynomial model but, insofar as they assume common income effects, also on the more flexible specifications discussed in the previous section.

4.3. "Spurious" Regressions

Another important issue that still remains unsolved is that of possible "spurious" EKC relationships. The early literature completely neglects the fact that the EKC regressions involve potentially non-stationary variables such as emissions and GDP.[§] We can only rely on results from regressions that contain non-stationary variables if these variables exhibit the co-integration property, that is, if there is a long-run equilibrium relationship between them.

The econometrics literature has extended non-stationarity (unit root) tests to panel data. Let x_{it} denote the variable on which we want to test for a unit root; in our case, emissions or income. In general, the panel unit root tests consider the following regression model

$$x_{it} = \mu_i + \varphi_t + \rho_i x_{it-1} + u_{it}$$

where u_{it} is a stationary process.[*] Under the null hypothesis, there is a unit root in x_{it}, i.e., $H_0 : \rho_i = 1$ for all $i = 1, ..., N$. On the other hand, the alternative hypothesis can take two forms depending on whether there are restrictions on the autoregressive coefficients ρ_i across cross-sections (regions). First, one can assume that the autoregressive coefficients are common across cross-sections. This gives rise to the homogeneous alternative of stationarity $H_a^{Homo} : \rho_i = \rho < 1$ for all i. A popular unit root test with homogenous alternative is the test of Levin et al. (2002) (LL), which is a modified augmented Dickey-Fuller (ADF) test. Alternatively, one can allow ρ_i to vary across cross-sections. This gives rises to the heterogeneous alternative of stationarity

$$H_a^{Hetero} : \begin{cases} \rho_i < 1 & \text{for } i = 1, ..., N_1 \\ \rho_i = 1 & \text{for } i = N_1 + 1, ..., N \end{cases}$$

for some N_1 such that $\lim_{N \to \infty} N_1 / N > 0$. The heterogeneous alternative is more flexible than the homogeneous one in two ways. First, it allows for some cross-sections to be non-stationary also under the alternative and, second, it does not restrict the autoregressive coefficient to be identical under the alternative hypothesis. Popular unit root tests with heterogeneous alternative are the two tests developed by Im et al. (2003). One of these tests is essentially a group-mean of individual ADF statistics (IPS) test and the other is a group-mean Lagrange multiplier (IPS-LM) test. Another popular panel unit root test with a heterogeneous alternative is the test proposed by Maddala and Wu (1999) (MW). The idea is based on

---

[§] In the macroeconometrics literature there is a lot of evidence that GDP series in particular are non-stationary.

Fisher's results to derive tests that combine the p-values from individual unit root tests. The MW test is flexible in that it can be applied to any type of unit root test.

If the null hypothesis of non-stationarity of emissions and GDP is not rejected, the next step is to test whether these variables are co-integrated using the recently developed co-integration tests for panel data.[+] Pedroni (2004) proposes seven co-integration tests which have become very popular in empirical work. All these tests are unit root tests performed on the residuals of the EKC regression. If carbon emissions and GDP are co-integrated, the residual process will be stationary. As in unit root tests, the co-integration tests can take two forms depending on whether there are restrictions on the autoregressive coefficients across cross-sections.

Wagner and Müller-Fürstenberger (2004) use the aforementioned panel unit root and co-integration tests to study the polynomial EKC. Their analysis is based on carbon emissions and GDP data for 107 countries over the period 1986–1998. Because of the short time span, they resort to both classical as well as bootstrap inference. Their results are mixed. Although, for carbon emissions there is clear evidence for non-stationarity, the test for GDP is not clear-cut. As for co-integration, results are not conclusive either. They depend upon the choice of the unit root and co-integration test, and also on whether one uses bootstrap or classical inference.[§]

The above panel integration and co-integration tests assume that the order of integration of a stochastic process can take on only integer values. This knife-edge distinction between, say, a stationary I(0) (integrated of order 0) and a non-stationary I(1) (integrated of order 1) process is too restrictive. Galeotti et al. (2009) challenge this assumption and consider tests of fractional integration and co-integration for panels. Fractionally differenced processes are flexible as the order of integration does not need to be an integer but can take any value between zero and one. Also, the order of integration is allowed to differ across cross sections. This framework gives flexibility to the EKC model as it allows for more possibilities for emissions and income to be co-integrated if they are non-stationary. The authors use a panel of 24 OECD countries over the period 1960–2002. The fractional integration tests find evidence of non-stationarity for the carbon emissions and GDP processes. Regarding co-integration, using a value of the (estimated) integration parameter of 0.5 as a threshold for fractional co-integration, the EKC hypothesis is supported in only five out of 24 countries. Overall, their results cast doubt on the validity of the EKC.

4.4. Spatial Dependence

Most papers estimating the EKC implicitly assume that regions' (countries') emissions are unaffected by the emissions in neighbouring regions. This assumption has recently been challenged in papers using spatial econometric techniques (Maddison, 2006, Auffhammer and Carson, 2009).[*] There are several reasons why spatial relationships may be present in the

---

* Note that region-specific time trends instead of the time-specific fixed effects can be included.
+ A comprehensive survey is given in Breitung and Pesaran (2008).
§ Similar mixed results are also reported by Müller-Fürstenberger and Wagner (2007). They use the same data but focus on the IPS and IPS-LM tests.
* Maddison (2006) use a country-panel of sulphur dioxide, nitrogen oxides, volatile organic compounds and carbon monoxide emissions for only two years of data (1990 and 1995). His methodology consists in a standard

income-pollution relationship. First, according to the "pollution haven hypothesis", and given that distance and common land borders may be important factors in increasing trade and investment, poor regions close to rich ones would be more likely to host the dirty activities of firms of developed countries and thus to have higher emissions. Second, the literature on the international diffusion of technology suggests that this is geographically localized, so that the R&D spillovers decline with geographical distance (Keller, 2004). Therefore, if there is technological progress that reduces emissions, it is reasonable to consider spatial relationships in emissions. Third, CO_2 emissions are strongly correlated with industrial activity. As economies are becoming increasingly linked over time so do their industrial activities, which in turn implies a stronger spatial relationship in emissions. Finally, governments often mimic each other environmental policies in order to reduce the costs of decision making and to legitimize their actions (Fredriksson and Millimet, 2002).

Auffhammer and Carson (2009) explore spatial econometric models to provide out-of-sample forecasts of China's aggregate emissions. Their analysis is based on province-level panel data of carbon emissions for 30 Chinese provinces over the period 1985–2004. The spatial econometric model considered by the authors is the following

$$p_{it} = \mu_i + \varphi_t + G(y_{it}) + G(y_{it-1}) + \pi\, p_{it-1} + \rho\left(\sum_{j=1}^{k} w_{ij} p_{jt-1}\right) + u_{it}$$

where $\left(\sum_{j=1}^{k} w_{ij} p_{jt-1}\right)$ are spatial lags which capture spillover effects across provinces and w_{ij} are the spatial weights given to previous year's CO_2 emissions by its k neighbouring provinces. In words, carbon emissions at a particular Chinese province are partially determined by a spatially weighted average of emissions of the neighbouring provinces. In principle, the model is semi-parametric as $G(.)$ has an unknown functional form which models the (possibly) non-linear relationship between emissions and GDP.[*] Moreover, the authors propose a dynamic model in order to take into account the partial adjustment of capital due to technological progress. For this, they include lagged emissions, p_{it-1}, as a regressor. In its most general form, the model allows for different speeds of adjustment across provinces π_i and this makes the technique even more flexible.

Their results support the use of the spatial model. In particular, the fit improves substantially with the inclusion of spatial lags. Moreover, the model clearly outperforms the static quadratic EKC specification on the basis of in-sample evaluation criteria. As for forecasting, the results point to a notable increase in carbon emissions in China during the current decade.

Summing up, these findings are encouraging for the use of spatial econometrics techniques and it rests for future research to see whether they can provide similar results for other datasets as well as for other types of pollutants.

---

quadratic model augmented by spatial dependence. The results do not give support to the existence of an EKC while reveal significant spatial effects across countries.

[*] In practice, Auffhammer and Carson search over three functional forms, that is, polynomial, spline and non-parametric, and finally settle with the spline model.

5. POLICY IMPLICATIONS

The shape of the relationship between carbon emissions and income has critical policy implications. An inverse U-shaped relationship seems to suggest that as countries experience economic growth, environmental deterioration eventually decelerates and thus becomes less of an issue. Therefore, taking these results for their face value would imply that growth is the "cause" and the "cure" of enviromental degradation. The problem would then be how to best accelerate growth to surpass the income threshold (turning point) as soon as possible. However, the survey carried out here shows that there are reasons to question this conclusion.

First, the EKC is not a structural model capturing the interrelations between technology, the composition of economic output, environmental policy and their effects on emissions, but a reduced form model. As such, it has the advantage that it is easily estimated. However, the observed relation between income and pollution reflects a correlation rather than a causal relationship. Furthermore, the EKC does not answer the question whether the reduction in emissions is achieved by more ambitious environmental policies (that may even be unrelated to economic growth) or by exogenous structural and technological changes. But, more fundamentally, the evidence presented in this survey suggests that the econometric foundations of the EKC are, in fact, weak and cast doubt on the generalization of the EKC to the majority of countries.

The failure to accept the EKC gives rise to radically different policy implications regarding environmental policy, with particularly dramatic consequences for developing countries. In effect, the environmental conditions in which the less advanced economies are developing today are much different from the ones faced by the developed countries in the past. The stock of greenhouse gases inherited by today's developing countries is certainly higher than that encountered by the developed countries in the early stages of their development. It is this stock, rather than the current flow of carbon emissions, that contributes mostly to global warming and its damages. For this reason, a policy of "accelerating growth in order to surpass the income threshold" based on a naïve interpretation of the EKC may have serious negative effects on the environment in the future.

This argument affects particularly the developing countries currently on the upward part of the curve. There is a good reason to believe that these countries may not be able to follow the same path as the developed countries. For instance, according to the "pollution heaven hypothesis" the EKC may be the result of environmental effects being displaced from developed countries (with stricter environmental regulations) to developing countries (with weaker environmental regulations), rather than reduced overall emissions. This implies that, without the implementation of the appropriate environmental policies, developing countries would not be able to find in turn some other countries to which "export" their pollution-intensive industries.

6. CONCLUSION

The empirical research on the relationhship between CO_2 emissions (a major greenhouse gas) and economic growth is continuously spurred by the renewed attention of scientists, policy-makers and the public opinion to the issue of climate change. A remarkably large number of recent contributions have investigated this relationship, correcting for some of the drawbacks of the early studies using the baseline polynomial model. In this survey we highlight the econometric issues related to functional forms, heterogeneity of income effects across countries, "spurious" EKC regressions and spatial dependence in emissions across regions.

With respect to functional forms, new parametric (e.g., spline, Weibull and smooth transition regression) and the non-parametric forms have been proposed as alternative and more flexible specifications to the baseline polynomial model. Despite these more sophisticated approaches, there is still no clear-cut evidence supporting or rejecting the existence of the EKC for carbon emissions. As for the assumption of homogeneous income effects across regions (countries), there is an aggreement in the literature rejecting such assumption. This is particularly clear when developed and developing countries are compared, with the EKC holding for some developed countries only.

With regard to the possible "spurious" EKC relationship, we reviewed studies adopting the recently-developed unit root and co-integration tests for panel data. Overall, they find that carbon emissions and GDP per capita are integrated variables, although not always co-integrated, what casts doubt on the validity of the EKC. Finally, some recent studies have allowed for spatial dependence in emissions across regions, which is intuitively appealing as regions' emissions are likely to be affected by emissions in neighbouring regions. The results, so far, are encouraging in the sense that the spatial econometric models clearly outperform the baseline polynomial EKC specification.

Other issues that, in our view, remain unresolved are the possible structural breaks in the EKC and contemporaneous feedback effects from emissions to GDP.[*] So far little work has addressed these issues. Azomahou et al. (2006) looks at the first issue and find no evidence of structural shifts in the (monotonic) relationship between CO_2 emissions and GDP. As for simultaneity, the results in Holtz-Eakin and Selden (1995) reject the existence of contemporaneous feedback effects. However, the evidence is still sparse and more work needs to be done in this direction.[+]

REFERENCES

Aslanidis, N., & Xepapadeas, A. (2006). Smooth transition pollution-income paths. *Ecological Economics, 57* (2), 182-189.

Aslanidis, N., & Iranzo, S. (2009). Environment and development: Is there a Kuznets curve for *CO2* emissions? *Applied Economics,* in press.

Auffhammer, M., & Carson, R.T. (2009). Forecasting the path of China's CO2 emissions using province level information. *Journal of Environmental Economics and Management,* in press.

---

[*] Regarding the latter, it is worth mentioning that the environment is a major factor of production as many countries heavily rely on natural resources to grow. At the same time, environmental degradation (e.g., high pollution levels) may reduce worker productivity as well as compromise potential growth.

[+] For instance, one could investigate a VAR-type model for CO_2 emissions and GDP, and to analyse the long-run and short-run effects of GDP.

Azomahou, T., Laisney, F., & Nguyen Van, P. (2006). Economic development and CO2 emissions: A nonparametric panel approach. *Journal of Public Economics, 90* (6-7), 1347-1363.

Bertinelli, L., & Strobl, E. (2005). The environmental Kuznets curve semi-parametrically revisited. *Economics Letters, 88* (3), 350-357.

Bousquet, A., & Favard, P. (2005). Does S. Kuznets's belief question the Environmental Kuznets Curves? *Canadian Journal of Economics, 38* (2), 604-614.

Breitung, J., & Pesaran, M.H. (2008). Unit roots and cointegration in panels. Forthcoming in L. Matyas, & P. Sevestre (Eds.), *The Econometrics of Panel Data: Fundamentals and Recent Developments in Theory and Practice* (Third Edition). Berlin: Kluwer Academic Publishers.

Cole, M.A. (2003). Development, trade and the environment: How robust is the environmental Kuznets curve? *Environment and Development Economics, 8* (4), 557–580.

Cole, M.A. (2004). Trade, the pollution haven hypothesis and the environmental Kuznets curve: Examining the linkages. *Ecological Economics, 48* (1), 71–81.

Cole, M.A. (2007). Corruption, income and the environment: An empirical analysis. *Ecological Economics, 62* (3-4), 637-647.

Cole, M.A., & Elliott, R.J.R. (2003). Determining the trade-environment composition effect: the role of capital, labour and environmental regulations. *Journal of Environmental Economics and Management, 46* (3), 363–383.

Cole, M.A., Rayner, A.J., & Bates, J.M. (1997). The environmental Kuznets curve: An empirical analysis. *Environment and Development Economics, 2* (4), 401–416.

Copeland, B.R., & Taylor, M.S. (2004). Trade, growth, and the environment. *Journal of Economic Literature, 42* (1), 7–71.

Dasgupta, S., Laplante, B., Wang, H., & Wheeler, D. (2002). Confronting the environmental Kuznets curve. *Journal of Economic Perspectives, 16* (1), 147-168.

de Bruyn, S.M., van den Bergh, J.C.J.M., & Opschoor, J.B. (1998). Economic growth and emissions: Reconsidering the empirical basis of environmental Kuznets curves. *Ecological Economics, 25* (2), 161-175.

Dinda, S. (2004). Environmental Kuznets curve hypothesis: A survey. *Ecological Economics, 49* (4), 431–455.

Dijkgraaf, E., & Vollebergh, H.R.J. (2005). A test for parameter homogeneity in CO2 panel EKC estimations. *Environmental and Resource Economics, 32* (2), 229–239

Fredriksson, P., & Millimet, D.L. (2002). Strategic interaction and the determination of environmental policy across U.S. States. *Journal of Urban Economics, 51* (1), 101–122.

Fredriksson, P., Vollebergh, H.R.J., & Dijkgraaf, E. (2004). Corruption and energy efficiency in OECD countries: Theory and evidence. *Journal of Environmental Economics and Management, 47* (2), 207-231.

Friedl, B., & Getzner, M. (2003). Determinants of CO2 emissions in a small open economy. *Ecological Economics, 45* (1), 133–148.

Galeotti, M., Lanza, A., & Pauli, F. (2006). Reassessing the environmental Kuznets curve for *CO2* emissions: A robustness exercise. *Ecological Economics, 57* (1), 152-163.

Galeotti, M., Manera, M., & Lanza, A. (2009). On the robustness of robustness checks of the environmental Kuznets curve. *Environment and Development Economics*, forthcoming.

Grossman, G.M., & Krueger, A.B. (1991). Environmental impacts of a North American Free Trade Agreement. National Bureau of Economic Research Working Paper 3914. NBER. Cambridge, MA.

Grossman, G.M., & Krueger, A.B. (1993). Environmental impacts of a North American Free Trade Agreement. In P. Garber (Eds.), *The Mexico-US Free Trade Agreement* (pp. 13-56). Cambridge, MA: MIT Press.

Grossman, G.M., & Krueger, A.B. (1995). Economic growth and the environment. *Quarterly Journal of Economics, 110* (2), 353–377.

Hill, R.J., & Magnani, E. (2002). An exploration of the conceptual and empirical basis of the environmental Kuznets curve. *Australian Economic Papers, 41* (2), 239–254.

Holtz-Eakin, D., & Selden, T.M. (1995). Stoking and fires? CO2 emissions and economic growth. *Journal of Public Economics, 57* (1), 85-101.

Im, K.S., Pesaran, M.H., & Shin, Y. (2003). Testing for unit roots in heterogeneous panels. *Journal of Econometrics, 115* (1), 53–74.

Jaffe, A.B., Peterson, S.R., Portney, P.R., & Stavins, R.N. (1995). Environmental regulation and the competitiveness of US manufacturing: What does the evidence tell us? *Journal of Economic Literature, 33* (1), 132-163.

Janicke, M., Binder, M., & Monch, H. (1997). 'Dirty industries': Patterns of change in industrial countries. *Environmental and Resource Economics, 9* (4), 467-491.

Jones, L., & Manueli, R.E. (2001). Endogenous policy choice: The case of pollution and growth. *Review of Economic Dynamics, 4* (2), 369-405.

Keller, W. (2004). International technology diffusion. *Journal of Economic Literature, 42* (3), 752-782.

Kuznets, S. (1955). Economic growth and income inequality. *American Economic Review, 45* (1), 1-28.

Lantz, V., & Feng, Q. (2006). Assessing income, population and technology impacts on CO2 emissions in Canada: Where's the EKC? *Ecological Economics, 57* (2), 229-238.

Levin, A., Lin, C.F., & Chu, C-S.J. (2002). Unit root tests in panel data: Asymptotic and finite sample properties. *Journal of Econometrics, 108* (1), 1-22.

Levinson, A. (2002). The ups and downs of the environmental Kuznets curve. In J. List, & A. de Zeeuw (Eds.), *Recent Advances in Environmental Economics*. London, Edward Elgar Publishing.

List, J.A., & Gallet, C.A. (1999). The environmental Kuznets curve: Does one size fit all? *Ecological Economics, 31* (3), 409–423.

Lopez, R., & Mitra, S. (2000). Corruption, pollution and the environmental Kuznets curve. *Journal of Environmental Economics and Management, 40* (2), 137-150.

Maddala, G.S., & Wu, S. (1999). A comparative study of unit root tests with panel data and a new simple test. *Oxford Bulletin of Economics and Statistics, 61* (S1), 631-652.

Maddison, D. (2006). Environmental Kuznets curves: A spatial econometric approach. *Journal of Environmental Economics and Management, 51* (2), 218–230.

Magnani, E. (2000). The environmental Kuznets curve, environmental protection policy and income distribution. *Ecological Economics, 32* (3), 431-443.

Martinez-Zarzoso, I., & Bengochea-Morancho, A. (2004). Pooled mean group estimation for an environmental Kuznets curve for CO2. *Economics Letters, 82* (1), 121-126.

Millimet, D.L., List, J.A., & Stengos, T. (2003). The environmental Kuznets curve: Real progress or misspecified models? *Review of Economics and Statistics, 85* (4), 1038-1047.

Müller-Fürstenberger, G., & Wagner M. (2007). Exploring the environmental Kuznets hypothesis: Theoretical and econometric problems. *Ecological Economics, 62* (3-4), 648-660.

Panayotou, T. (1993). Empirical tests and policy analysis of environmental degradation at different stages of economic development. International Labour Office Working Paper WP238. Geneva.

Panayotou, T. (2000). Economic growth and the environment. CID Working Paper No. 56. Harvard University.

Pedroni, P. (2004). Panel cointegration: Asymptotic and finite sample properties of pooled time series tests with an application to the PPP hypothesis. *Econometric Theory, 20* (3), 597-625.

Schmalensee, R., Stoker, T.M., & Judson, R.A. (1998). World carbon dioxide emissions: 1950-2050. *Review of Economics and Statistics, 80* (1), 15-27.

Shafik, N. (1994). Economic development and environmental quality: An econometric analysis. *Oxford Economic Papers, 46,* 757-773.

Shafik, N., & Bandyopadhyay, S. (1992). Economic growth and environmental quality. Background Paper for the 1992 World Development Report. The World Bank. Washington D.C..

Stern, D.I. (1998). Progress on the environmental Kuznets curve? *Environment and Development Economics, 3* (2), 173-196.

Stern, D.I. (2004). The rise and fall of the environmental Kuznets curve. *World Development, 32* (8), 1419-1439.

Stokey, N.L. (1998). Are there limits to growth? *International Economic Review, 39* (1), 1-31.

Suri, V., & Chapman, D. (1998). Economic growth, trade and energy: Implications for the environmental Kuznets curve. *Ecological Economics, 25* (2), 195-208.

Torras, M., & Boyce, J.K. (1998). Income inequality and pollution: A reassessment of the environmental Kuznets curve. *Ecological Economics, 25* (2), 147–60.

Taskin, F., & Zaim, O. (2000). Searching for a Kuznets curve in environmental efficiency using Kernel estimation. *Economic Letters, 68* (2), 217-223.

Tucker, M. (1995). Carbon dioxide emissions and global GDP. *Ecological Economics, 15* (3), 215–223.

Wagner, M., & Müller-Fürstenberger, G. (2004). *The carbon Kuznets curve: A cloudy picture emitted by bad econometrics?* University of Bern, Department of Economics Discussion Paper No. 04-18.

In: Handbook of Environmental Policy
Editors: Johannes Meijer and Arjan der Berg

ISBN 978-1-60741-635-7
© 2010 Nova Science Publishers, Inc.

Chapter 9

ENVIRONMENTAL CONSEQUENCES OF AGRICULTURAL DEVELOPMENT IN BANGLADESH: EMPIRICAL EVIDENCE, FARMERS' PERCEPTIONS AND THEIR DETERMINANTS[+]

Sanzidur Rahman[*]

School of Geography, Earth and Environmental Sciences, University of Plymouth, Plymouth, UK

ABSTRACT

Concern about the environmental consequences of agricultural development, and studies exploring farmers' awareness of this issue are few. The present paper provides an insight into the environmental consequences of Green Revolution technology diffusion in Bangladesh using selected material evidences, such as, loss of soil fertility and trends in fertilizer and pesticide productivity at the national level, as well as examines farmers' awareness of these adverse environmental impacts and their determinants using a survey data of 406 households from 21 villages in three agro-ecological regions. Results reveal that Bangladesh has lost soil fertility in 11 out of its 30 agro-ecological zones to the tune of 10–70% between 1968 and 1998 due to intensive crop cultivation practices. The intensive HYV rice cultivation pattern (i.e., three rice crops a year: *Boro* rice–Transplanted *Aus* rice–Transplanted *Aman* rice) depletes approximately 333 kg of N, P, K per ha per year. Also, the partial productivity measures clearly demonstrate that productivity from fertilizers and pesticides were declining steadily at a rate of 4.5 % and 7.0 % per year (p<0.01) between 1977 and 2002. Farmers are well aware of the adverse environmental consequences of Green Revolution technology, although their awareness remains confined within visible impacts, such as, loss of soil fertility, fish catches, and health effects. Their perception of intangible impacts, such as, toxicity in water and soils

---

+ This chapter heavily draws on materials published in Rahman and Thapa (1999), Rahman (2003) and Rahman (2005).
* Dr Sanzidur Rahman, Senior Lecturer in Rural Development, School of Geography, Earth and Environmental Sciences, University of Plymouth, Drake Circus, Plymouth, PL4 8AA, England, UK, Phone:+44-1752-585911, Fax:+44-1752-585998, E-mail: srahman@plymouth.ac.uk.

is weak. Among the determinants of such awareness, the level and duration of Green Revolution technology adoption directly influence awareness of its adverse effects. Education and extension contacts also play an important role in raising awareness. Awareness is higher among farmers in developed regions, fertile locations and those with access to off-farm income sources. Policy implications include investment in farmers' education, agricultural extension services, rural infrastructure and soil fertility improvements.

JEL classification: O33; Q18; C21.

Keywords: Environmental impacts, Green Revolution, farmers' perceptions, multivariate analysis, Bangladesh.

1. INTRODUCTION

Agriculture accounts for 23.5% of national income and employs 62% of the labour force (MoA, 2008). The dominant sector is the field crop agriculture accounting for more than 60% of agricultural value added. Among the field crops, rice is the major staple crop, occupying 73% of the gross cropped area (MoA, 2008). Being one of the most densely populated nations of the world, the land-man ratio is highly unfavourable resulting in a lack of food security and widespread hunger (Ahmed and Sampath, 1992). As such continued agricultural growth is deemed pivotal in alleviating poverty and raising standard of living of the population. Consequently, over the past four decades, the major thrust of national policies was directed towards transforming agriculture through rapid technological progress to keep up with the growth in population. This has led to a widespread diffusion of 'Green Revolution (GR)' technology with corresponding support in the provision of modern inputs, such as, chemical fertilizers, pesticides, irrigation equipment, institutional credit, product procurement, storage and marketing facilities (Rahman and Tapa, 1999). As a result food production grew at an estimated annual rate of about 3.3% during the period 1968/69 – 1993/94 with corresponding increase in area under irrigation and high yielding varieties (HYV) of rice, and use rates of fertilizers and pesticides per unit of land (Rahman, 2002).

Delayed consequences of GR technology on the environment and the question of sustainability of agricultural growth received priority only recently (e.g., Singh, 2000; Shiva, 1991; Alauddin and Tisdell, 1991; and Redclift, 1989). Singh (2000) identified widespread adoption of GR technologies as a cause of significant soil degradation in Haryana state of India. Shiva (1991), in her analysis of agricultural transformation in Indian Punjab, concluded that GR has produced scarcity and not abundance by reducing availability of fertile land and genetic diversity of crops. Redclift (1989), examining the issues of environmental degradation in rural areas of Latin America, noted that it is closely related to agricultural modernization. Similarly, in Bangladesh, historical analysis revealed that the productivity from GR is declining and these technologies now pose a threat to sustainability of economic development (Alauddin and Tisdell, 1991). The adoption rate of HYV rice varieties seemed to be stagnated around 47% (MoA, 2008) and there are claims that the ceiling level of adoption has been reached (Bera and Kelly, 1990). Such stagnation in the diffusion of HYV rice varieties was attributed primarily to slower expansion of modern irrigation facilities, susceptibility to pest and disease attack, and the requirement of heavy capital investment (Rahman and Thapa,

1999). Furthermore, soil analysis of 460 samples from 43 profiles from the same locations between 1967 and 1995 revealed a decline in fertility (Ali et al., 1997) although this decline has not been explicitly linked to GR technology.

Given this backdrop, the aims of the present chapter are three fold: (a) to examine empirical evidence of environmental consequences of GR technology in Bangladesh; (b) to examine whether farmers are aware of such environmental consequences; and finally (c) to identify the socio-economic factors responsible for raising their awareness. This is because sustainability of an agricultural production system depends largely on the actions of the farmers and their ability to make decisions given their level of knowledge and information available to them.

The chapter is organized into following sections. Section 2 presents a summary of the selected indicators of technological progress over the past five decades (1949/50 to 2001/03) and selected empirical evidence of environmental consequences of GR diffusion in Bangladesh. Section 3 then examines whether farmers at the farm level are aware of such environmental consequences of GR technology based on a farm-level survey data collected from 406 housheolds in 21 villages in three agro-ecological regions. Section 4 identifies the determinants of such environmental awareness of the farmers using a multivariate analysis. Finally, section 5 concludes and draws policy implications.

2. Environmental Consequences of Green Revolution in Bangladesh

The technological breakthrough in Bangladesh agriculture has been primarily in the foodgrain sector, and relates to the introduction of a rice-based 'Green Revolution' technology, followed by a gradual introduction of wheat based technology (Rahman and Thapa, 1999). Table 1 presents selected indicators of technological change over the past five decades from 1949/50 to 2001/03, using triennium averages of five periods.

Land area under agricultural production in Bangladesh has been operating up to its limits since the 1980s, with a declining net-cropped area owing to transfer of land for other uses. The total rice area also reached its upper limit, and is making way for expansion of area under wheat. It is interesting to note that the wheat acreage, which picked up in early 1980s, primarily represents HYV varieties, while the area under HYV rice varieties, introduced since 1963, accounted for only 46.7% of the total in 2003 (Table 1). Such stagnation in the diffusion of HYV rice varieties is attributed primarily to slower expansion of modern irrigation facilities, susceptibility to pest and disease attack, and requirement of heavy capital investment (Rahman and Thapa, 1999). On one hand, fertiliser use rates per hectare of gross cropped area increased about 10 folds over the past four decades in response to an increase in the areas of foodgrain under HYVs. Also pesticide use, negligible until the 1970s, recorded a dramatic increase in recent years, as it became an integral part of this GR technology. Rahman (2008) noted that pesticide use is almost perfectly correlated with area under HYV rice ($r=0.96$, $p<0.01$) as well as fertilizer use ($r=0.96$, $p<0.01$). On the other hand, the yield rates of HYV rice fell sharply from the 1970 levels due to a decline in soil fertility while the yield rates of local rice varieties is on the rise probably due to the use of modern inputs and variety screening. Rahman (2008) reported that while fertilizer and pesticide use grew at an astonishing annual rate of 6.5% and 9.0% per year during 1977–2002, the corresponding growth in yield rate of HYV rice and HYV wheat are less than 1% per year during the same period.

Table 1 Selected indicators of technological change in Bangladesh agriculture, 1949/50 – 2001/03.

	Indicators	Period 1950-52	Period 1968-70	Period 1980-82	Period 1992-94	Period 2001-03
1	Total cropped area (TCA) ('000 ha)	10,614	12,871	13,103	13,753	14,840
2	Net cropped area ('000 ha)	8,274	8,787	8,531	7,812	8,070
3	Cropping intensity (%)	128.3	146.5	153.6	176.0	176.7
4	Total rice area ('000 ha)	8,071	10,049	10,310	10,135	10,745
5	Rice as % of total cropped area (%)	76.0	78.1	78.7	73.7	72.4
6	HYV rice as % of total rice area (%)	nil	1.5	20.3	49.0	46.7
7	Total wheat area ('000 ha)	39	105	520	609	740.4
8	HYV wheat as % of total wheat (%)	nil	6.1	96.2	98.0	100.0
9	Total irrigated area (IA) ('000 ha)	< 1	1,057[b]	1,865	3,257	4,580
10	Irrigated area as % of TCA (%)	na	8.2[b]	14.2	23.7	30.9
11	Foodgrain irrigated area as % of IA (%)	na	85.8[b]	78.4	91.2	90.2
12	Irrigation methods by Modern (%)	na	31.5[b]	67.2	70.9	72.1
	Traditional (%)	na	68.5[b]	32.8	29.1	27.9
13	Total fertiliser used ('000 mt of nutrients)	< 1	113.1	380.8	664.8	1,462.9
14	Fertilizer use rate per TCA (kg of nutrients/ha)	na	8.8	29.1	48.3	98.58
15	Pesticide use ('000 mt)	na	na	2.2	6.5	14.4
16	Rice production ('000 mt)	7,367	11,504	13,417	18,211	24,858
17	Rice yield (kg/gross HYV variety ha)	nil	3,809	2,297	2,409	2,816
	Local variety	913	1,103	1,048	1,208	1,403
18	Wheat production ('000 mt)	22	86	932	1,124	1,625
19	Wheat yield (kg/gross ha)	564	819	1,792	1,846	2,195

Note: Periods are three-year averages. Information in the last column is computed from MoA (2008) and the remaining has been reproduced from Table 1 of Rahman and Thapa (1999).
Source: After Rahman and Thapa (1999), MoA (2008).

2.1. Impact on Soil Fertility

Serious loss of soil fertility has been documented in Bangladesh only recently, although Ali et al., (1997) noted a decline of soil fertility between 1967 and 1995. The MoA (2008) reported that 11 out of a total of 30 agro-ecological zones* of Bangladesh have lost soil fertility between 10–70% due to intensified crop cultivation over a 30 year period from 1968 to 1998 (Table 2). Barind Tract and Old Brahmaputra Floodplain areas seem to be the hardest hit areas in terms of soil fertility decline. Baanante et al., (1993) also noted that the present

---

* The Land Resources Appraisal of 1988 classified Bangladesh into 30 distinct agro-ecological regions (88 including sub-regions) based on information relevant for land use and assessment of agricultural potential (UNDP/FAO, 1988).

level of food crop production in Bangladesh takes up an estimated 0.93 million tons of nutrients (N, P, K and S) from the soil annually.

Table 3 presents the ranking of cropping system according to the rate of soil-fertility decline. It is clear from Table 3 that the most intensive cultivation system spurred by the diffusion of GR technology, i.e., three rice crops a year (*Boro* rice–Transplanted *Aus* rice–Transplanted *Aman* rice) ranks first and depletes approximately 333 kg of N, P, K per ha per year, which is alarming. However, adding 'green manure' in the system and keeping two crops of rice dramatically reduces the depletion rate to 121 kg of nutrients/ha/year. The least amount of soil nutrient depletion of 112 kg of nutrients/ha/year is associated with the system comprising of Wheat–Mungbean–Transplanted *Aman*. The results are not surprising although depressing. Widespread adoption of GR technology was identified as a cause of significant soil degradation and declining crop yields in India (Singh, 2000; Yadav et al., 2000). Pimentel (1996) indicated that extensive use of fertilizers and pesticides to support GR has caused serious public health and environmental damages worldwide, particularly in developing countries. Furthermore, it has been noted that continued, intensive production of rice has led to yield reductions in some countries in Asia, explained in part by soil nutrient exhaustion (Doberman et al., 2002).

Table 2. Losses of fertility of soil by intensified crop cultivation, 1967/68 to 1997/98.

Agro Ecological zone (Number)	Types of land	Increase of cropping intensity (%)	Losses of Soil fertility (%)
Old Himalayan Piedmont Plain (1)	HL	100	25-45
Tista Floodplain (2)	HL	100	10-35
Tista Meander Floodplain (3)	HL	100	10-40
Old Brahmaputra Floodplain (9)	MHL	100	25-65
High Ganges River Floodplain (11)	HL	100	20-45
Middle Meghna River Floodplain (16)	MLL	100	15-40
Surma Kushiyara River Floodplain (20)	MLL	100	20-40
North Eastern Piedmont Plain (22)	HL	100	20-70
Chittagong Coastal Plain (23)	HL	100	10-30
Barind Tract (26)	HL	100	30-60
Madhupur Tract (28)	HL	100	40-65

Note: HL = High land, MHL = Medium high land, MLL = Medium low land.
The land type classification in Bangladesh is based on flooding depth. HL = no flooding, MHL = flooding depth of 0.01 – 0.90 m, MLL = flooding depth of 0.91 – 1.83 m, LL (Low land) = flooding depth of 1.83 – 3.05 m, VLL (Very low land) = flooding depth >3.05 m (Source: Land Resources Information database, Bangladesh Agricultural Research Council).
Source: Adapted from MoA, 2008 (Table 4.04: Committee report for losses of soil fertility, 2004, BARI)

2.2. Trends in Fertilizer and Pesticide Productivity

Given the evidence that GR technology is associated with soil fertility loss, we now examine the contribution of two key production inputs, namely, the inorganic fertilizers and pesticides, to crop productivity growth. The key assumption was that the application of inorganic fertilizers will compensate for soil fertility decline from higher uptake of nutrients in the form of crop harvest. However, pesticides were not originally conceived as an integral part of the GR technology package. Nevertheless, there is a widespread acceptance that the expansion of modern agricultural technologies has led to a sharp increase in pesticide use (Roger and Bhuiyan, 1995; Pingali and Rola, 1995).

Table 3. Estimation of nutrient depletion in major cropping pattern in Bangladesh.

Major cropping pattern	Total yield (ton/ha/year)	Input (kg/ha)			Output (kg/ha)			Balance (kg/ha)			Approximate total depletion (kg/ha/year)
		N	P	K	N	P	K	N	P	K	
Boro-T.Aus-T.Aman	11.5	350	60	151	469	57	368	-119	+3	-217	333
Mustard-Jute-T.Aman	7.5	340	75	205	430	79	429	-90	-4	-224	318
Potato-Jute-T.Aman	36	380	70	240	385	55	496	-5	+15	-256	246
Potato-T.Aus-T.Aman	38	386	67	220	430	53	435	-44	+14	-215	245
Wheat-T.Aus-T.Aman	10	335	65	166	420	64	292	-85	+1	-126	210
Sugarcane+Potato intercropping	100	190	55	150	210	60	320	-20	-5	-170	195
Mustard-Boro-T.Aman	9.5	378	73	183	404	95	326	-26	-22	-143	191
Boro-Fallow-T.Aman	8	248	49	118	324	32	234	-76	+17	-116	175
Boro-GM-T.Aman	8	285	0	135	324	32	240	-39	+28	-105	121
Wheat-Mung bean-T.Aman	8	275	64	190	305	52	284	-30	+12	-94	112

Note: Input: Fertilizer, manure, fixation (BNF), deposition (rain), sedimentation (flood) and irrigation; Output: Harvested product, residues removed, leaching, dentrification, volatilization and erosion (Source: Information based on research conducted by Bangladesh Agricultural Research Institute (BARI)).
Source: Adapted from MoA (2008, Table 4.04a).

Since information on fertilizer and pesticide use in specific crops over time is not available, only a partial measure of productivity on total crop production was computed. Figure 1 presents the trend in pesticide productivity measured as the gross value added of total crop output at constant prices (i.e., in US dollars) contributed by per kg of fertilizer nutrients and active ingredients of pesticides, respectively. It is clear from Figure 1 that although pesticide productivity has increased during the first few years of application, the overall trend for both pesticides and fertilizers is negative. The average annual growth rate is estimated at -4.5% and -7.0% ($p < 0.01$) for fertilizers and pesticides, respectively over a 25 year period from 1977 to 2002. The implication is that increased dose of fertilizers and pesticides are required just to maintain existing yield levels of the major crops.

In short, the empirical evidences clearly demonstrate that the GR technology has resulted in adverse environmental consequences, particularly, leading to a depletion of soil fertility over time as revealed in Tables 2 and 3. We next examine whether the farmers, who are the actual practitioners of this GR technology, are aware of such adverse environmental awareness.

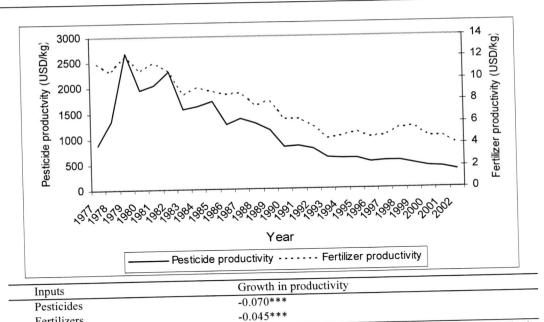

Inputs	Growth in productivity
Pesticides	-0.070***
Fertilizers	-0.045***

Note: Growth rates are computed using semi-logarithmic trend function: $lnY = \alpha + \beta T$, where Y is the target variable, T is time, ln is natural logarithm, and β is the growth rate.

*** = significant at 1% level (p<0.01)

Source: Adapted from Rahman (2008).

Figure 1. Trends in pesticide and fertilizer productivity.

3. Farmers' Awareness of the Environmental Consequences of Green Revolution

Prior to analysing farmers' awareness of the environmental consequences of GR technology, we first describe the study regions, its agro-ecological characteristics and the data.

3.1. Study regions and its agro-ecological characteristics*

Farm-level cross section data for crop year 1996 were collected from three agro-ecological regions of Bangladesh. The survey was conducted from February to April 1997. The specific selected regions were Jamalpur (representing wet agroecology), Jessore (representing dry agroecology), and Comilla (representing both wet agroecology and an agriculturally developed area). The specific selected areas were Jamalpur Sadar Thana (central sub-district), Manirampur Thana (sub-district) of Jessore region and Matlab Thana (sub-district) of Comilla region

---

* Descriptions of the agro-ecological characteristics of the study regions were selected from Land Resources Appraisal report (UNDP/FAO, 1988).

The Jamalpur study region falls under *Agro-ecological Region 9* defined as *Old Brahmaputra Floodplain*. This region occupies a large area of Brahmaputra sediments, which were deposited since the time that the river has shifted to present Jamuna channel about 200 years ago. This region encompasses large areas of Sherpur, Jamalpur, Tangail, Mymensingh, Netrakona, Kishoreganj, Narshingdi, and Narayanganj districts. Small areas of east of Dhaka and Gazipur districts are also included. The region covers a total area of 7,230 sq km. The entire region has broad ridges and basins with irregular relief. The elevation between ridge tops and basin centers range between 2-3 m. There are five sub-regions in this region with transitional delineations across sub-regions. The study area specifically falls under sub-region 9b characterized as medium high land and is mainly shallowly flooded during peak monsoon. A total of 17 soil series are identified in Jamalpur Sadar Thana. However, for the specific study area, the dominant soil series are, *Sonatala*, and *Silmandi*. The agricultural system is mainly rainfed. However, a large land area of Jamalpur Sadar Thana is irrigated with Shallow Tube Wells and Deep Tube Wells. Two rice crops of *Aman* and *Boro* dominate the cropping pattern (UNDP/FAO, 1988).

The Jessore study region falls under *Agro-ecological Region 11* defined as *High Ganges River Floodplain*. This region includes the western part of the *Ganges River Floodplain*, which is predominantly composed of highland and medium highland. This region encompasses Chapai Nawabganj, Rajshahi, southern Pabna, Kushtia, Meherpur, Chuadanga, Jhenaidah, Magura, Jessore, and northern parts of Satkhira and Khulna districts. The region covers a total area 13,205 sq km. Most areas have complex relief of broad and narrow ridges and inter-ridge depressions. There are three sub-regions in this region. The study area specifically falls under sub-region 11a with smooth ridge and basin relief crossed by broad and narrow belts of irregular relief adjoining old river channels. Lower ridges and basins are shallowly flooded by either rainwater or by raised groundwater table during heavy rainfall. A total of 7 soil series are identified in Manirampur Thana. However, for the specific study area, the dominant soil series are, *Gopalpur, Ishwardi*, and *Gheor*. The agricultural system is mainly rainfed. A large land area of Manirampur Thana is irrigated with Shallow Tube Wells and Deep Tube Wells. Two rice crops, Aman and Boro, dominate the cropping pattern (UNDP/FAO, 1988).

The Comilla study region falls under *Agro-ecological Region 16* defined as *Middle Meghna River Floodplain*. This region occupies the abandoned channel of river *Brahmaputra* on the border of the Dhaka and Comilla regions. This region occurs between the southern part of the Sylhet basin and the confluence of river Meghna with Dhaleshwari and Ganges rivers. The region encompasses parts of Kishoreganj, Brahmanbaria, Comilla, Chandpur, Narsingdi, and Narayanganj. This region covers a total area of 1,555 sq km. No sub-region is recognized. Most soils are deeply flooded, except on high floodplain ridges. Three main kinds of soils occur in this region. The *grey loams* and *clay* in areas of Meghna *alluvium* occupies greater part of the region. Ample surface water resource exists in the Meghna channels to irrigate the agricultural crops throughout the area. A Flood Control, Drainage and Irrigation (FCD/I) project is constructed with embankment on only one side of the Matlab Thana in 1987. This led to increase in cropping intensity inside the embankment, with two or three rice crops grown in a year (UNDP/FAO, 1988).

3.2. The Data

A multistage random sampling technique was employed to locate the districts, then the subdistricts, and then the villages in each of the three subdistricts and finally the sample households. A total of 406 households from 21 villages (175 households from eight villages of Jamalpur Sadar Thana, 105 households from six villages of Manirampur Thana and 126 households from seven villages of Matlab Thana) form the sample for the study. Detailed crop input-output data were collected for 10 groups of crops[*]. The dataset also includes information on the level of soil fertility[+] determined from soil samples collected from representative locations and information on the level of infrastructure[§] development in the study villages.

3.2. Construction of the Farmers' Environmental Awareness Index

In order to determine farmers' awareness of the environmental consequences of GR technology, we construct an environmental awareness index. The procedure is summarized in Figure 2 (Rahman, 2003). Farmers' perception on the environmental impacts of GR technology is elicited in two steps. First, a set of 12 specific environmental impacts was read to the respondents who were asked to reveal their opinion on each of these impacts (E_j). A value of 1 is assigned for each of the impact indicators where the farmer recognises the impact and 0 otherwise. Selection of the list of indicators was based on the Focus Group Discussions (FGD) with the farmers during a pre-testing stage prior to the administration of the structured questionnaire. In the next step, farmers were then asked to reveal the relative importance of each impact indicator on a five-point scale (R_m). A score of 1 is assigned for least importance and 5 for very high importance. These ranks are then converted into weighted scores (W_q). A weight of 0.2 is assigned for lowest rank of 1 and a weight of 1 is assigned for the highest rank of 5. A zero weight is assigned for indicators where the farmer does not recognise the impact. Then the overall environmental awareness index (EAI) for each farmer is computed by summing up the weighted scores of each impact indicator and then dividing by the total number of impacts (Figure 2).

---

* The crop groups are: traditional rice varieties (Aus – pre-monsoon, Aman – monsoon, and Boro – dry seasons), HYV rice varieties (Aus, Aman, and Boro seasons), HYV wheat varieties, jute, potato, pulses, spices, oilseeds, vegetables, and cotton. Pulses in turn include lentil, mungbean, and gram. Spices include onion, garlic, chilly, ginger, and turmeric. Oilseeds include sesame, mustard, and groundnut. Vegetables include eggplant, cauliflower, cabbage, arum, beans, gourds, radish, and leafy vegetables.

+ Information on physical and chemical properties of soil from the selected farmers' fields was collected to evaluate the general fertility status of the soil and to examine inter-regional differences (if any) between the study areas. Ten soil-fertility parameters were tested. These were: (1) soil pH, (2) available nitrogen, (3) available potassium, (4) available phosphorus, (5) available sulphur, (6) available zinc, (7) soil texture, (8) cation exchange capacity (CEC) of soil, (9) soil organic matter content, and (10) electrical conductivity of soil. The soil fertility index was constructed from test results of these soil samples. High index value refers to better soil fertility (for details, see Rahman and Parkinson, 2007).

§ The index of infrastructure was constructed using the cost of access approach. A total of 13 elements were considered for its construction. These are, (1) primary market, (2) secondary market, (3) storage facility, (4) rice mill, (5) paved road, (6) bus stop, (7) bank, (8) union office, (9) agricultural extension office, (10) high school, (11) college, (12) thana (sub-district) headquarter, and (13) post office. High index value refers to high under developed infrastructure (for details of construction procedure, see Ahmed and Hossain, 1990).

Opinion on the jth impact	Disagree	Agree				
Impact value (E_j)	0	1				
Rank of the importance of jth impact on a five-point scale (R_m)	0	1	2	3	4	5
Rank interpretation	None	Very low	Low	Medium	High	Very high
Weights (W_q)	0	0.2	0.4	0.6	0.8	1.0
Aggregate awareness index of the ith farmer (AAI_i)	$\sum_{j=1}^{12}\sum_{m=0}^{5}\sum_{q=0}^{1} ERW_{jmq}, \quad \forall\; j=1,2,...,12;\; m=0,1,.....,5;\; q=0,0.2,...1.$ (1)					
Overall environmental awareness index of the ith farmer (EAI_i)	$\dfrac{AAI_i}{N}, \; where\; N = 12\,(total\; number\; of\; impacts).$ (2)					

Source: Adapted from Rahman (2003).

Figure 2. Construction of the environmental awareness index.

Table 4. Ranking of farmers' perception on environmental impacts of GR technology.

Sl. no.	Environmental impacts of HYV agricultural technology	Index weighted by rank of responses[a]										
		Jamalpur region			Jessore region			Comilla region			All region		
		Agree (%)	Index	Rank	Agree (%)	Index	Rank	Agree (%)	Index	Rank	Agree (%)	Index	Rank
1	Reduces soil fertility	85	0.82	1	98	0.94	1	76	0.63	1	86	0.79	1
2	Affects human health	74	0.60	2	87	0.79	2	70	0.45	4	76	0.60	2
3	Reduces fish catch	65	0.55	3	65	0.59	4	91	0.57	2	73	0.56	3
4	Increases disease in crops	66	0.51	4	74	0.61	3	66	0.45	3	68	0.52	4
5	Compacts/ hardens soil	51	0.36	7	71	0.57	6	49	0.37	5	56	0.42	5
6	Increases insect/ pest attack	53	0.43	5	68	0.58	5	21	0.12	9	47	0.37	6
7	Increases soil erosion	53	0.39	6	67	0.49	7	18	0.11	10	46	0.33	7
8	Increases soil salinity	41	0.28	8	56	0.43	8	36	0.24	6	43	0.30	8
9	Contaminates water source	35	0.26	9	34	0.24	9	16	0.08	11	29	0.20	9
10	Increases toxicity in soil	22	0.14	11	20	0.16	11	21	0.13	7	21	0.15	10
11	Creates water logging	22	0.14	10	22	0.18	10	6	0.05	12	17	0.13	11
12	Increases toxicity in water	17	0.12	12	10	0.07	12	20	0.13	8	16	0.11	12
	All impacts	49	0.38[b]	2	56	0.47[b]	1	41	0.28[b]	3	48	0.37	

Rank correlation among regions (Spearman rank correlation coefficient r_s)				
Jamalpur region	1.00			
Jessore region	0.99***	1.00		
Comilla region	0.70***	0.73***	1.00	
All region	0.97***	0.98***	0.80***	1.00

Note: The higher the index the stronger the perception.
a Ranking done by weighting individual responses by their ranks.
b Ranking done across three regions.
*** Significant at 1% level (p<0.01).
Source: Adapted from Rahman (2003) .

3.3. Level of Farmers' Environmental Awareness

Table 4 presents the results of this exercise. 'Decline in soil fertility' featured at the top of the list of perceived adverse environmental impacts of GR technology diffusion, followed by 'health effects', 'decline in fish catch', 'increase in crop disease', 'soil compaction', 'increase in insect/pest attack', 'soil erosion' and 'soil salinity'. The perception of the adverse impact of HYV technology on water resources is, however, very weak, as evident from the sharp decline in index values. This implies that though farmers are aware of the adverse environmental impacts of HYV agricultural technology, their awareness of the extent remains confined to the visible impacts evident from farm fields and crop production on which their livelihoods depend. The awareness of indirect impacts such as 'contamination of soil and water bodies' is poor as indicated by low index values. This may well be due primarily to high levels of illiteracy amongst the farmers (see Table 5) and poor exposure to messages on health and hygiene. All relative rankings of impacts across regions are significantly ($p<0.01$) and positively related, with the value of rank-correlation coefficient varying within a range of 0.70 to 0.99 (see lower section of Table 4). The implication is that farmers from all three regions have similar level of awareness, although they are located in different agro-ecological as well as socio-economic settings (Rahman, 2003).

4. DETERMINANTS OF FARMERS' ENVIRONMENTAL AWARENESS: A MULTIVARIATE ANALYSIS

Having identified that the farmers are well aware of the environmental consequences of GR technology, next we attempt to examine the determinants of such awareness by using a multivariate analysis. We undertake this task by adopting an analytical framework similar to those used in the adoption-perception literature to determine the indicators of technology adoption and/or farmers' decision making processes (for details, see Rahman, 2003).

4.1. The Econometric Model

Among the limited dependent variable models widely used to analyse farmers' decision making processes, Tobit analysis has gained importance since it uses all observations, both those are at the limit, usually zero (e.g., non-adopters), and those above the limit (e.g., adopters), to estimate a regression line, as opposed to other techniques that uses observations which are only above the limit value (McDonald and Moffit, 1980). In our case, farmers could be unaware of any environmental impacts of GR technology even after adoption. Therefore, there are a number of farmers with zero environmental awareness at the limit. In such case, the application of Tobit analysis is most suited because of the censored nature of the data. The stochastic model underlying Tobit may be expressed as follows (McDonald and Moffit, 1980):

$$y_i \quad = X_i\beta + u_i \quad if \ X_i\beta + u_i > 0$$
$$= 0 \quad if \ X_i\beta + u_i \leq 0,$$
$$i = 1, 2, \ldots \ldots n, \quad (1)$$

where n is the number of observations, y_i is the dependent variable (farmers' environmental awareness), X_i is a vector of independent variables representing technology attributes and farm and farmer specific socio-economic characteristics, β is a vector of parameters to be estimated, and u_i is an independently distributed error term assumed to be normal with zero mean and constant variance σ^2. The model assumes that there is an underlying stochastic index equal to $(X_i\beta + u_i)$, which is observed when it is positive, and hence qualifies as an unobserved latent variable. The relationship between the expected value of all observations, E_y and the expected conditional value above the limit E_y^* is given by:

$$E_y = F(z) \ E_y^*$$

where $F(z)$ is the cumulative density normal distribution function and $z = X\beta/\sigma$.

4.2. The Empirical Model

The estimated empirical model uses a set of technological attributes, farm-specific socio-economic characteristics and regional characteristics as explanatory variables that are assumed to influence farmers' environmental awareness. Choice of the explanatory variables is based on the adoption–perception literature with similar justification thereof. Table 5 presents the description, measure and summary statistics of the variables used in the tobit model.

Two principal technology attributes, the 'level' and 'duration' of GR technology adoption, are hypothesized as the major determinants in raising farmers' environmental awareness since perception comes from experience of adoption (Negatu and Parikh, 1999). The variable 'area under HYV of rice and/or wheat' reflects the level and extent of GR technology adoption by these farmers[*] and 'years of actually growing HYV rice' reflects duration of involvement with this technology and are expected to insist the farmer to identify reasons for variation in output level and/or declining productivity over time, if any. Access to modern irrigation facilities is an important pre-requisite for growing HYV rice, particularly, for the HYV *Boro* rice grown in dry season. Lack of access to modern irrigation facilities has been identified as one of the principal reasons for stagnation in the expansion of HYV rice which currently accounts for a little over 50% of total rice area (Rahman and Thapa, 1999; Hossain, et al., 1990, and Hossain, 1989). Nevertheless, farmers choose to grow HYV rice during the main monsoon season (*Aman* season) with heavy reliance on monsoon rain as it still yields twice that of traditional rice varieties if managed with proper supplementary irrigation and water control. Hence, the irrigation variable is incorporated to account for its influence in raising awareness.

---

[*] In cross-section data, this is a standard proxy for specifying a technology variable, particularly in Bangladesh (see Ahmed and Hossain, 1990, Hossain et al., 1990, and Hossain, 1989).

The other variables included in the analysis are: education of the farmer, age of the farmer, subsistence pressure, type of tenant, extension contact over the past one year, share of off-farm income in total household income, the index of underdevelopment of infrastructure, and the soil fertility index (for details of justification for the inclusion of these variables, see Rahman, 2003). The dependent variable is the constructed environmental awareness index (EAI) of each farmer.

4.3. Determinants of Farmers' Environmental Awareness

Table 6 presents the parameter estimates of the Tobit model[+]. Except for the age and family size variables, the coefficients for the remaining nine variables representing farmers' socio-economic characteristics and production circumstances were significantly different from zero at 10% level at least, indicating that the inclusion of these variables were correctly justified in explaining farmers' overall environmental awareness. The Likelihood Ratio test result, presented at the bottom of Table 6, confirms that these variables contribute significantly as a group to the explanation of the environmental awareness level of the farmers. The parameters of the estimated Tobit model cannot directly reveal the actual magnitude of the effects. Therefore, the marginal effects are presented in columns 4 and 5 of Table 6.

'Level' and 'duration' of involvement with GR technology are the two most important determinants, which directly influence farmers' awareness of its ill effects thereby, supporting the maintained hypotheses. The marginal effects of 'level' and 'duration' of involvement are 0.04 and 0.01. This suggests that a 1% increase in the area under HYV rice and/or a change in the duration of growing HYV rice are expected to increase the awareness probability by 0.04% and 0.01%, respectively. Lack of access to modern irrigation also raises awareness. Lack of this important input, which is a pre-requisite, result in poor yield performance and perhaps induces higher incidence of pest and disease infestations, thereby, enabling farmers to realize the ill effects of HYV technology. The marginal effect estimate shows that a 1% reduction in irrigated area is expected to result in about 0.05% increase in the awareness probability.

Both education and extension contact significantly increase awareness, as expected. The marginal effects are estimated at 0.01 and 0.05 for education and extension, implying that a 1% increase either in education level or extension contact is expected to result in about 0.01% and 0.05% increase in the awareness probability, respectively. These findings conform to the results of other adoption–perception studies (e.g., Neupane et al.; 2002; Mbaga-Semgalawe, 2000; Baidu-Forson, 1999; and Hossain et al., 1990). Next, owner operators, who are presumably relatively large farmers as well, are relatively more aware than the tenants. One of the pathways to trigger awareness among owner operators might be through receipt of lesser amount of earning in the form of land rent wherein the popular arrangement (also set by law) is 33% of the total produce with selective sharing of input costs. Those who earn their livelihood substantially from off-farm sources are also more aware. Probably, these are the households who eventually turned towards off-farm activities, provided opportunities exist, after realizing that GR technologies are not paying off over time.

+ NLOGIT Version 4 was used for the analyses (ESI, 2007).

Table 5. Description and summary statistics of the variables.

Description	Measurement	Mean	Standard deviation
Technology characteristics			
Level or extent of involvement with GR technology	Hectare	0.74	0.79
Duration of involvement with GR technology	Years	9.71	5.71
Irrigation index	Proportion	0.62	0.30
Socio-economic characteristics			
Education level of the farmer	Completed years of schooling	3.74	4.26
Age of the farmer	Years	46.88	14.46
Subsistence pressure	Persons per household	6.02	2.53
Tenurial status	Value is 1 if owner operator, 0 otherwise	0.58	0.49
Extension contact	Value is 1 if had extension contact in the past one year, 0 otherwise	0.13	0.33
Off-farm income share in the household	Proportion	0.22	0.31
Regional characteristics			
Infrastructure	Number	33.32	14.95
Soil fertility level	Number	1.68	0.19

Table 6. Determinants of farmers' environmental awareness.

Variables	Tobit model Coefficient	t-ratio	Marginal effects Coefficient	t-ratio
Constant	0.1300	1.537	0.1289	1.537
Technological characteristics				
Extent of technology adoption	0.0354***	3.095	0.0351***	3.095
Duration of adoption	0.0045***	3.121	0.0045***	3.121
Irrigation facilities	-0.0495*	-1.861	-0.0491*	-1.861
Socio-economic characteristics				
Education of farmer	0.0049**	2.351	0.0048**	2.351
Age of farmer	0.0002	0.409	0.0002	0.409
Subsistence pressure	-0.0024	-0.719	-0.0023	-0.719
Owner operator	0.0299*	1.796	0.0297*	1.796
Extension contact	0.0464*	1.881	0.0467*	1.881
Off-farm income share	-0.0017***	-3.190	-0.0017***	-3.190
Regional characteristics				
Infrastructure	0.1249***	2.953	0.1239***	2.953
Soil fertility	0.0533**	2.041	0.0528**	2.041
Model diagnostics				
Likelihood Ratio test ($\chi^2_{11,0.95}$)	78.09***			
Total observations	406			

Note: Likelihood ratio (LR) test statistic is used to test the null hypothesis that there is no relationship between the farmers' environmental awareness level and the set of independent variables included in the model (i.e., H_0: $\beta_1 = \beta_2 = ... = \beta_{11} = 0$).

*** Significant at 1% level ($p<0.01$);

** Significant at 5% level ($p<0.05$);

* Significant at 10% level ($p<0.10$).

Source: After Rahman and Thapa (2003).

Farmers in developed regions[§] are more aware as it is probably endowed with better access to information and opportunities to exchange information. Negatu and Parikh (1999) concluded that proximity to town (a proxy of developed infrastructure) is an important explanatory variable affecting perception (of marketability of HYV crop). Also, awareness is significantly higher in areas with relatively better soil fertility status. The marginal effect of this variable is estimated at 0.05 indicating that a 1% improvement in the level of soil fertility will raise the probability of awareness by 0.05%.

5. CONCLUSION

The study deals with one of the least touched upon issues associated with the diffusion of GR technology, specifically its impact on the environment. The chapter first provides some selected empirical evidences of the adverse environmental consequences of GR technology in Bangladesh. Next it examines whether the farmers, who are actual practitioners of the GR technologies, are aware of such environmental consequences, and then identify the factors that raises such awareness. Results reveal that GR technology has exerted adverse environmental impacts, largely resulting in substantial loss of soil fertility in 11 out of a total of 30 agro-ecological regions of the country. The most intensive cropping system, i.e., three rice crops a year (*Boro*–Transplanted *Aus*–Transplanted *Aman*) depletes approximately 333 kg of soil nutrients per ha per year, which is alarming. The least two soil nutrient depleting cropping systems are those that have two cereals (rice or wheat) with a soil nitrogen fixing crop in the middle. These are: Wheat–Mungbean–*Boro* rice system depleting only 112 kg of soil nutrients/ha/year followed by *Boro* rice–Green Manure–Transplanted *Aman* rice system depleting 121 kg of soil nutrients/ha/year. Farmers are well aware of the adverse environmental impacts of GR technology. However, their awareness level remains confined within the visible impacts that are most closely related to their local experience, i.e., loss of soil fertility, fish catches, and health impacts.

All three technology attributes, the 'level' and 'duration' of GR technology adoption and 'lack of modern irrigation facilities' directly influence farmers' awareness of its ill effects. This has profound implications for agricultural sustainability because perception and/or awareness significantly condition adoption behaviour (Negatu and Parikh, 1999; Adesina and Baidu-Forson, 1995; and Adesina and Zinnah, 1993) and perhaps partly explains stagnation of HYV rice expansion after four decades of major thrust in its diffusion. Morris et al., (1996) reported that locations where facilities for mechanical irrigation are uncertain, farmers opt to choose HYV wheat and is one of the principal reasons for expansion in wheat acreage in recent years, although in financial terms, production of HYV *Boro* rice is far more profitable (Rahman, 1998). Also, such awareness may influence adoption of conservation measures, a proposition worth exploring. Mbaga-Semgalawe and Folmer (2000) found partial support in their empirical findings that perception of a soil-erosion problem as a first stage in the sequential household decision making process leads to the adoption of conservation measures and finally to the effort devoted to conservation. In fact, Rahman (2005) reported that these

---

[§] The index reflects the underdevelopment of infrastructure, and therefore, a negative sign indicates positive effect
   on the dependent variable.

environmentally aware farmers use a significantly less amount of all inputs (including chemicals) in order to avoid further environmental damage.

Among the socio-economic factors, education and extension contacts play an important role in raising awareness. This clearly provides an opportunity to design and strategise information dissemination process through existing educational institutions and agricultural extension system. Several studies highlighted the use of extension education to promote conservation (e.g., Neupane et al., 2002; Mbaga-Semgalawe and Folmer, 2000; and Baidu-Forson, 1999).

Regional characteristics (state of rural infrastructure and soil fertility status) also influence environmental awareness. This may very well justify improvement in rural infrastructure, as it seems to facilitate access to resources vis-à-vis improved information. Poor rural infrastructure has been identified as one of the major impediments to agricultural development in Bangladesh (Ahmed and Hossain, 1990). Promotion of soil fertility status, however, would require considerable effort in disseminating important conservation information as well as crop production practices and crop-mixes to suit specific agro-ecological niches. In fact, areas that are fertile are also home to relatively higher levels of HYV wheat acreage as well as legume crops (that fix soil nitrogen), particularly, the survey villages in Jessore region (Rahman, 1998). The empirical evidence presented in Table 3 also confirms that such a system is the least soil nutrient depleting system.

The policy implications are clear. Promotion of education and strengthening extension services both in terms of its quality and coverage would boost farmers' environmental awareness. Also, investment in the development of rural infrastructure and measures to replenish depleting soil fertility will play a positive role in raising awareness. It is hoped that the results of this study could be used to develop a comprehensive agricultural development strategy conducive to maintaining or even increasing agricultural production without affecting environmental quality.

REFERENCES

Adesina, A.A., Baidu-Forson, J. 1995. Farmers' perception and adoption of new agricultural technology: evidence from analysis in Burkina Faso and Guinea, West Africa. *Agricultural Economics*, 13: 1 – 9.

Adesina. A.A., Zinnah, M.M. 1993. Technology characteristics, farmers' perceptions and adoption decisions: a tobit model application in Sierra Leone. *Agricultural Economics*, 9: 297 – 311.

Ahmed, A.U., Sampath, R.K. 1992. Effects of irrigation induced technological change in Bangladesh rice production. *American Journal of Agricultural Economics*, 74: 144 – 157.

Ahmed, R., Hossain, M. 1990. *Developmental Impact of Rural Infrastructure in Bangladesh*. Research Report. No.83. Washington, D.C. :International Food Policy Research Institute.

Alauddin, M., Tisdell, C. 1991. *The Green Revolution and Economic Development: The Process and its Impact in Bangladesh*. London: Macmillan.

Ali, M.M., Shaheed, S.M., Kubota, D., Masunaga, T., Wakatsuki, T. 1997. Soil degradation during the period 1967 – 1995 in Bangladesh, 2: selected chemical characters. *Soil Science and Plant Nutrition*, 43: 879 – 890.

Baanante, C.A., Henao, J., Wan, X. 1993. *Fertilizer Subsidy Removal in Bangladesh: An Assessment of the Impact of Fertilizer Use, Crop Yields, and Profits of Farmers.* International Fertilizer Development Centre, Muscle Shoals, Alabama.

Baidu-Forson, J. 1999. Factors influencing adoption of land-enhancing technology in the Sahel: lessons from a case study in Niger. *Agricultural Economics*, 20: 231 – 239.

Bera, A.K., Kelly, T.G. 1990. Adoption of high yielding rice varieties in Bangladesh. *Journal of Development Economics*, 33: 263 – 285.

Coelli, T., Rahman, S., Thirtle, C. 2002. Technical, allocative, cost and scale efficiencies in Bangladesh rice cultivation: a non-parametric approach. *Journal of Agricultural Economics*, 53: 607 – 626.

Dobermann, A., Witt, C., Dawe, D., 2002. "Site-specific nutrient management for intensive rice cropping systems in Asia". *Field Crops Research*, 74: 37-66.

ESI, 2007. NLOGIT-4, Econometric Software, Inc. New York.

Hossain, M. 1989. *Green Revolution in Bangladesh: Impact on Growth and Distribution of Income.* Dhaka: University Press Limited.

Hossain, M., Quasem, M.A., Akash, M.M., Jabber, M.A. 1990. *Differential Impact of HYV Rice Technology: The Bangladesh Case.* Dhaka: Bangladesh Institute of Development Studies.

Mbaga-Semgalawe, Z., Folmer, H. 2000. Household adoption behaviour of improved soil conservation: the case of the North Pare and West Usambara Mountains of Tanzania. *Land Use Policy*, 17: 321 – 336.

McDonald, J.F., Moffit, R.A. 1980. The uses of tobit analysis. *Review of Economics and Statistics*, 61: 318 – 321.

MoA, 2008. *Handbook of Agricultural Statistics, 2007.* Ministry of Agriculture, Dhaka, Bangladesh. Available @ http://www.moa.gov.bd/statistics/statistics.htm

Morris, M., Chowdhury, N., Meisner, C. 1996. Economics of wheat production in Bangladesh. *Food Policy*, 21: 541 – 560.

Negatu, W., Parikh, A. 1999. The impact of perception and other factors on the adoption of agricultural technology in the Moret and Jiru Woreda (district) of Ethiopia. *Agricultural Economics*, 21: 205 – 216.

Neupane, R.P., Sharma, K.R., Thapa, G.B. 2002. Adoption of agroforestry in the hills of Nepal: a logistic regression analysis. *Agricultural Systems*, 72: 177 – 196.

Pimentel, D. 1996. Green Revolution and chemical hazards. *The Science of the Total Environment.* 188 (Supplement, 1): S86 – S98.

Pingali, P.L., Rola, A.C., 1995. Public regulatory roles in developing markets: the case of Philippines. In: Pingali, P.L., Roger, P. (Eds.), Impact of Pesticides on Farmer Health and the Rice Environment. Kluwer Academic Publishers, Boston.

Rahman, S. 1998. Socio-economic and environmental impacts of technological change in Bangladesh agriculture. Unpublished Ph.D. Dissertation. Pathumthani: Asian Institute of Technology, Thailand.

Rahman, S. 2002. Technological change and food production sustainability in Bangladesh agriculture. *Asian Profile,* 30: 233 – 246.

Rahman, S. 2003. Environmental impacts of modern agricultural technology diffusion in Bangladesh: an analysis of farmers' perceptions and their determinants. *Journal of Environmental Management*, 68: 183 – 191.

Rahman, S. 2005. Environmental impacts of technological change in Bangladesh agriculture: farmers' perceptions, determinants, and effects on resource allocation decisions. *Agricultural Economics*, 33: 107 – 116.

Rahman, S. 2008. Pesticide use in Bangladesh: trends in consumption and productivity. *Outlook on Agriculture* (in press).

Rahman, S., Parkinson, R. J. 2007. Productivity and soil fertility relationships in rice production systems, Bangladesh. *Agricultural Systems*, 92: 318 – 333.

Rahman, S., Thapa, G.B. 1999. Environmental impacts of technological change in Bangladesh agriculture: farmers' perceptions and empirical evidence. *Outlook on Agriculture*, 28: 233 – 238.

Redclift, M. 1989. The environmental consequences of Latin America's agricultural development: some thoughts on the Brundtland Commission report. *World Development*, 17: 365 – 377.

Roger, P.A., Bhuiyan, S.I., 1995. Behaviour of pesticides in rice-based agro-ecosystems: a review. In: Pingali, P.L., Roger, P. (Eds.) Impact of Pesticides on Farmer Health and the Rice Environment. Kluwer Academic Publishers, Boston.

Shiva, V. 1991. *The Violence of the Green Revolution: Third World Agriculture, Ecology and Politics*. London: Zed Books.

Singh, R.B. 2000. Environmental consequences of agricultural development: a case study from the Green Revolution state of Haryana, India. *Agriculture, Ecosystem and Environment*. 82: 97 – 103.

UNDP/FAO, 1988. *Land Resources Appraisal of Bangladesh for Agricultural Development. Report #2: Agroecological Regions of Bangladesh*. Rome: United Nations Development Programme/ Food and Agricultural Organization of the United Nations.

Yadav, R.L, Dwivedi, B.S. Prasad, K., Tomar, O.K. Shurpali, N.J. Pandey, P.S. 2000. Yield trends, and changes in soil organic-C and available NPK in a long-term rice-wheat system under integrated use of manures and fertilizers. *Field Crops Research*, 68: 219-246.

In: Handbook of Environmental Policy
Editors: Johannes Meijer and Arjan der Berg

ISBN 978-1-60741-635-7
© 2010 Nova Science Publishers, Inc.

Chapter 10

Informal Waste Recycling* and Urban Governance in Nigeria: Some Experiences and Policy Implications

*Thaddeus Chidi Nzeadibe[1] * and Chukwuedozie K. Ajaero[2]*

[1]Waste Management & Recycling Research Unit, Department of Geography,
University of Nigeria, Nsukka, Nigeria.
[2]Department of Geography,
University of Nigeria, Nsukka, Nigeria.

Abstract

A contemporary paradigm shift in waste management is to regard wastes as resources. One approach to exploiting the resource value in waste is recycling. Waste recycling in low- and middle- income countries is being driven by the informal sector, often with minimal if any input from institutions of the state. The focus of this chapter is on the often unrecognized and unacknowledged recycling activities of the urban informal sector in Nigeria. Recent experience on Nigerian waste recycling system is reviewed using insights from authors' fieldwork in various cities of Nigeria. The chapter also includes re-analysis of secondary data from published works on informal recycling in Nigeria, and first-hand field experiences. This chapter focuses primarily on the cities of Enugu and Onitsha in southeastern Nigeria, Lagos and Ilorin in the southwest, and Abuja in northern Nigeria. It explores the linkages and contributions to urban governance by the informal recycling sector. The chapter draws attention to some key characteristics of Nigerian recycling systems that have enabled them to cope with the vicissitudes of the recycle trade. Trends and commonalities in research on informal waste recycling in Nigeria are identified. Contributions of ordinary citizens involved in waste recovery and recycling to urban sustainable development, the need to acknowledge and support these contributions through reform of solid waste management, and implications of *integration* of the informal recycling sector into formal solid waste policy and practice in Nigeria are

---

* The term 'recycling' is used here primarily to refer to the recovery of materials by the informal sector, and to a lesser extent, their processing, sale or use.

* E-mail: chidinzeadibe@yahoo.com Tel: +234-803-772-7927

examined. Conclusions of the chapter form the basis of generalization on waste recycling by the informal sector in Nigeria.

Keywords: Governance reforms, informal sector, Nigeria, Millennium Development Goals, recycling, solid waste management, urban sustainable development.

INTRODUCTION

Varied numbers of people in developing world cities make a living through recovery and recycling of materials from municipal solid waste (MSW). Such activities constitute the so-called informal recycling sector. In general, informal recycling refers to the waste recycling activities of waste pickers and other groups such as itinerant waste buyers (IWBs), middlemen and recycling companies involved in the processing/transformation and trade of materials recovered from waste (Wilson *et al*, 2006; Nzeadibe & Iwuoha, 2008; Wilson *et al*, 2009).

In the developing world and countries in transition, governance problems and also a tangible shortfall of financial resources at the local government level, have contributed to producing and maintaining urban environmental service deficits. But at the same time, these deficits have created opportunities for an informal sector to contribute to urban governance in a way that statutory municipal authorities are unable to do.

Furthermore, as a result of an increasing recognition of the complex realities of poverty and disguised unemployment, hence the "discovery" of the informal sector (Hart, 1973; ILO 1972), and a secondary and slower recognition that waste recovery and recycling forms one such sub-sector within the informal sphere, the informal recycling sector, it has been argued, is making significant contributions to urban governance in many cities around the world, sometimes propelled by global structures and processes of change (Medina, 2007).

Waste picking has been identified as a major driver of informal sector recycling in developing countries (Wilson, 2007). Unfortunately, early research on informal recycling systems considered waste picking as a social problem to be eliminated and waste pickers as marginal groups and poor subjects who needed help ostensibly to exit the recuperative activity (see for example, Birkbeck, 1978; Furedy, 1984; Tevera, 1994). Oftentimes, such research contained recommendations oriented towards modernization of the waste management system. In addition, development interventions have often been instituted to improve the working and living conditions of the informal recycling sector and in relieving their daily needs and problems, and sometimes to empower them (Anschütz *et al.*, 2004; Scheinberg *et al.*, 2006).

More recently, however, the informal waste management sector, in general, and waste pickers in particular are gradually being recognized as active participants in urban socio-economic and environmental governance (Van de Klundert & Anschütz, 2001). Under this later conception, the place of scavenging and informal recycling in modern waste management and contributions of scavengers to the economic aspects of the recycle trade, poverty reduction, job creation and social inclusion of people often excluded in society, is emphasized over and above the health, social and economic problems associated with waste picking (Nas & Jaffe, 2004; Wilson *et al.*, 2006; Medina, 2007; Gutberlet, 2008). In this approach, attempt is often made at exploring ways to improve the livelihoods and recycling capacity of the informal sector. And in doing so, recommendations are often oriented towards making use of their skills, experience and expertise in urban management (Muller & Scheinberg, 2003; Wilson *et al.*, 2006; Rouse, 2006).

While activities of scavenger groups in some countries of Africa, Asia and Latin America would appear to have gained prominence in cross-cultural development literature over the past few years (Medina, 2000, 2007; Nas & Jaffe, 2004; Fahmi & Sutton, 2007; Chintan Environmental Research and Action Group, 2007; Waste Pickers Without Frontiers, 2008), research on scavenger and informal recycling systems in Nigeria has tended to be particularistic, focusing on single city phenomena and generally lacking cross-cultural comparative perspectives (Adeyemi *et al.*, 2001; Olufayo & Omotosho, 2007[*]; Imam *et al*, 2007; Nzeadibe & Eziuzor, 2006; Nzeadibe & Iwuoha, 2008; Nzeadibe, 2009 a, b). Even where there appears to be a semblance of cross-cultural efforts, emphasis has often been placed on the economic and environmental aspects of the recycle trade (Agunwamba, 2003).

More importantly, recent research on Nigerian waste management appears to have given scant attention to linkages between waste management and urban governance (Onibokun and Kumuyi, 1999; Whiteman *et al.*, 2006).Consequently, research on the informal recycling sector in Nigeria has only considered the economic, environmental, socio-political and spatial aspects of waste recycling (Adeyemi *et al.*, 2001; Agunwamba, 2003; Nzeadibe and Eziuzor, 2006; Nzeadibe and Iwuoha, 2008; Kofoworola, 2007; Wilson *et al.*, 2009), with little attention paid to contributions of the sector to urban governance (Nzeadibe, 2009a).

This chapter presents some first-hand experiences and a re-analysis of earlier published works on recycling by the informal sector in Nigeria. It places particular focus on the cities of Enugu and Onitsha in southeastern Nigeria, Lagos and Ilorin in southwestern Nigeria, and Abuja in northern Nigeria. However, reference is made to available information in other Nigerian cities for purposes of comparison and completeness. The chapter draws attention to some of the characteristics of scavenger systems that have enabled them to cope with the vicissitudes of the recycle trade. Contributions of the informal recycling system to urban sustainable development in Nigeria and policy implications of integration of the informal recycling sector into formal solid waste policy and practice are also discussed.

METHODOLOGY

This chapter adopts cross-cultural comparative perspectives for analysis. The use of cross-cultural comparative perspectives in research on informal recycling systems have been advocated to be holistic, and having the potential to isolate important features that characterize these systems for detailed analysis and generalization (Nas & Jaffe 2004; Medina, 2007; Sluka & Robben, 2007).

The chapter is essentially based on authors' insights on the topic and on fieldwork on scavenger and informal recycling systems in different cities of Nigeria[+]. Data from authors' participant observations and ethnographic interviews in Enugu, Lagos and Onitsha were

---

[*] As methodologically and epistemologically unsound as this work may be, it is, however, mentioned for the sake of reference and not for its contribution to the phenomenon of this discourse.

[+] Lead author has recently conducted fieldwork in the cities of Aba, Enugu, Lagos, Nsukka and Onitsha. Similar fieldwork has also been planned for other cities of southeastern Nigeria.

"reused". First-hand experiences on Abuja recycling system, and limited reliance on secondary sources (2 of which until 2006 appeared to be the only peer-reviewed publications on informal recycling in Nigeria), also form part of this analysis.

Nigerian population is put at about 140 million (Federal Republic of Nigeria, 2007), made up of about 250 ethno-cultural groups, which fall under 3 major languages namely, Igbo, Hausa and Yoruba found predominantly in southeastern, northern and southwestern regions respectively. Because of availability of secondary data on scavenger systems in Igbo-speaking southeastern Nigeria (Enugu and Onitsha) and Yoruba-speaking southwestern Nigeria (Ilorin and Lagos), as well as dearth of published data on northern Nigeria*, this chapter has been limited to re-analysis of southeastern and southwestern Nigeria recycling systems, while Abuja is taken to represent northern Nigerian scavenger system. These varied insights are, thus, combined to bear on this chapter. The analysis is taken as a basis for generalization about scavenging and informal recycling systems in Nigeria.

SYNOPSIS OF SOLID WASTE MANAGEMENT PRACTICE IN URBAN NIGERIA

Informal waste recycling is a sub-sector of the solid waste management (SWM) sector in Nigeria as in other cities of the developing world. It is, therefore, only proper in this discourse, to place it within the context of SWM practice in Nigeria in order to have a better understanding of the political, economic and socio-environmental issues involved.

In Nigerian cities, the increasing rate of consumption and urbanization leading to an increase in the rate of waste generation is taking a toll on urban environmental quality. Incomplete collection of waste by authorities is a significant problem, and uncollected piles of waste materials dot the urban landscape causing health, aesthetic and environmental problems (see for example, Osibanjo, 1999; Afon, 2007b; Afon & Okewole, 2007; Sha'Ato et al., 2007).

Although municipal waste management in Nigeria is the constitutional responsibility of the local government councils (Federal Republic of Nigeria, 1999; Federal Ministry of Environment, 2000), they have not been able to effectively perform this task largely due to a variety of factors some of which are related to governance (Agunwamba, 1998). Consequently, the federal and state governments have often had to initiate intervention measures such as setting up specialized agencies or taskforces to tackle the menace of MSW especially in the larger cities.

Solid waste management in most Nigerian cities is both formal and informal. In the city of Lagos, for example, two formal outfits namely the Lagos State Waste Management Authority (LAWMA) and the private sector service providers (PSSPs) registered by the Authority undertake solid waste management. Informal sector cart pushers (locally known as *kole-kole* or *barro' boys*) estimated to be over 5000 in number are also active in collection

---

* Kwaire (2008) observes that Tuareg smiths recycled wastes from motor radiators, water taps, television and radio plastic cases to produce ornamental rings, dagger and sword handles. This author further stated that the rope weavers in the production of hanging ropes, animal ropes and water ropes have used polythene sacks. Dauda and Osita (2003), on the other, hand found that scavengers' activities reduced the total amount of waste in Maiduguri by 11.4%. However, these works are not considered as representative of detailed analyses of scavenger systems of northern Nigeria.

and disposal of waste (Adebola, 2006). This informal sector initiative is said to be popular among the residents of this city (Afon, 2007a).

On the other hand, the formal structure for solid waste management in Enugu is the Enugu State Waste Management Authority (ESWAMA) and its PSSPs , while in Nigeria's capital city, Abuja, management of urban solid waste is the responsibility of the Abuja Environmental Protection Board (AEPB) (Adama, 2007; Imam et al., 2008). In most of Nigerian cities, informal sector operatives such as waste pickers, itinerant waste buyers (IWBs), informal collectors & cart pushers, middlemen and micro and small enterprises (MSEs) are also involved in SWM. This group often consists of self-employed or family-owned enterprises working for a livelihood and doing so without social security and the benefit of official recognition in SWM policy. As a result of indiscriminate dumping of residual waste, which is sometimes practiced by some informal sector operatives, the entire informal waste management sector is being persecuted in some of Nigerian cities (LAWMA Bulletin, 2006; Afon, 2007a; Imam et al., 2007; Aderibigbe, 2008; Nzeadibe and Iwuoha, 2008).

The seeming prevailing intolerance of activities of the informal solid waste sector is exemplified by Lagos State. While inaugurating 50 waste compactors newly acquired by the LAWMA, the Governor of this State ordered all cart pushers in the metropolis to leave the State by December 31, 2008, stating that a law was already in place prohibiting cart pushers and their activities (Aderibigbe, 2008). It should be recalled that the State Government had during the previous administration, threatened to ban cart pushers (LAWMA Bulletin, 2006; The Guardian Newspapers, 2006). The current proscription of cart pushers' activities would appear to suggest that environmental policy of Lagos State Government has consistently been hostile to the informal SWM sector, despite the recent observation that cart pushers are critical to driving the urban waste management and recycle trade chain, accounting in some cases for 71.3% of waste disposal in the city (Afon, 2007a).

On the other hand, the action of Lagos State Government would tend to support the view that current thinking in SWM by authorities in most cities of Nigeria revolves around making huge investments in importation of waste collection and disposal equipment as though ipso facto, a panacea for the solid waste problem would be found. A case was recently reported in one of the states in southeastern Nigeria where such investments were made. In one sweep, the government of Abia State spent ₦700 (about US$5.5 million) to purchase equipment and vehicles for refuse collection and disposal (Federal Radio Corporation of Nigeria, 2006). This equipment all imported from developed countries, would appear to have limited application to local conditions because of the characteristics of wastes and the nature of the area[*].

It is also pertinent to note that the predominant waste disposal method in Nigerian cities is open dumping and burning at designated or illegal sites. Little attention is currently paid by governments to recycling, as there are no statutory structures or requirements for recycling of waste, no designated official(s) responsible for recycling and no incentive for the public to recycle. Recycling, therefore, remains squarely a private informal sector-driven activity, which is often held in contempt and antagonized by government officials and the public.

A noteworthy later development in Nigerian SWM system is the consideration being given to generation of energy fronm waste (EFW) in some cities. For example, in the past few

---

[*] Ezeah (2008) points out that about 58% of MSW in Abuja is organic material which would seem to suggest that any approach short of composting would be inappropriate for managing such waste.

months, Memorandums of Understanding (MOU) have reportedly been signed between SWM authorities in the city of Enugu and some foreign investors for partnering in waste-to energy program*. A close observation appears to indicate a situation in which the investors have shown little zeal in following up such partnerships beyond putting pen to paper. Although there may be a cornucopia of waste materials to support EFW projects, the apparent cold feet of investors would appear to have resulted from the seeming unwillingness of government to guarantee the security of investments and personnel of these firms. The overall effect is that EFW projects are yet to get off the ground. And the garbage glut continues unabated in Nigerian cities, in the process, creating opportunities for scavenging and informal recycling activities to thrive.

In recent times, attempts have been made at reforming the SWM systems of some cities in Nigeria, supposedly to make them more effective (Whiteman *et al.*, 2006; Nzeadibe, 2009a). Unfortunately in these interventions, little or no regard is paid to the informal waste collection and recycling services already in existence in these cities. This has often led to disruption of activities of the existing informal system. In general, some progress has been made in managing solid waste in some areas in Nigeria: a composting plant was recently set up in Lagos; plastic, metal scrap, abattoir waste and organic waste recycling industries have also taken off in the cities of Ibadan, Port Harcourt, Kaduna, all based on indigenous technology (Ugwuh, in press).

Similarly, the Ondo State Government took the initiative of establishing an integrated waste recycling project in which local producers used indigenous technologies to convert organic matter into organic and organo-mineral fertilizers and a number of other products (Olarewaju and Ilemobade, 2009). These are commendable initiatives. However, sustainability of the initiatives would require adequate funding, less political interference and development of capacity of municipal staff to undertake effective solid waste planning and implementation.

MAJOR FEATURES OF NIGERIAN INFORMAL RECYCLING SYSTEM

Availability of waste as resources, harsh economic situation, income inequalities and the need to survive in Nigeria all combine to sustain the informal recycling sector in Nigerian cities. The sector shows a wide range of locally formed and adapted characteristics and social systems. Outlined below are some salient features of this system.

Spatial Locations of Scavenger Systems

This relates to the points in space where scavengers harvest waste. Waste picking in Nigeria is common in waste dumps in all cities in Nigeria, particularly the major cities. In addition, scavengers operate at designated waste disposal sites (landfill sites). In addition to the landfill and dumpsite scavengers, there also exist itinerant scavengers who recover

* Assistance from Greg Anyaegbudike, Managing Director of Enugu State Waste Management Authority (ESWAMA) on this section is gratefully acknowledged.

materials from public places and itinerant waste buyers (IWBs) who trade cash for trash mostly with householders and businesses.

In Lagos, waste recovery activities are carried out at the Ojota landfill and other legal open waste dumps in Solus and Abule-Egba. These dumpsites are legal because they were so designated by LAWMA, the government agency in charge of waste management in the city. Some waste pickers however work at illegal dumps scattered around the city (Nzeadibe & Iwuoha, 2008).

In Enugu, the Ugwuaji landfill is currently the only designated waste disposal site in the entire city, and all the PSSPs tip waste they collect there (Nzeadibe, 2009a). In the commercial city of Onitsha, Nkwelle Ezunaka landfill site is an important center of waste recovery by the pickers although scavenging activities may also be found at major roadside dumps.

In Ilorin, material recovery from refuse takes place along the streets, 3 major illegal dumpsites and within neighborhoods. Only the dump at Sobi Army Barracks Road is officially recognized and controlled by authorities. On the other hand, waste recovery activities in Abuja take place at Gosa and Mpape , the officially designated waste disposal sites in the city, and recycling by scavengers at this site is recognized by the AEPB. However, authorities usually frown at activities of the informal recycling sector elsewhere in the city.

Organization of the System

Contrary to popular perception, there appears to be some form of order in scavenger and informal recycling systems in some cities of the developing world. Although this organization may vary between cities (see, for example, Birkbeck (1979) for Cali, Colombia; Agarwal *et al* (2005) & Chintan Environmental Research and Action Group (2007) for Indian cities; Wilson *et al*, (2006[*]), a hierarchical arrangement may often be visualized. Figure 1 illustrates the hierarchy of informal recycling typical of Nigerian cities.

In this hierarchical structure, there is direct relationship between increasing influence, affluence and specialization on the one hand, and an inverse relationship in the population of people involved on the other as we move up the pyramid (Nzeadibe & Iwuoha, 2008). At the bottom are four different scavenger groups as follows:

- scavengers that operate at the landfill site;
- scavengers at the legal and illegal waste dumps;
- itinerant bottle and can collectors; and,
- cart pushers

Mention must be made that the scavenger groups operate at the lowest end of the waste economy. The pyramid, which shows many waste pickers at the base, tapers off upwards to the micro- and -small enterprises (MSEs).

---

[*] Wilson et al (2006) shows a generalized model of the hierarchy of actors in informal sector recycling system in Third World cities.

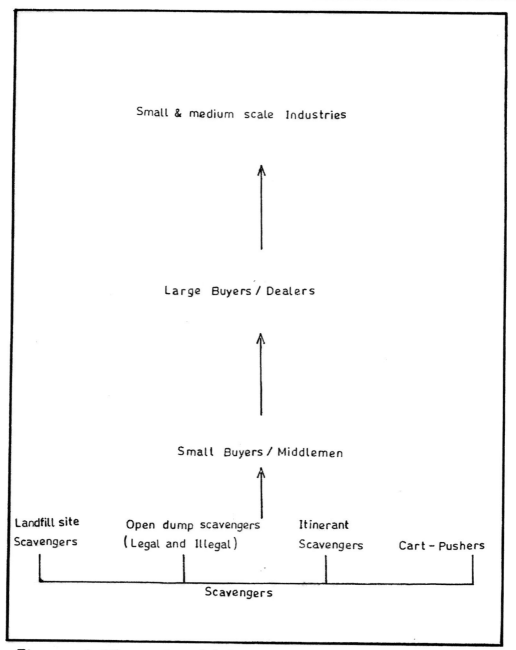

Figure₁ : Hierarchy of informal recycling in Lagos ,Nigeria

Source: Nzeadibe & Iwuoha, 2008

Data and illustrations reproduced by kind permission of IWM Business Services Ltd and CIWM

Figure 1: Hierarchy of informal recycling in Lagos, Nigeria;

Socio-Political Context

Relationship of scavengers with authority varies among cities. Public policy and attitudes towards informal waste workers in Lagos is largely negative and is characterized by *repression* of the landfill scavengers (The Guardian Newspapers, 2006; Nzeadibe & Iwuoha, 2008). However, in Enugu, official attitudes towards scavengers may be characterized as *neglect* - a practice in which authorities simply ignore scavengers and their operations, leaving them alone, without persecuting or helping them (Anyaegbudike[+], personal communication, 2008).

Authorities in the capital city of Abuja appear to have a more sympathetic attitude towards waste pickers at the Gosa landfill site. In this connection, 150 waste pickers have purportedly been registered by the AEPB to recover materials at the site. The pickers are said to belong to a recognized association. Regrettably, other initiatives aimed at improving the lives and recycling capacity of the informal sector in Abuja such as creation of a separate area at the Gosa site where pickers could operate safely, without interfering with vehicle movements or with waste placement at the landfill face section, and provision of protective clothing and medical services to the pickers, were reportedly stopped when a Director of the Board whose brainchild they were was removed from office[*].

Similarly, Agunwamba (2003) reports that although scavengers at waste disposal sites in Onitsha operate at no cost, landfill officers may sometimes collect small amounts of money from them and this may be characterized as *collusion* (Medina, 2000). On the whole, public attitude and policy towards informal recycling in Nigeria could vary from open hostility in Lagos, through neglect in Enugu to collusion in Onitsha and "qualified" support in Abuja. Public policy and action to stimulate the informal recycling sector given its contribution to urban environmental management and development would, therefore, seem imperative if the working conditions and efficiency of recycling by the informal sector in Nigeria is to be improved.

Noteworthy is the fact that the typical scavenger system in Nigeria is fiercely independent, having a closely–knit network of socio-political relations which are employed to advantage in the consumption and production process (Nzeadibe & Iwuoha, 2008). This form of organization has been examined in Lagos. Although not fully investigated, similar observations have also been made in other cities of Nigeria particularly Aba in the southeast[+].

Furthermore, in order to cope with the vicissitudes of the recycle trade, particularly exploitation by middlemen and harassment by authorities, some scavenger groups in Nigeria have attempted to band together to form and operate a co-operative. For example, the Ojota landfill scavenger's community in Lagos inhabits an area known as Olususun has an influential cooperative, which they refer to as Union. A powerful chairman and executive

---

+ Oral interview with Anyaegbudike on 10th June 2008. Authors are grateful for the interesting insight into Enugu Waste Management System he provided.

* Some of these issues were examined at a Focus Group Discussion (FGD) organized by University of Wolverhampton, UK on "Analysis of barriers and success factors affecting sustainable management of MSW in Abuja, Nigeria", held at Day Spring Hotel, Abuja on 15th September 2008. This FGD is significant because it included 2 representatives of Abuja Scavengers' Association, officials of AEPB & FCDA as well as researchers on informal waste management. Authors are particularly grateful to Amos Odunfa and Hassan Abubakar Dogon Daji of AEPB for providing an overview of MSWM in Abuja.

+ Field notes and transcriptions from lead author's fieldwork in Aba, Nigeria in 2006, unpublished. Grateful thanks are due to Chinenye Obodozie for insight on Aba scavenger system.

committee regulate activities of the pickers while also protecting the interest of the scavengers. All waste pickers at the site are said to belong to the cooperative and have been reported to pay weekly dues to it (Nzeadibe & Iwuoha, 2008). This cooperative is probably the most active in the country at the moment.

Methods of Resource Recovery

This relates to the techniques they adopt in waste recovery and their system of resource exploitation. The most common is direct picking of waste from the streets and municipal refuse dumps. In this method, the scavengers would dip their hands into the mass of commingled waste to pull out materials they perceive as recyclable. They sometimes use sticks, hooks and rods to facilitate this recovery process. They also recover materials from municipal trucks and push carts tipping waste at the dumps.

Another method involves setting fire to the dumps. The fire burns over combustible components such as wood, plastics etc, leaving non-combustible parts such as metal scraps (Rouse, 2006). In fact, this is a method of choice for the reduction of the volume of other wastes in order to reach the scraps. Using this method, metal scraps may be recovered from materials with outer plastic casings and an internal metallic framework.

Economic Aspects of Waste Recycling in Nigeria

The economic aspects of the recycle trade relate to the remuneration received by the informal sector for their efforts at waste recovery and recycling. Often, this amount varies but could sometimes be significant .In the city of Port Harcourt, for example, it has been reported that annual earnings in the informal recycling sector range from US$4248 for middlemen through US$3058 for large itinerant buyers to US$550 for waste pickers (Wilson et al, 2009). It could therefore be argued that the informal sector provides a source of livelihood for a large proportion of the urban poor in Nigeria.

Table 1. Minimum wage in Lagos compared with recyclers' income (₦).

Recycler groups	Average daily income	Average monthly income	Min. wage	US$ Equivalent (Min. wage)
Ojota landfill site	700	21, 000	7,500	65
Solus/Abule-Egba legal & other dumps	550	16,500	7,500	65
IWBs/Cart pushers.	350	10, 500	7,500	65

Source: Nzeadibe and Iwuoha (2008).

Sometimes, earnings could be high in comparison to the money earned by the least paid formal sector workers in some states. Table 1 compares the amount of money made by the informal sector in Lagos with the state's minimum wage, while Table 2 illustrates the situation in other Nigerian cities. When viewed in the context of the Millennium Development Goal of eradication of extreme hunger and poverty, this observation would seem to be significant although the sector remains unrecognized in government economic and development programs (National Planning Commission, 2004, 2007).

Table 2: Comparison of waste pickers' monthly income and the minimum wage in some Nigerian cities (US$)

City	Income	Minimum wage
Port Harcourt	153.0	75
Nsukka	48.3	35
Enugu	99.3	35*
Onitsha	84.9	35

Source: adapted from Agunwamba (2003) and Nzeadibe (2009a).
* Current rate is $50
Data and illustrations reproduced by kind permission of IWM Business Services Ltd and CIWM

Environmental Aspects

This relates to the amount of recovered materials, hence savings in landfill or disposal space as a result of waste picking, and on the role played in the informal sector by groups such as the itinerant waste buyers (IWBs), who collect clean, source separated materials directly from households and businesses. Table 3 indicates the average amount of various materials recovered by waste pickers in Lagos while Table 4 shows the major recovered materials in Enugu. Table 5 presents data on Nsukka. Although Nsukka is not one of our focus cities, the data are intended to accentuate the contribution of waste recycling to diversion of materials away from disposal.

Similarly, in a study on the city of Port Harcourt, Nigeria, Wilson *et al* (2009) focused on quantifying the recycling rates achieved by the informal sector. This study estimated recycling rates in Port Harcourt to range from 5-15% for paper and cardboard; 10-40% for scrap metal; 20-40% for plastics and 25-70% for bottles and glass, giving overall recycling rates of 8-22% with nearly half contributed by IWBs (Wilson *et al*, 2009).

It is pertinent to note at this point that the informal sector in Nigeria is achieving significant recycling rates with little or no support from the formal sector, development agencies and NGOs. There is also little attempt at strengthening the capacity of the informal sector to improve their recycling rates and livelihoods. Authors of this chapter support the view that building the capacity of the informal sector is critical if living and working conditions as well as efficiency of informal waste workers in SWM and recycling are to improve in Nigeria.

Table 3. Average quantities of recovered materials in Lagos (in tonnes).

Scavenger group	ferrous metals	glass bottles	non-ferrous metals	plastics	tyre	Paper	nylon	Total
Ojota landfill site	81.82	71.12	60.96	91.44	50.8	60.96	91.44	508.54
Solus/Abule-Egba dumps	50.8	60.96	50.8	81.28	30.48	40.64	60.96	375.92
IWBs/Cart pushers	40.64	91.44	81.28	40.64	5.08	20.32	5.08	284.48
Total	173.26	223.52	193.04	213.36	86.36	121.92	157.48	1168.94

Source: Nzeadibe & Iwuoha (2008).
Data and illustrations reproduced by kind permission of IWM Business Services Ltd and CIWM

Table 4. Major recovered and recycled materials in Enugu (Kg).

Material	Quantity
Plastics	14,400
Non-ferrous metals	550
Ferrous metals	18, 120
Glass bottles	8,005
Total	41,075

Source: Nzeadibe, 2009a

Table 5. Quantities and Prices of materials stocked by the waste dealers in Nsukka.

Type of material	Quantity in stock	Unit price (₦)	*US$ rate
Plastics	4,500kg	30	0.25
Scrap Metals	10,000kg	120	1.03
Aluminium &Brass	2,500kg	120	1.03
Bones	150kg	5000	42.73
Bottles	5,000kg	10	0.08
Slippers	2,000kg	25	0.21
Automobile batteries	300 units	1000	0.18
Motorcycle batteries	120 units	250	2.13
Tyres	150 units	200	1.70

*Exchange rate used is ₦ 117 to US$1.
Source: Nzeadibe (2009b)

CONTRIBUTIONS TO URBAN GOVERNANCE AND DEVELOPMENT

While earlier studies on the Nigerian informal recycling sector were concerned with economic and environmental aspects of the recycle trade, some of the more recent works have attempted to explore the linkages between activities of the sector, the Millennium

Development Goals (MDGs) and urban governance (Nzeadibe, 2009a,b). MDGs of interest are those related to poverty reduction, universal primary education, gender equality and empowerment, health-related MDGs , environmental sustainability and job creation.

Waste recovery and recycling is an activity that prevents a significant segment of the urban population from going into starvation. For example, it has been reported that about 3000 waste pickers operate in Lagos (Nzeadibe & Iwuoha, 2008). For individuals faced with the choice of either going hungry or eking out a living from the recovery and recycling of waste materials, the later option would appear to make economic sense, the social and occupational health implications notwithstanding. As noted earlier, studies in Lagos, Ilorin , Onitsha and Enugu suggest that scavengers often earn much more than the benchmark 1US$ per day, and in some cases, their earnings may be more than the statutory minimum wage (Adeyemi *et al.* 2001; Agunwamba, 2003; Nzeadibe, 2009a; Nzeadibe and Iwuoha, 2008). It could then be argued that the sector is making some contribution towards achieving MDG 1.

Similarly, child labor appears to characterize waste recycling in most Nigerian cities. Children of school age are often involved in the recycle trade either as child waste pickers operating alone or with their parents, or as helpers in their parents' junk shops. Such activities are often at the expense of their education. Involvement of children in the recycle trade could therefore inhibit attainment of MDG 2 on achieving universal primary education.

In some Nigerian cities, a third of the scavenger population is women, mostly widows. In Nsukka, for example, widowed female pickers reported that they took to waste picking after the death of their husbands (Nzeadibe, 2009b). The occupation therefore provided the widows a chance to earn an independent income. In addition, waste picking also provides an income opportunity for recent migrants from the rural areas (Nzeadibe and Eziuzor, 2006). The activity therefore serves as a refuge occupation for vulnerable groups and is inadvertently contributing to MDGs 4 and 8.

On the other hand, the environmental benefit of informal recycling is evident in the amount of landfill space saved by recovery of the materials shown in Tables 3, 4 and 5. With the absence of statutorily prescribed recycling targets and the non-involvement of municipal authorities in recycling, the only form of waste recycling in Nigerian urban areas is that practiced by the informal sector. Recycling rates achieved by the sector could often be high and at no cost to municipalities. Informal recycling would therefore seem to have a strong linkage with MDG 7.

A downside to the waste recycle trade, however, is the health impact. Scavengers and informal recyclers in Nigerian cities often suffer from health consequences of their occupation. The waste disposal sites are a breeding ground for mosquitoes, flies and other disease pathogens. Because they seldom work with personal protective devices, waste pickers clearly face many serious occupational health risks.

The main complaints of the scavengers concerned bites from mosquitoes, cuts and bruises, body aches, general weakness, and frequent fever. The pickers are often unaware of the dangers they face but are economically tied to continuing the occupation when they do (Nguyen *et al*, 2003; Cointreau, 2006). The nature of their waste recovery and recycling activities would therefore seem to negatively correlate with MDGs 6 and 7 related to sanitation, malaria and other diseases (Nzeadibe, 2009a).

In sum, although SWM is not specifically listed among the MDGs, this chapter argues that strong linkages exist between the recycling activities of the informal sector in Nigeria and the MDGs. It should , however , be acknowledged that some activities in the informal recycling sector tend to retard, instead of promote achievement of the MDGs. Public policy should therefore aim at harnessing the capacity of the informal recycling sector in Nigeria to make contributions to development while striving to reduce or eliminate the aspects that tend to inhibit attainment of the goals.

TRENDS IN RESEARCH IN INFORMAL RECYCLING IN NIGERIA

The informal recycling sector in Nigeria comprises men and women carrying out legitimate activities for a livelihood. Although public policy and attitudes may not be favorable to their activities, it has been shown to make significant contributions to urban governance. Whether or not society chooses to recognize that fact, scavengers perform a task which, however marginalized, is both useful and helpful for the environment and society (Berthier, 2003).

Although substantial work has been done on scavenging and informal recycling in Nigerian urban areas in recent years, studies on the phenomenon are still evolving. However, from what has been done so far, a pattern to findings of these studies could be discerned. First is the development of human capital and institutional capacity for environmental management, job creation and poverty reductions, which are some of the Millennium Development Goals (MDGs) set for 2015 (Nzeadibe, 2009a,b; Nzeadibe & Eziuzor, 2006). Significantly, some of these publications have paid particular attention to exploration of the linkages between the MDGs and recycling activities of the informal sector. Reform of governance to achieve the MDGs by the informal sector has also been emphasized.

Another theme that appears to run through some publications on Nigerian recycling system is integration (Adeyemi, *et al*, 2001; Agunwamba, 2003). This concept would appear to have different meanings to different people. However, one could argue that integrating informal waste recyclers into *formal waste management programs* is fundamentally different from integrating them into *formal waste management policies* (Gozenbach & Coad, 2007). In the latter case, they are considered as stakeholders who could contribute towards the SWM planning rather than as people who are made to work for formal waste management systems as a result of lack of opportunities in informal sector resulting from modernization, as is the former case.

Support for and organization of informal recycling systems, particularly the socio-political aspects, hierarchical structure and power relations of actors in the system, and the need to support the formation of waste picker co-operatives has also received some attention in recent literature on Nigerian scavenger system (Onibokun & Kumuyi, 1999; Nzeadibe, 2009a,b; Nzeadibe and Iwuoha, 2008).

Lastly, vulnerability and gender issues in the Nigerian informal waste economy including the role that women (particularly widows), migrants and children play, and constraints (socio-economic, demographic and cultural pressures) that they are often confronted with in waste recycling have also been reported (Adeyemi, *et al*, 2001; Agunwamba, 2003; Nzeadibe, 2009b). In addition to the above common threads, contributions of the sector to urban livelihoods and environmental governance have also been noted albeit with varying levels of emphasis in some of the studies.

NEED TO SUPPORT THE INFORMAL RECYCLING SECTOR IN NIGERIA

It could hardly be argued that informal recycling is a phenomenon prevalent in most, if not, all developing country cities. Reasons for the ubiquity of waste recycling in cities of the developing world and countries in transition are often economic and socio-cultural. If indeed informal recycling is part of the social system of these countries, *the need to accept this reality in Nigeria and view it not as an impediment to development but instead as a basis of further development,* could not be over-emphasized (Nas and Jaffe, 2004).

Furedy (1997) having examined the social and cultural issues relating to waste recycling by the informal sector in Asian countries especially India and China came to the conclusion that waste scavenging is an inevitable phenomenon in Third World cities. She observes that the role of the informal sector in reducing wastes, creation of livelihood opportunities and provision of resources for manufacturing, cannot be dismissed as unimportant arguing further that we cannot afford to ignore the status and needs of sector (Nguyen *et al*, 2003).

In spite of the contributions of informal recycling activities to economic growth, poverty reduction and environmental conservation in Nigerian cities, the sector does not appear to have the benefit of official recognition and has not been given the attention it deserves in the context of the solid waste management policy and practice. In order to reap the full benefits of the informal recycling sector in solid waste management in the country therefore, active support and recognition extended to them by government, development agencies and NGOs would seem to be imperative (Onibokun and Kumuyi , 1999). Supporting the formation of co-operatives and provision of grants and strategic advice for the development of competitive markets for recyclables will go along way in improving the lives and recycling capacity of the informal sector, while further enhancing their contribution to urban governance in Nigeria.

Evidently, concerns about the occupational and environmental health aspects of informal recycling activities in Nigerian cities and the social challenges confronting the sector are not misplaced. These authors are, however, of the view that while informal waste recycling may not be a development goal *per se*, the government of Nigeria should accept the reality of informal waste management and work towards improvement of lives, livelihoods and working conditions of those involved. Recognition of the role of informal recycling in SWM and development as important, and effecting a change in perception and public attitudes towards the informal recycling sector are challenges that NGOs and donor agencies can help in overcoming. This has been done in other parts of the world and could be replicated in Nigeria.

CONCLUSION AND POLICY IMPLICATIONS

Although it is not operating at its utmost at the moment, the informal waste recycling system in Nigeria has shown some remarkable resilience to withstand shocks and stresses associated with the recycle trade. Reform of governance to recognize the contributions of the sector for improved SWM and livelihoods would seem to be a corollary of the drive of government to reduce poverty, improve lives and quality of the environment (National Planning Commission, 2004, 2007).

This chapter is, perhaps, a pioneer attempt at examining in a holistic and comparative manner scavenger and informal recycling systems in Nigeria. Major threads within the system have been identified, while characterization of the system is also made. The overall intent is to have an insight into the characteristics of Nigerian informal waste management system such that generalizations could be made and in order to see how well Nigerian informal recycling systems compare with those of other cultures. It is also aimed at highlighting the significant contributions the sector is making to urban governance and sustainable development, arguing that these be recognized and supported.

An aspect of the recycle trade that could inhibit progress towards the MDGs is the involvement of school age children in waste recycling. Public policy should give priority to taking children away from dumps and getting them back to school, as education is the bedrock for achieving most of the other MDGs. The Brazilian experience provides an interesting example of how this could be done (Dias, 2006).

The informal sector in Nigeria, at present, has received no support from the formal waste management sector, development agencies or NGOs but has been noted to be achieving significant recycling rates. Public policy should , therefore, be geared towards harnessing the potentials of the system for improved solid waste management and urban livelihoods as a large proportion of urban residents are involved in the trade. Integrating the informal sector into solid waste policy is a sure way to increase recycling rates at no cost to municipalities, with the additional benefits of creating sustainable jobs, social inclusion and poverty reduction.

Integration proposed in this chapter would involve giving the waste pickers access to recyclables at disposal sites and support for the development of secondary materials market outlets. Opening up channels of communication with formal stakeholders and decision-makers in the waste planning process, as is the case in Abuja, could also provide platforms for the recyclers' concerns to be heard and addressed in a proactive manner.

If the government decides to replace the current lip service being paid to poverty reduction programs with more sustainable approaches, one sure way of giving a bite to the poverty reduction efforts will be to involve the informal sector in SWM and recycling policy, giving them a distinct role in the development process. Although this change might not occur overnight, courageous and determined political leadership will be key to achieving the goal. Doing this could be a veritable tool for fighting poverty in the informal sector.

ACKNOWLEDGMENTS

The authors would like to thank Emmanuel T. Osu of Stockholm University, Sweden for assistance received during the preparation of this chapter. The illuminating opportunity presented by the Focus Group Discussion, organized by University of Wolverhampton, UK on Abuja MSWM in September 2008 is gratefully acknowledged. We also like to thank representatives of Gosa Scavengers' Association, Abuja, and officials of AEPB and ESWAMA for insights on SWM in Abuja and Enugu. Finally, the authors are indebted to IWM Business Services Ltd and CIWM, UK (www.ciwm.co.uk), and Habitat International Journal for permission to reuse data and illustration.

REFERENCES

[1] Adama, O. (2007).*Governing from above: solid waste management in Nigeria's new capital city of Abuja*.Stockholm: Stockholm University press. ISBN 978-91-85445-67-7

[2] Adebola, O.O. (2006). *New approaches of solid waste management in Lagos*. In: Solid waste, health and the Millennium Development Goals, CWG-WASH workshop, 1–5 February, 2006, Kolkata, India.

[3] Aderibigbe,Y. (2008). Cart pushers get Dec 31 deadline to quit Lagos. *The Nation*, Friday 5 September, 2 (0777), 10.

[4] Adeyemi A.S.,Olorunfemi J.F., & Adewoye T.O (2001). Waste scavenging in Third World cities: A case study in Ilorin, Nigeria.*The Environmentalist* , *21 (2)*, 93-96.

[5] Afon, A.O. (2007a). Informal sector initiative in the primary sub-system of urban solid waste management in Lagos, Nigeria. *Habitat International, 31 (2),*193-204.

[6] Afon, A. (2007b). An analysis of solid waste generation in a traditional African city: the example of Ogbomoso, Nigeria. *Environment and Urbanization, 19 (2)*, 527-537 .

[7] Afon, A.O., & Okewole , A. (2007).Estimating the quantity of solid waste generation in Oyo, Nigeria. *Waste Management & Research, 25 (4)*, 371-379.

[8] Agarwal, A., Singhmar, A., Kulshrestha, M., & Mittal, A.K. (2005).Municipal solid waste recycling and associated markets in Delhi, India. *Resources, Conservation and Recycling, 44 (1)*, 73-90.

[9] Agunwamba, J.C. (1998). Solid waste management in Nigeria: problems and issues. *Environmental Management, 22(6)*, 849–856.

[10] Agunwamba, J.C. (2003). Analysis of scavengers' activities and recycling in some cities of Nigeria. *Environmental Management, 32(1)*,116-127.

[11] Anschütz, J., Scheinberg, A.,& Van de Klundert, A. (2004). *Addressing the exploitation of children in scavenging (waste picking): A thematic evaluation on action on child labour .A global report for the International Labour Organization*. Geneva, Switzerland: International Labour Organization.

[12] Berthier, H.C. (2003). Garbage, work and society. *Resources, Conservation & Recycling, 39(3)*,193 – 210.

[13] Birkbeck, C.(1978). Self-employed proletarians in an informal factory: The case of Cali's garbage dump. *World Development 6 (9-10)*, 1173-1185.

[14] Birkbeck, C. (1979). Garbage, industry and the "vultures" of Cali, Colombia. In R. Bromley & C. Gerry, C. (eds) *Casual work and poverty in third world cities* (161-183). New York: John Wiley.

[15] Chintan Environmental Research and Action Group (2007).*Wasting our local resources: the need for inclusive waste management policy in India*. New Delhi: Chintan.

[16] Cointreau, S. (2006). *Occupational and environmental health issues of solid waste management: special emphasis on middle and lower-income countries*. Washington, D.C: World Bank.

[17] Dauda, M., & Osita, O.O (2003). *Solid waste management and re-use in Maiduguri, Nigeria*. Proceedings of the 29th WEDC International Conference-Towards the Millennium Development Goals, Abuja, 20-23.

[18] Dias, S.M. (2006).*Waste and citizenship forums- achievements and limitations*. In: Solid waste, health and the Millennium Development Goals, CWG-WASH workshop 2006, 1 –5 February, Kolkata, India.

[19] Ezeah, C.(2008). *Municipal solid waste management in Abuja, Nigeria*. Paper presented at Focus Group Discussion on Analysis of barriers and success factors affecting sustainable management of MSW in Abuja, Nigeria, held at Day Spring Hotel, Abuja on 15th September 2008

[20] Fahmi, W.S., & Sutton, K. (2006). Cairo's Zabaleen garbage recyclers: multi-nationals' takeover and state relocation plans. *Habitat International* , *30 (4,)* 809-837.

[21] Federal Ministry of Environment. (2000).*Blueprint on municipal solid waste management in Nigeria*. Abuja: FMENV.

[22] Federal Radio Corporation of Nigeria. (2006). *Network News*, 3rd August.

[23] Federal Republic of Nigeria. (1999). *The constitution of the Federal Republic of Nigeria*. Abuja: Federal Government printer.

[24] Federal Republic of Nigeria. (2007). *Details of the Breakdown of the National and State Provisional Population Totals 2006*. Official Gazette vol.94, no.24.

[25] Furedy, C. (1984). Survival strategies of the urban poor- scavenging and recuperation in Calcutta. *GeoJournal, 8 (2),* 129-136.

[26] Furedy, C. (1997). *Reflections on some dilemmas concerning waste pickers and waste recovery*. Gouda, the Netherlands: WASTE.

[27] Gonzenbach, B., & Coad , A .(2007).*Solid waste management and the Millennium Development Goals: Links that inspire action*. St. Gallen, Switzerland: Collaborative Working Group on Solid Waste Management in Low -and   Middle-income Countries.

[28] Gutberlet, J. (2008). Empowering collective recycling initiatives: Video documentation and action research with a recycling co-op in Brazil. *Resources, Conservation & Recycling, 52*, 659–670.

[29] Hart, K. (1973).Informal income opportunities and urban employment in Ghana. *Journal of Modern African Studies, 11 (1)*, 61-89.

[30] Imam, A., Mohammed, B., Wilson, D.C., & Cheeseman, C.R.(2008). Solid waste management in Abuja, Nigeria. *Waste Management, 28*, 468 –472.

[31] International Labour Office. (1972). *Employment, incomes and equality: a strategy for increasing productive employment in Kenya*. Geneva: ILO.

[32] Kofoworola, O.F. (2007). Recovery and recycling practices in municipal solid waste management in Lagos, Nigeria. *Waste Management, 27 (9)*, 1139-1143.

[33] Kwaire, M. (2008) .*Tuareg Migration from Niger Republic to Sokoto Metropolis: 1900 – 1985*. Paper presented at the Nigerian Universities Doctoral Thesis Award Scheme, Usmanu Danfodiyo University, Sokoto, 8-9 July.

[34] LAWMA Bulletin. (2006). LAWMA carries out enforcement, shuts Itire, Ikate, Ojuelegba, Iddo and Tejuosho market/shops. *LAWMA Bulletin, 1 (3)*, 4, October/November Edition.

[35] Medina, M. (2000). Scavenger cooperatives in Asia and Latin America. *Resources, Conservation & Recycling, 31 (1)*,51-69.

[36] Medina M. (2007).*The world's scavengers: salvaging for sustainable consumption and production*. Lanham: AltaMira Press.

[37] Muller, M., & Scheinberg, A. (2003). Gender-linked livelihoods from modernising the waste management and recycling sector: a framework for analysis and decision-making.

In V. Maclaren & N.T.A Thu (Eds), *Gender and the waste economy: Vietnamese and International experiences* (15-39) Hanoi: National political publisher.

[38] Nas, P J M., & Jaffe, R. (2004) .Informal waste management: Shifting the focus from problem to potential. *Environment, Development & Sustainability, 6*, 337-353.

[39] Nigerian National Planning Commission. (2004). *Meeting Everyone's Needs: National Economic Empowerment and Development Strategy*. Abuja:

[40] Nigerian National Planning Commission. Nigerian National Planning Commission (2007). *Draft National Economic Empowerment and Development Strategy - NEEDS2*. Abuja: Nigerian National Planning Commission.

[41] Nguyen, H., Chalin, C. G., Lam, T.M., &. Maclaren, V.W. (2003). Health and social needs of waste pickers in Vietnam. *Research paper, Canadian International Development Agency (CIDA) WASTE-ECON programme South East Asia.*

[42] Nzeadibe,T.C.(2006).Cash for Trash.*CIWM Journal for Waste & Resource Management Professionals*, October 32-33.

[43] Nzeadibe, T.C. (2009a). Solid waste reforms and informal recycling in Enugu urban area, Nigeria. *Habitat International, 33* (1), 93-99.

[44] Nzeadibe, T.C. (2009b). Development drivers of waste recycling in Nsukka urban area, Southeastern Nigeria. *Theoretical and Empirical Researches in Urban Management 12(3)*, 137-149.

[45] Nzeadibe, T.C., & Eziuzor, O.J.(2006). Waste scavenging and recycling in Onitsha urban area, Nigeria. *CIWM Scientific &Technical Review, 7 (1)*, 26-31.

[46] Nzeadibe, T.C.,& Iwuoha, H.C. (2008).Informal waste recycling in Lagos, Nigeria. *Communications in Waste & Resource Management (CWRM), 9(1)*, 24-30.

[47] Olarewaju, O.O., & Ilemobade, A.A.(2009). Waste to wealth : a case study of Ondo State integrated waste recycling and treatment project, Nigeria. *European Jounal of Social Sciences, 8 (1)*, 7-16.

[48] Olufayo, O.,& Omotosho, B.J (2007). Waste scavenging as a means of livelihood in southwestern Nigeria. *Pakistan Journal of Social Sciences 4 (1)* ,141-146.

[49] Onibokun, A. G., & Kumuyi, A.J.(1999).Governance and waste management in Africa. In A.G Onibokun (Ed) *Managing the monster: Urban waste and governance in Africa.* Ottawa, Canada: International Development Research Centre (IDRC).

[50] Osibanjo, O. (1999).*Municipal and industrial wastes management.* An invited paper presented at The Impact Assessment Workshop 1999, October 21-22, Port Harcourt.

[51] Rouse, J.R. (2006). Seeking common ground for people: livelihoods, governance and waste. *Habitat International, 30 (4,)* 741-753.

[52] Scheinberg, A., Anschütz, J., & Van de Klundert, A. (2006). *Waste pickers: poor victims or waste management professionals?* In: Solid waste, health and the Millennium Development Goals, CWG-WASH workshop 2006, 1 –5 February,Kolkata, India.

[53] Sha 'Ato, R., Aboho, S.Y., Oketunde, F.O., Eneji, I.S., Unazi,G., & Agwa, S. (2007). Survey of solid waste generation and composition in a rapidly growing urban area in Central Nigeria. *Waste Management, 27,* 352 –358

[54] Sluka, J.A.,& Robben, A.C.G.M. (2007). Fieldwork in cultural anthropology: An introduction. In: Robben, A.C.G.M & Sluka, J.A (eds) *Ethnographic fieldwork: An anthropological reader* (1-28). Oxford: Blackwell.

[55] Tevera, D. S.(1994) .Dump scavenging in Gaborone, Botswana: anachronism or refuge

occupation of the poor? *Geografiska Annaler Series B, Human Geography* 76 (1) 21-32.

[56] The Guardian Newspapers. (2006).Lagos plan six new refuse stations. Lagos: Guardian Newspapers Limited, August 15,p.5.

[57] Ugwuh, U.S. The state of solid waste management in Nigeria. *Waste Management*, in Press, doi:10.1016/j.wasman.2009.06.030.

[58] Van de Klundert, A., & Anschütz, J. (2001).*Integrated sustainable waste management-the concept: tools for decision-makers*. Experiences from the Urban Waste Expertise Programme (1995-2001). Gouda, The Netherlands: WASTE.

[59] Waste pickers without frontiers. (2008) .The 3rd Latinamerican congress and 1st world – Conference of waste pickers http://www.recicladores.net/index.php?lang=english (Accessed on 03:06:08).

[60] Whiteman, A., Barratt, L., &Westlake, K.(2006) . *Solid waste management as a catalyst for governance reforms: micro-licensing for private sector participation in Nigeria*. In: Solid waste, health and the Millennium Development Goals, CWG-WASH workshop 2006, 1 –5 February, Kolkata, India.

[61] Wilson, D.C.(2007).Development drivers in waste management. *Waste management & Research, 25,* 198-207.

[62] Wilson, D.C., Velis, C., &. Cheeseman, C.(2006). Role of informal sector recycling in waste management in developing countries. *Habitat International, 30(4)*,797-808.

[63] Wilson, D.C., Araba, A.O., Chinwah, K., and Cheeseman, C.R. (2009).Building recycling rates through the informal sector. *Waste Management, 29(2)*,629-635.

In: Handbook of Environmental Policy
Editors: Johannes Meijer and Arjan der Berg

ISBN 978-1-60741-635-7
© 2010 Nova Science Publishers, Inc.

Chapter 11

THE ECONOMIC AND ENVIRONMENTAL EFFECTS OF WATER PRICING POLICY IN CHINA: AN ANALYSIS OF THREE IRRIGATION DISTRICTS

Han Hongyun[1] and Zhao Liange[2]*

[1]Center for Agricultural and Rural Development, Hangzhou , Zhejiang, China
[2]College of Economics, Zhejiang Gongshang University, Hangzhou Zhejiang, China

ABSTRACT

Although pricing mechanism—especially raising the price of water—has become a high priority in dealing with the problem of water scarcity and the inefficiency of agricultural water management, it is a controversial issue both in developed and developing countries. Based on a linear programming model (LPM), this chapter analyzes the impact of water pricing policy on economic and environmental effects. The examination indicates that farmers will cut their rice planting as a direct response to rising water prices. Due to large capital costs and the labor-intensive characteristics of rice production, the reduction in rice area leads to an increase in farmers' agricultural income and a decrease in agricultural employment and revenue of irrigation districts (IDs). Moreover, together with the reduction of surface water consumption, the effects of changing fertilizers and pesticides resulting from rising water prices on the local environment are negative. In contrast to the findings of other researchers, in China the rising water price will result in an increase of farmers' agricultural income rather than a decrease; water pricing as a single instrument is not a valid means of significantly reducing agricultural water consumption due to the substitution of groundwater for surface water under the current water management institutions.

Keywords: water pricing policy, irrigation district, economic and environmental effects, water allocation.

* Corresponding author: 268 Kaixuan Road, HuaJiaChi Campus of Zhejiang University, Hangzhou, Zhejiang Province, China 310029. Fax: 86-571-86971646; Telephone: 13575786597 , 86-571-88210030; E-mail: hongyunhan@zju.edu.cn.

Literature Review and Outline

The huge population and socio-economic development are continuously posing a great pressure on China's limited water resources. As the biggest consumer of water, "agriculture is becoming the focus to which all analysts are pointing as the culprit of the nation's water problems as well as capturing most of the attention to introduce better policies aiming at increasing water use efficiency" (Varela-Ortega et al., 1998, p.194). Although inappropriate institutional arrangements for water management have contributed to the low efficiency of water utilization, the low-price policy for irrigation is considered the origin of the waste of agricultural water consumption (Dinar, 2000). Hence, the rising price of water is considered to be a useful instrument for encouraging the conservation and efficient allocation of water resources (Qubáa et al., 2002).

Although the efficiency of water pricing is an important concern, the rising price of water is a controversial issue all over the world. Even in California, the purpose of agricultural water management is different from the traditional assumption of profit maximization, which depends on political power and exerts significant effects on rural employment, social welfare and the environment (Dinar, 2000). In Spain, farmers have responded to the increase in water prices by reducing their water consumption (Gómz-Limón and Riesgo, 2004, p.48). "The revenue collected by the government agency tends to decrease as water saving increases" (Varela-Ortega et al., 1998, p. 200). In Mexico, the differentiating pricing policy has a catastrophic effect on water allocation, especially in regions where agricultural water occupies a high proportion of the water supply (Diao and Roe, 2001). In Spain, price increases for water are not a suitable policy because of the resulting high negative impacts on the agricultural sector, including decreasing farmer income, a reduction in the number of crops available for farming, and a significant low level of employment. The pricing of water has a negative effect on farm income as a result of two factors that reinforce each other, including income transfer from the private farming sector to the public sector and a reduction in valuable water-demanding crops (Berbel and Gómez-Limón, 2000; Gómez-Limón and Riesgo, 2004).

The responsiveness of farmers' water use to different prices is influenced by many factors concerning water management systems, availability of substitute crops and the overall rural and urban economic development. Even in most advanced countries, "where sound governance, participation, institutions, and skills do exist", much work in improving water use efficiency needs to be done, and agricultural water use is still heavily subsidized (Azevedo and Balter, 2005, p. 21). A mix of polices is necessary to render pricing an effective tool. Farmers shift to new high-valued and more water-intensive crops as water prices are raised. "As such, the efficacy of water pricing for water conservation cannot be taken for granted and needs to be assessed in the context of the particular geographical area, cropping pattern, existing institutional arrangements, and complementary agricultural polices" (Dinar and Mody, 2004, p. 120).

Pricing mechanism, especially raising agricultural water prices, has become a high priority in dealing with the problem of water scarcity worldwide and the inefficiency of water management. It is crucial for policy makers to know the exact effects of raising water charges to illuminate the decision making process of policy options to attain the desired goals of reducing water demand and alleviating water scarcity (Varela-Ortega et al., 1998). Although the reform of water pricing has attracted much attention from researchers and policy-makers, there are few empirical research studies on the likely economic, employment and environmental effects of water pricing policy in China due to the difficulty in accessing data; it is unclear how farmers will respond to and what is the effect of rising water prices on rural development. Using panel data of households in northern China, the chapter intends to analyze the effect of water pricing policy on rural development based on a simple LPM, which should be helpful in decision making regarding water management not only in China but throughout the world.

The chapter proceeds as follows: firstly, a model is given, including data acquisition and methodology; secondly, an analysis of the impacts of water pricing policy on rural development is conducted, including economic, employment, and environmental effects; finally, a brief conclusion is given. We focus in this chapter on the effect of water pricing on rural development under a given institutional framework within which these policies can be implemented. By simulating land use allocation under varying price scenarios, the model traces a variety of reactions to pricing changes subject to the constraints of land, water, and budget. All of these are very useful devices for advising policy makers on making strategic policy options.

"Small reductions in water consumption appear only when water prices are high, yields are sharply reduced, dry farming may take place and farm income decreases" (Varela-Ortega et al., 1998, p. 200). Water conserving polices are feasible where increasing prices push profit-maximizing farmers either to switch to low water intensive crops or to introduce water saving technology as a long-term measure, "pricing irrigation water is expected to enhance water conservation and thereby reduce demand" (Dinar and Mody, 2004, p. 113). By contrast, the article has two main empirical findings. Firstly, farmers' agricultural income will increase with the change of cropping pattern from rice to rainfed crops as a response to rising water prices; secondly, farmers have cut surface water consumption as a response to rising water prices, but single water pricing policy is not an useful instrument to approach the desired goal of reducing water consumption because it has induced farmers to switch from surface water to underground water instead of the change of irrigated technologies and management practices.

1. THE MODEL

1.1. Description of Study Area and Data Acquisition

There has been a trend in farm crops away from rice to field crops in China; 92.7% of the loss in rice crops has taken place in Heilongjiang, Liaoning and Jilin provinces (Zhang and Liu, 2003). Although rice production in northern China accounts for only 7% of the total national output, the desirable taste has earned its rice a high demand and, accordingly, a high price (Ministry of Agriculture, 2000). The rice area in Liaoning province dropped quickly. Liaoning province lies in the east of the Eurasian continent. It enjoys 130–200 frost-free days and 600–1100 mm of precipitation every year. In spite of favorable conditions relating to topography, soil and water availability, and high demand and high price for rice compared to other regions in China, farmers in Liaoning Province have reduced their rice area especially in the late 1990s..

In China, agriculture, the primary industry, was the biggest consumer of water and accounted for 65.0% of the total in 2005. More severely, the coefficient of irrigated water utilization was only 0.45, hence a water price policy was developed as an approach aiming to enhance the efficiency of water utilization and to alleviate the financial burden of operation and maintenance of IDs. A pricing mechanism has been implemented swiftly nationwide to encourage the efficient use of irrigated water. In some major IDs in the HHH (Huanghe, Haihe, and Huaihe) region, irrigation water prices doubled between 1998 and 2000 (Yang et al., 2003). How could farmers in China respond to the rising price of water under specific economic and political circumstances? What will be effects of water pricing on IDs?

To find out the potential impacts of rising water prices on the rural economy, Bayi, Hunpu, and Tiejia IDs in Liaoning Province were surveyed in 2003, where the rice area dropped dramatically and some irrigated rice area was converted into soybean, corn and vegetables. The crops occupying the three IDs are rice, vegetables including field and green house vegetables, corn and soybeans. In sum, 331 households were interviewed to collect the data for an analysis of farmers' land use change. Following is a description of IDs in detail. All sampled IDs are gravity IDs.

- Tiejia ID

Tiejia ID resides in the southeast of Dandong, Liaoning province. There are six reservoirs and 12 rivers. It covers 10 towns, 68 villages, and 80,000 rural households. Its average precipitation is 900 millimeters; the total arable land is 30,500 ha, of which field area is 9500 ha; irrigation area is 211,000 ha. The key crops there are rice, corn, and vegetables. The fixed cost of Tiejia ID was roughly 30 million Chinese Yuan[*]; the cost of operation and maintenance was 10 million Chinese Yuan on an average; the main source of income was revenue from agricultural use, about 10 million Chinese Yuan.

- Bayi ID

Bayi ID resides in the southwest of Shenyang, Liaoning province. Its projected irrigation area is 19, 333 ha, the real irrigation area is 8,527 ha. The key crops are rice, corn and soybeans. In 2003, the fixed cost for Bayi ID was roughly 14.04 million Chinese Yuan, the cost of operation and maintenance was 2.67 million Chinese Yuan, the main source of income was revenue from agricultural use, about 1.00 million Chinese Yuan.

- Hunpu ID

Hunpu ID resides in the southwest of Shenyang, Liaoning province, which covers 19 towns and two state-owned farms. The projected irrigation area is 29,333 ha, and the real irrigation area is 26, 040 ha. The key crops are rice, corn, soybeans and greenhouse vegetables. The water fee at the gate of Hunqu ID is 0.03 Yuan/m^3 and water chare for farmers is 0.05 Yuan/m^3. In 2003, the fixed cost infrastructure was roughly 14.04 million

---

* One US$ equals 8.27 Chinese Yuan.

Chinese Yuan, the cost of operation and maintenance was 2.02 million Chinese Yuan, the main source of income was revenue from agricultural use, about 1.87 million Chinese Yuan.

On-site observations and a series of interviews were conducted in May 2003. During the field surveys, the following variables were assembled concerned with farmers' benefits and costs: the crop price at farm gate, yield per hectare, seeds, fertilizers, pesticides, machinery, taxes, labor and charges for water and electricity, all these were measured in monetary terms. Machinery covered all cost of land preparation, fertilizing, seeding, cultivation, harvesting and transport; labor cost was only employed labor because of the difficulty in collecting data with family labor cost; cost of water including the fee paid to IDs and electricity fee for pumping groundwater.

1.2. Methodology

Given the existing subsidies, it is a difficult task to account for the cost of water supply systems. Many water distributors are supported by local taxes, receive water subsidized by state and federal taxes, and "the actual subsidies provided to California Agriculture from the Bureau of Reclamation are far beyond what even Congress intended, resulting in nearly $ 1.5 billion of hidden and illegal expenditures of public funds" (Brajer and Martin, 1989, p. 261). As a well-established technique for modeling agricultural systems, LPM is able to represent a major change and the components of LPM can be broken down into manageable segments (Hall, 2001). LPM has been widely used to solve the problems of company resource allocation, "the technique's ability to predict how companies will adjust to changes in a variety of exogenous factors is well known, and when used at company level, it enables us to avoid aggregation problems" (Berbel and Gómez-Limón, 2000, p. 219).

Although LPM has resolved the difficulty in the computation of the fixed costs by the calculation of gross margins, there exist aggregation bias (Gómz-limón and Riesgo, 2004). This aggregation bias can only be resolved with the preconditions of farms' homogeneity, including technological, pecuniary, and institutional proportionality. Hence, cluster analysis is an effective way of overcoming this difficulty of heterogeneity (Joseph and Bryson, 1998).

Utilizing a simple LPM that maximizing economic welfare of farmers, we will quantify the effects of rising water prices from a perspective of an irrigation district. To avoid aggregation bias from lumping together farmers, data of households is divided into different groups by IDs. "The amount of crops produced by agricultural farms may influence market prices. If they produce a crop in plenty, the market price of the crop lowers and they cannot make much profit. Conversely, if they produce in a small amount, the price may move up and consumers are struck" (Itoh et al., 2003, p. 555).Due to the characteristics of small-scale farms and existing land distribution system in China, farmers in a given ID are homogenous, including technological, pecuniary and institutional homogeneity.

Based on a simple LPM, we aim to measure the impacts of changes in water pricing on rural development. Simulation based on an aggregation of farmers is proposed as a suitable technique for a hypothetical implementation of cost recovery of water pricing as a proposed policy. To simulate varying water prices of irrigated water, following variables were assembled in our field survey of local water authorities: the costs of IDs (operation and maintenance, and fixed investment), revenue from agricultural water utilization, and water supply of IDs. The effects of water pricing polices are simulated at different scenarios of water prices.

1.3. Scenario Simulations of Water Charges

Historically, water supply facilities have been developed by Central Chinese Government, including the construction of reservoirs, distribution systems, and public works. As happened in most other countries, irrigators in China are heavily subsidized with water prices that correspond exclusively to the costs incurred in the construction and maintenance of conveyance and storage facilities. The price for water has been held at an artificially low level by the Government. The prevailing pattern of subsidies exerts a heavy burden on the Government.

Pushed by heavy financial burden of water use and pressure from water scarcity, Chinese government has increased water prices for agricultural use due to the fact of coexist of water shortage and low efficiency. Progressive pricing for over-quota water use was implemented, the average water tariff increased from 0.028 Yuan/m^3 in 2000 to 0.06 Yuan/m^3 in 2005 (Ministry of Water resources, 2005). The reform of water policy has proven to be difficult and the price of water was set without regard for the costs of supply. The knowledge of how water pricing potentially affects agriculture and the environment is not available. In China, "lack of both time-series and cross-sectional data has deterred such approaches, one direct reason is that irrigation charges did not vary significantly until recently"(Yang et al., 2003 p.151).

This chapter aims to contribute to the discussion of water pricing policy by simulating the impact that a policy based upon rising water prices could have. The simulations greatly simplify the actual physical and economic settings, and allow us to measure the potential impact of water pricing changes on rural development which had not been done in China. Various scenarios that simulate current conditions and varying water charges are examined to determine the optimal cropping pattern and consequent impacts on rural development. To allow for the simulation of different pricing policy, different levels of likely water charges based on the components of water application costs are calculated, they are actual water price, a price covering O&M costs, and a price recovering full costs, including the fixed cost and O&M costs, but no environmental cost.

1.4. The Model

We hypothesize the objective to be the maximization of profit estimated as the gross margin of the farmers. An underlying hypothesis is that farmers and IDs wish to maximize their profits. To overcome the difficulty in computation of general costs and depreciation, we assume that the gross margin is a good estimator of profit, and that the maximization of profit is equivalent to the maximization of gross margin, which is income less variable costs. The gross margin is calculated as the difference between the market value of produced crops and variable costs for production, such as seeds, fertilizers, pesticides, machinery, labor, water and electric fees(including the payments for surface water from ID and pumping costs of underground water), and tax for land. The separate handling of water costs allows us to examine farmers' rational responses in the optimal solution to changes in water prices and water supply of water irrigation institutions. The objective function is as follows:

$$Max \sum_i Z_i = \sum \sum A_{ij} X_{ij}$$

Here, i is the farmer, j the crop, Z_i the total net revenue of farmer i from all crops, X_{ij} the area of crop j of farmer i, n the total number of farmer households in a given irrigation district.

$B_{ij} = p_{ij} * T_{ij}$, B_{ij} is the return per hectare of farmer i, P_{ij} the gate price of crop j, T_{ij} the yield of crop j of farmer i; C_{ij} is the variable costs of farmer i, $A_{ij} = B_{ij} - C_{ij}$, A_{ij} is the net return per hectare.

The model is formulated to allocate the land area between various crops in order to maximize the net return from the available water and land resources, which is consisting of three parts: a linear objective function for maximization of gross margin; a set of linear constraints; and a set of non-negativity constraints.

Land Constraints: Land constraints consider the total available area for irrigation, the total cultivated area should be equal to or less than land available to each farmer i:

$$\sum X_{ij} \leq D_i \, (D_i \text{ is the total land available to farmer } i \,)$$

Non negativity constraints:

$$X_{ij} \geq 0$$

Capital constraints: In addition to land constraints, farmers' planting structure is constrained by budget available, it is farmer i's total income last year.

$$C_i \leq \sum A_{ij} X_{ij}$$

According to Berbel and Gómez-Limón(2000), farmers are facing the constraints of market and water resources. The marketing channels put an upper limit on short-run variations in areas planted; the projected consumption of water is the variable that policy makers wish to control via changes in water pricing policies. With specific consideration of market and institutional arrangements, there are no constraints of market and water resources in China due to the small scale of farms.

With regard to the constraint of water availability, in an attempt to improve water resource use efficiency, the Central Government implemented a license permit system in 1994. The presence of many small, fragmented farms makes it costly to measure the amount of water used by a farmer. These high transaction costs make it difficult for the license permit system to be implemented in agriculture. Hence, the water resource in a rural area remains an open access resource. Farmers can substitute surface water with groundwater, there is no limitation for farmers pumping of underground water. Water resources in rural areas are absolutely free.

2. RESULTS AND DISCUSSION

2.1. Crop Response to Rising Water Prices

Farmers adjust cropping structure under land, non-negative, and financial constraints. There are four kinds of response to higher water prices: they are typically "use of less water on a given crop, change in irrigation technology, shifting of water application to more water-efficient crops, and change in crop mix to higher valued crops" (Rosegrant et al., 1995 p.206). Various scenarios of water charges are simulated to find out the optimal cropping pattern as a result of rising water prices in China.

As a response to rising water prices, farmers have three options with their accessible land: rice, field crops, or fallow land. The optimal cropping pattern resulting from the LPM simulation at different levels of water pricing is shown in table 4. The optimal solution proposes that total cultivated rice areas of three IDs are different, the optimal rice areas are slightly smaller than have been at Tiejia and Hunpu, but it is a little bit greater at Bayi ID. The higher current water fee at Bayi ID is responsible for the difference between cultivated rice area and optimal rice area. Differences between real and optimal rice areas at current water price at three IDs means that water price provides a signal to farmers; however, rice is still the most important crop at three IDs, the high level of asset specificity, low level of commercialization, and the difficulty of labor transfer out of agricultural activities have constrained to the change of cropping structure.

The major difference in cropping patterns between the actual and optimal land allocation was an increase in vegetable and corn. "The insecurity of water deliveries to individual farmers has been shown to cause farmers to select crops which are more drought resistant" (Bromley et al., 1980, p.369). Farmers respond to the rising water prices with a modification of their cropping patterns by growing less water demanding crops, substituting dry crops for previously irrigated crops and in some areas of frequent part-time farming by quitting all farming activities (Varela-Ortega et al., 1998). In China, vegetable farmers have changed to groundwater due to the consideration of transaction costs for irrigated water from IDs.

2.2. The Indirect Result of Rising Water Prices

In addition to the reduction of rice area, the indirect result of rising water prices is the increase of agricultural income, the decreasing revenue of water supply agencies and rural employment, mixed demand for fertilizers and pesticides.

2.2.1. The Economic Effects of Rising Water Policies

"Water pricing has two key roles: (1) an economic role of signaling the scarcity value and opportunity cost of water to guide allocation decisions both within and across water subsectors; and (2) a financial role as the main mechanism for cost recovery" (Dinar and

Mody, 2004, p.114). Although the benefit of rice per hectare is greater than that of corn, the high capital cost of rice has led to an increase of farmers' agricultural income resulting from the adjustment of planting structure. The total gross margin at three IDs, as indicators of the farmers' income, will shrink from optimal allocation at current price to higher water prices because farmers' net income will decrease with further increase in water price; at that point, the increase of cost resulting from water price increase offset the effects of adjustment of planting structure, it is a transfer of net income from farmers to public works; with higher water price, the transfer of benefit will be furthered. Increases of water prices from actual price to the level covering O&M, and then the level covering full cost will reduce the amount of water supply and also the revenue of IDs from agricultural water consumption.

In sum, we can conclude that the effect of changing cropping structure is the increase of farmers' agricultural income. With the increase of water pricing, farmers will reduce the rice area, which in turn results in the decrease of water demand and decreased revenue of water supply agencies. However, the relatively high water price has led to the decreased real rice area and efforts to recover cost have led to a direct transfer of farm incomes to water authorities (Yang et al., 2003). Even though the model includes some simplifications, the difference in gross margins indicates that optimizing the allocation of irrigation water still has a reasonable potential to increase the farmers' agricultural income.

2.2.2. The Effects of Water Pricing on Rural Employment

The change of water pricing causes a serious reduction in farm labor since farmers respond to price increase by reducing water consumption through changes in crop planting, introducing less profitable crops to substitute for higher value and higher labor crops with high water demand. This implies the substitution of water-demanding crops by less water and more mechanized crops, rising water prices inevitably lead to the reduction of rural employment. This phenomenon can be evaluated via labor demand attribute, which is measured in agricultural labor units per ha. At three IDs, there are different effects of water pricing on rural employment in detail; however, it is clear that there is a negative effect of farmers' planting adjustment from rice to rainfed crops because of the labor intensive characteristics of rice production.

On an average, the hired labor for rice is 141 US$ per hectare, only 111US$ per hectare for maize. According to the sampled village heads, rice production incurs 150 working days per hectare de facto; by contrast, it is 60 days per hectare for soybean and maize production (Authors field survey). Although farmers' profit margins in rice farming are lower than that of vegetables, rice farming remains the major provider of rural employment because of the lack of opportunities out of agriculture (Yang et al., 2003). Rice is seen by many Asian governments as a strategic commodity since it is the single most important element in the diet of the poor and an important source of employment and income for farmers (Hossian and Narciso, 2004).

"Cost recovery through pricing strategies will require technological and institutional modifications that help farmers gain greater control over their water usage. However, raising irrigation water prices may hurt farm incomes and employment" (Dinar and Mody, 2004 p.112). Gómz-limón and Riesgo (2004) argue that a rise in water price would lead to a decrease in rural employment caused by substitution of most water intensive crops, which are normally more labor intensive. They believe that these conclusions could be generalized to other countries and regions, where irrigated agriculture is a strategic sector in rural areas.

2.2.3. The Effects of Water Pricing Policy on Local Environment

Besides the effect of water pricing policy on cropping patterns and rural employment, there are three-fold environmental impacts, including the drop of groundwater level, the effect of non-point pollution, and the negative impact on field irrigation facilities.

- The drop of groundwater level

Paddy field is a natural storage of precipitation during rainy seasons, irrigation has induced high groundwater levels due to its seepage, percolation and runoff from rice cultivation (Renault et al., 2001). Farmers respond to rising water price by reducing water consumption through changes in cropping patterns, the reduction of rice area has resulted in the decrease of water demand from IDs, which in turn results in the drop of groundwater level.

At the same time, rising water fees have induced farmers to make full use of underground water instead of surface water. In rural areas of China, farmers pay primarily the cost of power and equipment for pumping underground water, water itself is absolutely free. At three sampled IDs, 22 watt-hour can pump water of 210 m^3, the fee for electricity is 0.46 Yuan per watt-hour, farmers can pump underground water of 20.75 m^3 at the cost of 1 Yuan. In contrast, the charge for surface water is from 0.03 to 0.06 Yuan/m^3, the costs for surface and underground waters are nearly the same. If conveyance loss and transaction costs are taken into consideration, farmers are willing to transfer from surface water to underground water.

Over-exploitation of groundwater resources has intensified with the shift to higher value-added vegetables in Northern China (Yang et al., 2003). The groundwater table in our sampled IDs has dropped from 1 meter in 1980s to 3 or even 15 meters. Without the recharge of seepage and percolation from rice systems, groundwater levels will drop at a fast pace. The value of rice systems in China should involve not only the direct revenues form rice production, but also the indirect revenues and costs from man-made wetland systems.

- The effect of non-point pollution

Since 1966, fertilizers have been the most important contributors to yield increase in China, particularly nitrogen. "The average annual application rate of N in China was gradually increased from 38 Kg N ha-1 in 1975 to 130 Kg N ha-1in 1985, and rapidly increased to 236 Kg N ha-1 in 1995 and 262 Kg N ha-1 in 2001" (Liu et al., 2005, p.212). Agriculture has become extremely intensive by using more inputs of inorganic fertilizers and chemical pesticides (Li and Zhang, 1999). Since 1978 fertilizer consumption in China has been growing rapidly, increasing from 8.84 million tons of 1978 to 46.366 million tons of 2004, the application rate of chemical fertilizers per hectare has increased from 58.892 kg in 1978 to 301.954 kg in 200). As a result, China has emerged as the largest consumer, the second largest producer and a major importer of chemical fertilizers in the world (Wang et al., 1996).

No statistical data are available for pesticide consumption, "from 1949 on, the consumption of pesticides in China increased rapidly, 1920 tons in 1952, 537,000 tons in 1980, and 271, 000 tons in 1989 after the manufacture of organic chlorinated pesticides ceased at the beginning of the 1980s" (Li and Zhang, 1999, p. 29).

In recent decades, non-point pollution has become a major concern of water quality degradation in China. With rapidly increasing application of manufactured nitrogen since the early 1970s, the marginal response ratios to N have dropped because of the unbalanced provision of other crop nutrients, phosphates and potash has become a constraint in many areas. For annual crops, the N uptake efficiency is less than 50%, even under good management practices (Ju and Zhang, 2003). "33.3–73.6% of applied N fertilizer is lost each year with an average loss rate of 60 % in arable land, including 20% by nitrogen gas emission, 15 % by denitrification, 10% by soil leaching, and 15% by agricultural land drains and runoff" (Li and Zhang, 1999, p.27). It is generally believed intensive farming with high application rates of N would lead to more severe groundwater pollution.

Compared to field crops, the amounts of leached NO^3-N have been found to be 17 and 53% higher in low and high volume irrigated fields (Feng et al., 2005, p.132). "About 30–70% of fertilizer N applied to paddy fields and 20–50% of fertilizer N applied to uplands were lost, and about 5–10% of N was directly lost through leaching"(Liu et al., 2005, p.212). It means that the amounts of leached N have decreased with the change of cropping patterns from rice to corn production; however, nitrate pollution of groundwater for vegetable cropping systems was found to be worse than for cereal cropping systems (Liu et al., 2005). China's home-produced and important chemical fertilizers are mainly nitrogen fertilizers, the change of irrigated rice to upland crops means the mixed pollution of N fertilizers.

On one hand, the higher capital costs for rice production induce a high level of pesticide and fertilizer residuals. On the other hand, with the transfer of irrigated rice to irrigated vegetable production, this means more N fertilizer pollution is introduced to underground water. The increasing share of vegetables and decreasing ratio of rice to total irrigated area indicates that the reduction of rice area likely means an increase in residuals of pesticides and fertilizers. Gómez-Limón and Riesgo (2004) argue that a rise in water price would lead to a reduction of fertilizer application. Agriculture is a primary and important source of pollution. Agricultural pollution is predominantly non-point due to fertilizer runoff, pesticide run-off, and discharges from intensive animal production enterprises.

- The impact of rice area reduction on field irrigation facilities

Historically, many large irrigation schemes constructed in the 1950s and 1960s were designed hastily, constructed to low standards, and built with poor quality materials and equipment. Many were not completed or currently lack distribution and drainage networks at the tertiary and farm level. Most systems require major upgrading, rehabilitation and completion, but all of this cannot be done because of financial deficits. Under the Home Responsibility System, farmers hold relatively independent land use rights. However, with regard to water supply systems, rights are held by the states. The physical infrastructure again is open access property.

Although the management responsibility of irrigation districts (ID) is now being transferred to water user associations (WUAs) to establish financially autonomous enterprises and self-governing WUAs, due to the poor quality of the infrastructure the WUAs are unwilling to take over the responsibility for management. Furthermore, the authorities that have been responsible for the infrastructure are unwilling to give up their monopolistic position. Farmers residing in a wide area within the ID are responsible only for the operation and maintenance of field facilities. Low water prices inevitably result in low cost recovery and, consequently, inadequate funding for the maintenance and rehabilitation of the irrigation infrastructure (Yang et al., 2003). "The deterioration of water systems can be seen worldwide, particularly in developing countries" (Azevedo and Balter, 2005, p. 21). The ratio of on-farm infrastructure in good condition is only 35%, resulting in the low efficiency of water delivery in northern China. The average efficiency of the canal water delivery system is only 30 to 40%, compared with a rate of 70 to 90% in most developed countries (Han and Bennett, 2003).

With the change in farmers' land use from rice to rainfed crops, more and more infrastructure scattering along fields has been transferred into arable land because farmers want to make full use of limited arable land. More severely, this kind damage to water facilities is irreversible. The value of rice systems in China should involve not only direct revenues from rice production, but also the indirect revenues from man-made wetland systems. The production of rainfed crops is facing the threat of flooding during the rainy season because of the damage of drainage facilities in Liaoning province. The local water authority is also responsible for flood protection, and paddy fields provide natural storage of precipitation during the rainy season.

3. BRIEF CONCLUSION

This study has examined the impact of water pricing as a policy instrument on rural development taking selected three IDs in Liaoning province as a case study. Gains and losses of increasing water prices for different interest groups are elaborated. The results of LPM indicate that optimizing cropping patterns and water distribution may yield substantial economic gains from agricultural production. A comparison of the actual cropping pattern with the optimal patterns at current water prices shows that the actual decisions of farmers tend to focus on the rising cost resulting from increasing water prices. The change in water pricing will result in the increase of farmers' agricultural income; however, the increase in cost resulting from a water price increase offsets the increase of farmers' income from agricultural activities; it is only a transfer of net benefit from farmers to public works. At the same time, farmers' land use change will lead to negative effects on irrigation systems and rural employment, as well as environment.

Water demand from agriculture should react to increasing water prices in a positive way over a long time. In terms of the ecosystem, rice production systems have both positive and negative effects. The value of rice systems in China should involve not only the direct revenues form rice production, but also the indirect revenues from man-made wetland systems. If all likely effects of increasing water prices are considered, it is obvious that single water price cannot meet the goal of reducing water demand for agricultural production; by contrast, rising water price will result in direct decrease of ID revenue from agricultural utilization and the low level of rural employment, negative environmental effects, and fallow land. Water pricing as a single instrument for the control of water utilization is not a valid means of significantly reducing agricultural water consumption because the slight decrease of water demand for surface water is coming at the cost of increased groundwater pumping. "Implementing pricing mechanisms will have a large impact on the rural development but little impact on water-use behavior. For this reason, a full-cost recovery of irrigation price may be neither practically feasible nor effective as a tool for water conservation" (Yang et al., 2003, p.151).

ACKNOWLEDGMENTS

Special thanks are extended to National Natural Science Foundation of China in 2002 (70273023), in 2005(70573091) and The National Planning Office of Philosophy and Social Science Project in 2003 (03BJY038).

REFERENCES

[1]   Azevedo, T.DE., Balter, A.M. (2005). Water pricing reforms: issues and challenges of implementation. *Water Resources Development*, .21 (1), 19-29.Berbel, J, Gómez-Limón J.A. (2000). The impact of water-pricing policy in Spain: an analysis of three irrigated areas. *Agricultural Water Management*, 43 (2), 19-238.

[2]   Brajer, V., and Martin, W.E. (1989). Allocating a scarce resource, water in the west: more market-like incentives can extend supply, but constraints demand equitable policies. American Journal of Economics and Sociology, 48(3), 259-272.

[3]   Bromley, D.W., Taylor, D.C., Parker, D.E. (1980). Water reform and economic development: institutional aspects of water management in the developing countries. *Economic Development and Cultural Change*, 28(2), 365-387.

[4]   Dinar, A. (2000). *Political Economy of Water Pricing Reforms*, Oxford University Press, New York, N.Y.

[5]   Dinar, A., and Mody, J. (2004). Irrigation water management policies: allocation and pricing principles and implementation experience. *Natural Resources Forum 28*, 112-122.

[6]   Diao, X.SH., Roe, T. (2001). Can a Water Market Avert 'Double-whammy' of Trade reform and Lead to 'Win-win' Outcome? *Journal of Environmental Economics and Management 45* , 708-723.

[7]   Gómez-Limón, J.A., Berbel, J. (2000). Multicriteria analysis of derived water demand functions: a Spanish case study. *Agricultural systems 63*, 49-72.

[8]   Gómz-Limón, J.A., Riesgo, L. (2004). Irrigation water pricing: differential impacts on irrigated farms, *Agricultural Economics 31*, 47-66.

[9]   Hall, N. (2001). Linear and quadratic models of the southern Murray-Darling basin. *Environment international 27*, 219-223.

[10]   Han, H.Y., Bennett, J.(2003). Chinese Agricultural Water Resource Utilization in the 21st Century, *Environmental management and development Occasional Papers* No.2, 1447-6975.

[11]   Hossian, M.,and Narciso, J.(2004). Global Rice Economy: Long Term Perspective, FAO Rice Conference, Rome, Italy, 12-13 February 2004.

[12]   Itoh, T., Ishii, H., and Nansekt, T. (2003). A model of crop planning under uncertainty in agricultural management. *Int. J. production economics* 81-82, 555-558.

[13] Joseph, A., Bryson, N. (1998).   Theory and Methodology Parametric Linear Programming and Cluster analysis. *European Journal of Operational Research* 111, 582-588.

[14] Ju, X T, Zhang, FS. (2003). Nitrate accumulation and its implication to environment in north China. *Ecol. Environ. 12*, 24-28.

[15] Li, Y, Zhang, JB. (1999). Agricultural diffuse pollution from fertilizers and pesticides in China, *Wat. Sci.Tech. 39(.3)*, 25-32.

[16] Liu, G D, Wu, W L, and Zhang, J. (2005). Regional differentiation of non-point source pollution of agriculture-derived nitrate nitrogen in groundwater in northern China, *Agriculture, Ecosystems and Environment 59*, 211-220.

[17] Ministry of Water resources. 2005. 2005 Statistic Bulletin on China Water Activities.

[18] Ministry of Agriculture (2000). Tabulation of Production Cost and Return of Agricultural Products. Ministry of Agriculture, Beijing.

[19] Qubáa, R., El-fadel, M, and Darwish, M.R.(2002). Water pricing for Multi-sectoral allocation: A case Study, *Water Resource Development,* 18(4), 523-544.

[20] Renault, D., Hemakumara, M., and Molden D (2001). Importance of Water Consumption by Perennial Vegetation in Irrigated Areas of the Humid Tropics: Evidence from Sri Lanka", *Agricultural Water Management 46*, 215-230.

[21] *Rosegrant, M W, Schleyer, R G, and Yadav, S N.1995. Water policy for efficient agricultural diversification: market-based approaches, Food policy, 20(3), 203-223.*

[22] Yang, H., Zhang, X.H., and Zehnder, A.J.B. (2003). Water Scarcity, Pricing Mechanism and Institutional Reform In Northern China Irrigated Agriculture, *Agricultural Water Management*, 61, 143-161.

[23] Varela-Ortega, C., Sumpsi, J.M., Garrido, A., Blanco, M., and Iglesias, E. (1998). Water pricing Policies, Public decision making and farmers' response: implications for water policy. *Agricultural Economics 19*, 193-202.

[24] Zhang, G.P., and Liu, J.Y. (2003). Spatial-temporal Changes of Cropland in China for the Past 10 Years Based on Remote Sensing. *Journal of Geography*, 58(.3), http://www.igsnrr.ac.cn/geo/guestarticle.jsp?id=1122.

In: Handbook of Environmental Policy
Editors: Johannes Meijer and Arjan der Berg

ISBN 978-1-60741-635-7
© 2010 Nova Science Publishers, Inc.

Chapter 12

THE PRECAUTIONARY PRINCIPLE AND ENVIRONMENTAL PROTECTION: THE AUSTRALIAN EXPERIENCE

Gamini Herath and Tony Prato
Deakin University, Victoria, Australia

1. INTRODUCTION

For at least two decades, sustainable development (SD) has been an important concept that promotes development that meets the demands of present and future generations while maintaining essential ecological processes and support systems. Policy makers need to identify environmental activities that limit development of society, configure usable knowledge, and develop sustainable strategies. An important challenge is to manage interactions of people with the environment so as to sustain critical ecological processes. Environmental resources, such as water, forests, and soil, are generally scarce and need to be conserved. Furthermore, ecosystem processes are not completely understood and too uncertain to permit accurate predictions of the environmental impacts of anthropogenic disturbances. Avoiding irreversible losses of species and preserving biodiversity are deemed critical to SD and not readily substitutable by human-produced capital (Barbier and Markandya 1990). When one species becomes extinct, it cannot be replaced by another species.

The Precautionary Principle (PP) is an environmental management concept used since the 1980s. It is appropriate in decision-making situations in which there is considerable uncertainty about the environmental impacts of policy actions and such impacts are irreversible (Perrings 1991). Critical environmental policy decisions cannot be postponed until all scientific information is available (Francis 1996). The complexities of natural systems and irreversibility of human impacts justify a commitment of resources to safeguard against serious threats to the integrity of the environment. The PP focuses conservation policy away from the current generation to future generations and changes the burden of proof to the resource developers and users. The PP is criticised as being poorly defined leading to many difficulties in its implementation (Beckerman 1994).

The PP is an integral element in SD. Despite the virtues of this principle, the PP has not been widely used in environmental management in Australia, where its validity has been questionioned (Hohl and Tisdell 1993; Rogers and Sinden 1994; Wills 1997). This paper reviews the PP concept, the different formulations of the principle, and the issues that affect its successful use in managing the Australian environment. Specific objectives of this paper are to: (1) review the major features of the PP; (2) assess the major elements influencing its adoption; and (3) investigate the extent to which the PP is being used in Australia.

2. THE PRECAUTIONARY PRINCIPLE: DEFINITIONS

The PP has been popular in international law since 1987. It maintains that policy makers should take precautionary measures even when certain cause-and-effect relationships are not fully understood. In 1990, the U.N. Secretary General stated that the principle "... has been endorsed by all recent international forums." Several national/international laws and treaties are based on the PP, including the U.S. Endangered Species Act, Convention on International Trade in Endangered Species of Wild Fauna and Flora, Framework Convention on Climate Change, Kyoto Protocol, Montreal Protocol on Substances that Deplete the Ozone Layer, Law of the Sea Convention, and the U.S. National Environmental Policy Act (Prato 2005). Although there are many definitions of the PP, it continues to be a somewhat nebulous concept (McAllister 2005).

Principle 15 of the Rio Declaration on Environment and Development states the PP as follows:

> "In order to protect the environment, the precautionary approach shall be widely applied by States according to their capabilities. Where there are threats of serious irreversible damage, lack of full scientific certainty shall not be used as a reason for postponing cost-effective measures to prevent environmental degradation" (UN doc. A/CONF.151/5 1992).

There are other definitions of the PP that complicate its application. The main differences in the definitions relate to:

a. the threshold level of environmental damage that is acceptable.
b. the potential level of risk and the balance against other factors, such as economic loss, moral and cultural values, and value placed on the resource (e.g., existence value).
c. level of adherence required, such as whether it is obligatory or not.
d. burden of proof (i.e., is it on developers or environmentalists).
e. who bears the liability for resource damages.

Cooney (2005) distinguishes between strong, moderate, and weak versions of the PP. The weak version is the least restrictive and simply states, "lack of scientific certainty should not be used as a reason for postponing action." It requires no action even if the threshold level of environmental damage is exceeded. Under the weak formulation, scientific uncertainty alone may not be sufficient and more serious damage should occur before action is taken (i.e., damage must be very significant or irreversible) (Peterson 2006). The weak formulation requires the costs of precaution to be considered. Principle 15 of the Rio Declaration is a weaker formulation because it requires the cost effectiveness of actions to be considered. The burden of proof falls on those who advocate caution.

The moderate version of the PP requires action to address perceived threats to the environment. The Ministerial Declaration of the Third International Conference on the Protection of the North Sea states:

The participants … will continue to apply the PP (i.e., to take action to avoid potentially damaging impacts of substances that are persistent, toxic, and liable to bioaccumulate) even where there is no scientific evidence to prove a causal link between emissions and effects" (Peterson 2006). In this version of the PP, there is no requirement to consider the economic or social costs and the trigger for action is less definite. The burden of proof should be on those who advocate caution.

The strongest statement of the PP is that where there are threats of serious and irreversible environmental damage, lack of full scientific certainty should not be used as a reason for postponing measures to prevent environmental degradation. Implementation of the PP requires that all public and private decisions be based on a careful evaluation, avoidance, wherever practicable, of serious or irreversible damages to the environment, and an assessment of the risk-weighted consequences of various options (Peterson 2006).

An example of the strong version is the definition used by the Earth Charter:

"Prevent harm as the best method of environmental protection and when knowledge is limited, apply a precautionary approach. It places the burden of proof on those who argue that a proposed activity may not cause significant harm, and make the responsible parties liable for environmental harm"

Under the strong version, there is reversal of proof of damage. The proponent of any activity needs to prove that the activity does not do any significant harm. This version may result in bans and prohibitions of potentially threatening activities. Instead of prohibition, it can be temporarily suspended while research is being done on alternative activities.

The interpretation, use, and implications of different versions of the PP remain controversial. In particular, the economics and the legal professions have different interpretations. Lack of scientific data relating to the quality of the environment compounds the solution of environmental problems and the application of the PP. Collection of information, analysis, and interpretation and formulation of policy guided by such analysis has not occurred at a rapid enough pace.

3. FACTORS INFLUENCING THE PP

3.1. Moral and Ethical Dimension and the Environment

The ethical and moral implications of using the PP cannot be ignored. Anthropocentrism and biocentrism continue to polarise community opinion on the environment.

Anthropocentrism is the idea that human needs, interests, human goods, and human values are the focal point of any moral evaluation of environmental policy. Biocentrism assumes nature has moral standing in itself, and, as such, has intrinsic value. Many agree that nature is a prerequisite for social life, but the biocentric approach considers anthropocentric values to be subjective human creations. The above distinction can result in different interpretations of the PP. The "deep ecologists" or the biocentrists may require developers to meet more restrictive environmental standards (Wills 1997). In Australia, anthropocentrism seems to dominate most controversies associated with the use of the PP.

3.2. The Values of Natural Resources

Natural resources have multiple values, including use and non-use values. The different value concepts and their assessment are mired in considerable controversy. The methods used to measure use values remains controversial. Loomis (1999) found the values of the western wilderness in the lower 48 states to be around $415 per ha and a total value of $7 billion for the 17.3 million ha. People place value on the existence of a resource, which is a non-use value (Krutilla 1967).

The unique and irreplaceable natural features of the environment can result in large non-use values. Non use values are based on altruism and are truly divorced from self-interest. Public decision-making based on a complete accounting of non-use values of the environment could improve society's overall welfare. Unfortunately, non-use values are not easily incorporated into economic procedures for evaluating environmental change. Actions resulting from genuine altruism could reduce an individual's welfare, which makes choices "counter preferential."

Margolis (1982) suggests that individuals can have two utility functions: one for self-interest; and the other for group interest, which comes from one's sensitivity to community and social responsibilities. When there is group interest, citizens behave as collective of individuals rather than as individuals. Because of the presence of genuine altruism associated with ethical and moral imperatives, individuals may not accept the validity of trade-offs between standard consumer goods and public or citizen goods, such as the environment. Hanley and Spash (1995) provide evidence of lexicographic ordering where respondents were not willing to give any amount as a trade-off. They wanted the resource protected regardless of cost. Tradeoffs are ruled out when assigning intrinsic values because important sites of interest, such as important cultural or biological features, can be protected from economic exploitation. If society is committed to preserving these areas based on their intrinsic worth, then intrinsic value cannot be compared with use value, thereby making protection absolute. Under the PP, preservation is the only alternative.

3.3. Risk Perceptions and Risk Attitudes

Risk and uncertainty are important concepts in using the PP. Risk attitudes have been examined by eliciting utility functions. In a seminal analysis, Gollier et al. (2000) identified conditions under which scientific uncertainties justify an immediate reduction of consumption of a substance or the adoption of the PP. According to Gollier et al. (2000) "… the choice of abatement paths involves balancing the economic risks of rapid abatement now against the corresponding risk of delay." The second statement in the above expression is compatible with the PP because it supports delay in use of a resource. Gollier et al (2000) shows that tension between use and delay in use of a resource depends on the shape of the utility function. They show that for small risks, the PP is not justified. The consumer must be sufficiently prudent to reduce current consumption with better information. Working within a Bayesian framework, Gollier et al. (2000) argue that larger variability in beliefs should induce society to take stronger preventive measures today. In other words, if there is greater uncertainty, it should induce the decision-maker to take more conservative measures today to preserve the value of these options in the future.

The other related issue is risk perception. The risk perceptions of the public, the government; scientists; industry groups, and individuals can differ, which makes decision-making more difficult. Some of the complexities of varying risk perceptions in relation to genetically modified food are provided by Love (1992). He noted that the nearly 140 submissions received on the risk of GM foods indicated differing perceptions to risk. To the environmentalist, risk meant the risk to the environment caused by GM organisms, but to the industrialist, risk meant commercial risk. The researchers identified commercial risk above others. It is also possible that the government and the judiciary can have different perceptions and interpretations of risk. This difference can lead to conflicts in decisions made by the two institutions. In the context of the PP, the full complexity of the issues and environmental risk safety nets must be viewed more broadly in a social policy framework.

3.4. Specific PP Methods: Safe Minimum Standard

The Safe Minimum Standard (SMS) is a collective choice rule that places considerable emphasis on the protection of the environment wherever thresholds of irreversible damage are likely to be exceeded (Crowards 1998). It implies that a minimum stock or threshold level of any important resource should be maintained. The basic idea is illustrated in Table 1 (Bishop 1978; Ready and Bishop 1991; Herath 2002). The problem is cast as a two-person zero sum game with the opponents being nature and society. First, the cost of implementing SMS is viewed as an insurance premium paid against potential future losses. Society has two strategies: (1) no timber harvesting or SMS, which involves leaving the area intact if the damage due to timber harvesting (in terms of species extinction) is in excess of the SMS; and (2) timber harvesting, which allows timber harvesting to proceed, but poses a risk of species extinction. Table 1 represents the values in terms of losses to identify the strategy that minimizes the maximum possible loss. When there is no timber harvesting and a disease outbreak occurs, but the disease is cured due to the presence of the species, losses are zero. When there is no timber harvesting and no disease outbreak occurs, the losses are zero. If timber harvesting occurs and there is no outbreak of the disease, then the total net gain of logging is (expressed as losses) is $-B_d$. If timber harvesting occurs and the disease breaks out, then the net cost to society is the loss in value of the extinct species (L) less the benefits from logging. This net cost is $L - B_d$, assuming that the benefits from species preservation is greater than the benefits from logging, which is positive. In this game, the maximum possible loss is zero without timber harvesting or SMS, and the maximum possible loss is $L - B_d$ with timber harvesting. If society's goal is to minimize the maximum loss that can occur (i.e., minimax strategy), then the preferred strategy is no timber harvesting.

Table 1. Social loss matrix for timber harvesting strategies.

Strategy	State 1:disease outbreak	State 2: no disease outbreak	Max Loss
No timber harvesting (SMS)	0	0	0
Timber harvesting	$L - B_d$	$-B_d$	$L - B_d$

Source: Bishop and Ready (1991)

SMS reflects lexicography because it is only when the SMS is satisfied that further use of the resource is permitted. The SMS approach gives priority to observing the threshold values. If non-use values, such as existence value are considered, use of the PP may be difficult because individuals exhibit both citizen and consumer behaviours. The PP needs to consider both use and non-use values.

Several authors have applied the SMS to natural resource issues (Ciriacy-Wantrup 1968; Rogers and Sinden 1994). Berrens et al. (1998) used the SMS to examine the importance of maintaining minimum environmental flows in the Colorado River. The study estimated a present value of $1.29 billion in regional benefits of restoring environmental flows during the study period. If the SMS is formulated by the political process, the standard may be lower because a governmental institution can absorb risk and the risk of any single decision may be seen as low. This implies the SMS may differ depending on the decision-maker, which can cause additional confusion.

4. APPLICATION OF THE PP IN AUSTRALIA

4.1. The PP and Genetically Modified Organisms (GMOs)

Biotechnological developments in agriculture have raised the possibility of developing new forms of life to meet specific requirements, such as plant resistance to agrochemicals. GM crops have the potential to generate economic benefits to the producing regions. In 2004, 81 million ha of biotechnology crops were grown by more than 8 million farmers in 17 different countries. The development of varieties of cotton that are resistant to Heliothis in Australia is a good example of the potential of biotechnology. Heliothis is the major pest in cotton costing approximately $60 million annually in chemical use. The variety of cotton resistant to Heliothis is a significant advance in its control (Herath 1998). Several thousand hectares of this new cotton variety have been planted in NSW and Queensland. This development was achieved through the introduction of the "INWARD" gene by biotechnology methods. The "INWARD" gene was developed by Monsanto, a large agricultural company in the USA. Another significant biotechnological development is the introduction of nitrogen fixing bacteria into crops other than legumes, which reduces the amount of synthetic fertilizer required to grow those crops. These developments are extremely significant because they reduce chemical pollution and costs of production, and increase farm profitability (Herath 1998).

The production and consumption of GM crops has generated significant and contentious societal debate globally (Cocklin et al. 2008). The risks arise from the absence of credible information on the long term effects of GM crops on human health, the natural environment, and long-term agricultural productivity. Staunch opponents of biotechnology extol the virtues of alternative strategies, such as organic food and GM-free agriculture.

Modern capitalism is a powerful engine of biotechnological advancement. The private sector defines biotechnology as an approach to fine tune nature and preserve its diversity while reaping its bounty (Krismky and Wrubel 1996). The profit-motivated private industry may lack incentives to focus on the long-term risks, and hence policy makers need to mediate the debate. Governments face the dilemma of capturing the competitive advantage afforded by biotechnology and concomitantly addressing the risk concerns of the public. How governments will respond and mediate depends on the institutional and policy environment among other factors.

The global market system had a stronger impact on Australia after globalization because of the heavy dependence on agricultural exports. Around 400 biotechnology companies operate in Australia. Sixty-five companies are involved in agriculture and only a few GM crops have been approved for commercial use in Australia. Hindmarsh and Lawrence (2001) observe that "… a Byzantine web of formal contractual obligations and informal connections has emerged between public sector research agencies such as CSIRO and universities, small businesses and large multinationals."

The Australian government can either promote increasing Australia's competitive advantage by permitting the production of GM crops or adopt the PP for risk management by promoting an environmentally benign system of agriculture. However, there is a unique constitutional division of powers and responsibilities between the Commonwealth and the six states in Australia. The research, production, adoption, and use of GM crops are policies for the states. States can effectively resist national environmental policies at the sub-national level. This division of powers can create tension if Commonwealth policies on GM foods contravene that of the states.

The global initiative for the control of genetic technology is the Cartagena Protocol on Biosafety, which is part of the Convention on Biodiversity (Lipman 2005). The convention adopts a comprehensive approach to the conservation of biological diversity and the conservation of biological resources. It covers all living modified organisms (LMO). The Protocol is designed to address the risk of LMOs to biological diversity and human health. The PP has a central position in the Protocol (Lipman 2005). Australia has not signed the Protocol due to perceived cost considerations associated with potential losses of Australian agricultural exports. The food industry and the farm lobby believe the Protocol can adversely impact trade in Australian GM food products, perhaps reflecting the stronger version of the PP.

In 2002, the Gene Technology Act (GTA) of 2000 was introduced as a part of a scheme to control gene technology in Australia. It established the Office of Gene Technology Regulator (OGTR). The basic aim of the Act is to protect the health and safety of people, and to protect the environment by identifying risks posed by or as a result of gene technology and by managing those risks through regulating certain aspects of GMOs. The GTA favours biotechnology development while mediating public concern through the introduction of regulatory frameworks that prescribe the identification and management of risk and through dissemination of public information promoting the benefits of GM technology. The OGTR has approved commercial scale release of GM varieties of cotton, carnations, and two varieties of canola. Limited field research has been granted for GM varieties of grapewine, papaya, sugarcane, and wheat. The GTA does not incorporate the PP, as does the Protocol.

Advocates of the PP have criticised the GTA for narrowly focusing on the scientific assessments of the health and environmental risks of GM food products and the exclusion of potential adverse socioeconomic impacts (Hain et al. 2002; Lawson 2002). The wording of the GTA indicates it does not consider economic impacts and ethical concerns, employs a highly restrictive definition of environment, and fails to apply the PP discussed in one section of the GTA (Cocklin et al. 2008).

Recent studies suggest ways to ameliorate this omission. Lipman (2005) suggests that the PP be incorporated as a specific objective of the GTA. The problems have been further highlighted by Lawson (2001) who reviewed the risk assessment plan of the first licence granted under the GTA for the release of transgenic cotton. He states that the regulator used inadequate information and guess work in the absence of relevant information. The PP was not used and, in particular, it did not consider the impact of transgenic crops on ecosystems (Lipman 2005). Thus, Australia still has a long way to go in fully accepting the Protocol and ensuring that potential harm through gene technology is minimised by applying the PP.

Conflicts between Commonwealth government and the states and among the states in Australia in using the PP for GM crops can occur. Tasmania adopted a moratorium on commercial release of GM crops based on the PP. The Tasmanian government wanted to maintain the image as a producer of pure, high quality, green food products. In 2003, the Tasmanian government extended the moratorium for another five years and prohibited research trials on GM crops in the open environment. Thus, Tasmania adopted the weak version of the PP (i.e., lack of scientific certainty should not be used as a reason for postponing actions to prevent adverse impacts of GM crops). The state of Victoria actively supported and adopted GM food crops in agriculture. The Victorian government extolled the virtues of biotechnology and argued that the future success of agriculture is tied to biotechnology. However, the Victorian government reversed its position in mid 2004, by imposing a four-year ban on the commercial release of GM crops. Evidently, the Victorian government adopted the weaker or moderate version of the PP in which the threat of environmental damages justifies or requires appropriate action. In Victoria, deep divisions within the industry and regional communities about the impact of GM crops on markets led to an extended moratorium (Dowie 2004). Besides risk, the political differences between the Liberal Federal Government and state labour governments may be at play (Cocklin et al. 2008).

Some of the complexities of varying risk perceptions in relation to GM food and use of the PP are provided by Love (1992). Love (1992) refers to the submissions made to the House of Representatives standing committee on Industry, Science and Technology regarding the release of GM organisms to the Australian environment. Nearly 140 submissions were received, which indicated differing perceptions to risk. Also the risk perceived is coloured by the persons special interest. For example, to the environmentalist, risk meant the risk to the environment through GM organisms, but to the industrialist, risk meant commercial risk. Such differences can also complicate the use of the PP in managing GM crops.

4.2. The PP and Forest Management in Australia

In 1992, Australia introduced a very comprehensive management plan referred to as the Regional Forest Agreement (RFA) Process for the sustainable use and management of forests. A major objective of the RFA is the conservation and management of areas of old-growth forests and wilderness areas in the reserve system (Dargavel 1998; Musselwhite and Herath 2005). Due to limited data and major gaps in knowledge, there are high levels of uncertainty about the impacts of forest management practices on species extinction. For this reason, the PP is appropriate for conservation of forest biodiversity in the context of the RFA process. To date, PP has not been explicitly used in the RFA Process. However, state governments are expected to undertake research and long-term monitoring to detect adverse biodiversity impacts of forest management practices and redress those impacts by revising codes of practice and management plans (Australian Government 1992).

For example, harvesting in the Wombat state forest in the West Victorian RFA has been unsustainable. It is estimated that logging rates in Victoria would have to be reduced by 20 percent to be sustainable. The Department of Natural Resources admitted that it got its figures wrong. If the PP was adopted, the appropriate action would have been to postpone harvesting until more information became available on sustainable harvest rates which could have avoided excessive harvest practices.

The clear-felling of old- growth forest behind Wye River in the Otways in Victoria despite the presence of the endangered spot tailed or tiger quolls further highlights the failures in implementing the PP in Australia's forestry sector. The tiger quoll numbers declined in the Otways during the past 30 years due to logging. The area has been earmarked as a special management zone under the RFA. The conservation values of the forest have yet to be considered and logging rates were increased before conservation legislation required a better assessment of the conservation values of the quoll. The PP would have stopped logging until a better assessment of the number of quolls is made. The protection of a rare species of Astelia species in Victoria state forests is another case where the government department responsible for legislation gave the benefit of the doubt to the developer, but not to the ecologists who manage the forests (Kirkpatrick 1991). The Chaelundi State Forest provides a safe haven for nearly 15 endangered species. Attempts to log this forest were stopped by the courts. However, the NSW government is considering changes to the law to allow logging. In all of the above cases, the SMS could serve as a guide to implement the PP.

Governments may view certain situations as less risky compared to the courts when deliberating on a given issue or vice versa. These differences in risk assessments mean that use of PP/SMS may differ according to the decision-maker, which can cause considerable confusion. In implementing PP, governments need to come to terms with the full complexity of the issues and environmental risk safety nets need to be viewed more broadly in a social policy framework.

Kirkpatrick (1991) favours using the SMS in forests in south eastern New South Wales. When the forests of Coolangubra and Tantawanglo were listed on the National Estate, a major conflict flared up between the developers and the conservationists. The Australian Heritage

Commission (AHC) favoured preservation to protect endangered species. The resulting political conflict between the Commonwealth government and the state government resulted in a joint scientific committee to study the issue. The committee report recommended establishing a system of small reserves along with appropriate changes to forest harvesting operations and concluded "there is no scientific evidence that current forestry management practices will result in the extinction of any organism plant or animal." This scenario places the burden of proof on the conservationists. Since logging can adversely impact endangered species, the burden of proof should have been placed on the developers. It should have been phrased as "... there is no evidence that development will not damage the ecosystem" (Kirkpatrick 1991). Clearly, the weaker version of the PP was implemented, but the SMS would have been more appropriate.

The PP supports wilderness conservation. According to the PP, although the exact ecological benefits of wilderness preservation remain uncertain, as a matter of fairness to present and future generations, developers of natural areas should demonstrate that their actions will not cause irreversible losses in wilderness quality (Habermas 1993). This is a strong version of the PP in which the liabilities are assigned to the developers. Accordingly, existing pristine wilderness should be maintained until we have better understanding of how to manage it in an ecologically sustainable manner. However, wilderness managers consider the PP to be legislatively weak in preserving wilderness. They state that "the moral and cultural arguments are easy to dispute and if there is no significant following in the community, it is a position that politicians can ignore. Wilderness is in such a position in Australia. The inundation of wilderness areas in Tasmania due to construction of a reservoir is a case in point But if political support exists for wilderness in Australia, it is easy to get the decision makers to agree to preserve wilderness" (Wright 1993). The political nature of the implementation of the PP is clearly demonstrated here. The use of the PP is conditional upon the presence of strong pressure groups.

In using the PP, the tradeoff between income and conservation objectives needs to be established. The SMS can be used to make decisions involving such tradeoffs. Rogers and Sinden (1994) used the SMS to determine tradeoffs between income and preservation of old-growth forests in New South Wales.

Rogers and Sinden (1994) applied the SMS to old-growth forests in Australia. They examined two strategies: (1) timber harvesting that results in extinction of species; and (2) no timber harvesting that results in retention of species. The game theory matrix is given in Table 2. This model was empirically applied to old-growth forests in New South Wales, Australia. The communities of Armidale and Dorrigo in northern New South Wales were surveyed to determine acceptable levels of economic and ecological tradeoffs. Respondents were asked to compare a loss from logging (a given 18 species) against a loss from preservation ($900,000 income and four jobs) and to nominate which was the greater loss. The Armidale and Dorrigo samples under both specific and general tradeoff scenarios, expressed preferences for species protection when the cost of doing so was at $900,000 regional income and four jobs. A smaller but statistically significant decrease was recorded in the number of respondents who preferred species over income when they were told the species could be found elsewhere (Armidale 2). It shows that $ 900,000 and four jobs is not an unacceptably large loss in preserving species. A majority of the samples were prepared to forgo some regional income to ensure survival of endangered species.

Rogers and Sinden (1994) also evaluated what an unacceptably large economic loss is from preservation using an iterative questioning procedure. The number of rare and endangered species which were threatened by logging was changed systematically to determine what was an acceptable level of environmental loss. Of the Armidale 1 sample, only 19.6 percent were prepared to accept some loss of species. Of the Armidale 2 sample, with the state-wide scenarios, only 11.2 percent were prepared to accept some loss. The study shows that in areas not directly affected by economic downturn, at least two thirds of the citizens regard any loss of species as unacceptable.

The implication for policy makers and those involved in forest management debates is that the community as a whole believes that there is a need to ensure species survival and is prepared to make significant economic sacrifice to achieve this end (Rogers and Sinden 1994).

4.3. Legislation and the PP in Australia

There is debate over the legal status of the PP and the definition of damages. The EU considers it as a general customary rule of international law, while the US considers it to be an approach. Despite the legal standing of the PP, it is being widely applied in decisions involving the environment and natural resource management. Lack of clarity may lead to legal challenges, and the potential for courts to rule in favour of adoption, contradicting policy makers. To leave it to the interpretation of the judiciary may not be helpful. There is a need to place the PP within a framework of good regulatory practice (Peterson 2006).

Table 2. Preferences for protection of species versus income and jobs.

Alternative	The loss	Armidale 1	Armidale 2	Dorrigo
Timber harvesting	18 rare species	95.0	86.3	58.0
No timber harvesting	$ 900,000 and four jobs	5.0	13.7	42.0

Source: Rogers and Sinden (1994)

The PP is an integral element in most environmental legislation. The PP is also an important element in international covenants that can support domestic legislation or implementation. Australian experience to date has been patchy and several reasons have been identified for this record. In many cases, the PP has been used only as a guide and not a legislative requirement. Its implementation remains conspicuously rare, while it continues to be incorporated in international legal instruments, national environmental strategies, and domestic legislation.

PP was judicially considered in a few specific cases in NSW, which highlight the multiplicity of subjective interpretations possible in the judicial system. One way to ensure effective use of the PP in decision-making is to integrate it legislatively in environmental impact assessment regimes (Gullett 1998). Several judicial decisions handed down by the judiciary involving the PP highlight the issues involved. The principle has had no strong legal status as evidenced by a number of cases. The Nyngan case, which failed to prosecute the council for not taking due care in building dams for flood protection, is a case in point. There is scope for inclusion of the PP in environmental impact statements (EIS) in development projects in Australia, but absence of legal status for the principle has precluded its use in important environmental decisions (Gullett 1998).

Often investments involving the use of environmental resources tend to occur rapidly outpacing the emergence of legislation. The Nature Conservancy Council (NCC) of the UK embarked on a ten-year programme to collect information to develop conservation strategies in the marine industry. However, the availability of scientific information was slow to emerge to address important environmental issues related to rapid pace of economic development.

Scale economies may be intertwined with allocation and oftentimes decision-makers believe that resources are being allocated to the human economy and away from nature. The decision rule is to increase scale if the net present value of benefits is greater than the costs. However, market allocative criteria and cost-benefit analysis that does not incorporate physical safe-minimum standards do not fully consider the important influences on scale decisions. Efficient resource allocation does not guarantee a sustainable scale. The argument that ecologically destructive policies are inefficient must in the end rely as much on the scale criterion as on the allocative criterion.

5. Concluding Remarks

The PP is grounded in sound reasoning, but considerable problems have been encountered in its application. Applications to date have been limited, but demonstrate that innovative use of the principle is possible. If a decision environment involves considerable uncertainty about environmental impacts of resource use and development, then the PP should be adopted in the absence of a better approach. There are many conflicting issues, such as the aims between the commonwealth and the state governments. Differing interpretations of the PP by the judiciary indicate that more headway needs to be made in using the PP to reduce environmental impacts of decisions. There is a need for further research to generate better guidelines for effectively employing the PP to control serious damage to the Australian environment.

References

Australian Government. 1992. Intergovernmental Agreement on the Environment (IGAE). http://www.deh.gov.au/esd/national/igae/

Barbier, E.B and Markandya, A. 1990. *The Conditions for Achieving Environmentally Sustainable Growth.* London, London Environmental Economics Centre,

Beckerman, W. 1994. Sustainable development: Is it a useful concept. *Environmental Values* 3:191-209.

Berrens, R.P., Brookshire, D., McKee, M., and Schmidt, C. 1998. Implementing the safe minimum standard approach: two case studies from the US endangered species Act. *Land Economics* 74:147-161.

Bishop, R.C. 1978. Endangered species and uncertainty: the economics of a safe minimum standard. *American Journal of Agricultural Economics* 60:10-18.

Ciriacy-Wantrup, S.V. 1968. *Resource Conservation: Economics and Politics* (3rd ed.).: California, University of California Press.

Cocklin, C., Dinden, J., and Gibbs, D. 2008. Competitiveness versus 'clean and green'? the regulation and governance of GMOS in Australia and the UK. *Environmental Economic Geography* 39:161-173.

Cooney, R. 2005. From promise to practicalities: the precautionary principle in biodiversity conservation and natural resource management, In Cooney, R and Dickson, B (eds.), *Biodiversity and the Precautionary Principle: Risk, Uncertainty and Practice in Conservation and Sustainable Use*. London: Earthscan.

Crowards,T.M. 1998. Safe minimum standards: costs and opportunities, *Ecological Economics* 25:303-314.

Dargavel, J. 1998. Regional forest agreement and the public interest. *Australian Journal of Environmental Management* 5:25-30.

Dowie, C. 2004. Industry split on GM bans. *Australian Dairy Farmer* May-June: 9.

Francis, J.M. 1996. Nature conservation and the precautionary principle. *Environmental Values* 5: 257-264.

Gollier, C., Julien, B., and Treich, N. 2000. Scientific uncertainty and irreversibility: an economic interpretation of the "precautionary principle." *Journal of Public Economics* 75:229-253.

Gullett, W. 1998. Environmental impact assessment and the precautionary principle: legislating caution in environmental protection. *Australian Journal of Environmental Management* 5:146-158.

Habermas, J. 1993. *Justification and Application: Remarks on Discourse Ethics*. Cambridge: Polity Press.

Hain, M., Cocklin, C., and Gibbs, D. 2002. Regulating bioscience: the Gene Technology Act of 2000. *Environmental and Planning Law Journal* 19:163-179.

Hanley, N., and Spash, C. 1995. Problems in valuing the benefits of biodiversity protection. *Environmental and Resource Economics* 5:249-272.

Herath, G. 1998 Agrochemical use and the environment in Australia: A resource economics perspective. *International Journal of Social Economics* 25:283-301.

Herath, G. 2002. Research methodologies for planning ecotourism and nature conservation. *Tourism Economics* 8:77-101.

Hindmarsh, R., and Lawrence, G. 2001. Bio-utopia: future natural?, In R. Hindmarsh and G. Lawrence (eds.), *Altered Genes II: The Future?* (2nd ed.). Melbourne: Scribe Publications.

Hohl, A., and C. Tisdell. 1993. How useful are safety standards in economics? The example of safe minimum standards for protection of species. *Biodiversity and Conservation* 2:168-181.

Kirkpatrick, J.B. 1991. The geography and politics of species endangerment in Australia. *Australian Geographical Studies* 29: 246-254.

Krimsky, S., and Wrubel, R. 1996. *Agricultural Biotechnology and the Environment Sciences, Policy and Social Issues.* Urbana and Chicago: University of Illinois Press.

Krutilla, J.V. 1967. Conservation reconsidered. *American Economic Review* 57:777-786.

Lawson, C. 2002. Risk assessment in the regulation of gene technology under the Gene Technology Act 2000 and the Gene Technology Regulation 2001. *Environment and Planning Law Journal* 19:195-216.

Lipman, Z. 2005. Gene technology regulation and the precautionary principle: how Australia measures up. *Journal of International Wildlife, Law and Policy* 8:63-89.

Loomis, J.B. 1999. Economic values of wilderness recreation and passive use: what we think we know at the turn of the century. Paper presented at the conference Wilderness Science in a Time of Change. Missoula, Montana USA.

Love, R. 1992. The public perception of risk. *Prometheus* 10:17-29.

Margolis, H. 1982. *Selfishness, Altruism and Rationality.* Cambridge: Cambridge University Press.

McAllister, L.K. 2005. Judging GMOs: judicial application of the precautionary principle. *Ecology Law Quarterly* 32:149-174.

Musselwhite, G., and Herath, G. 2005. Australia's regional forest agreement process: analysis of the potential and problems. *Forest Policy and Economics* 7:579-588.

Perrings, C. 1991. Reserved rationality and the precautionary principle: technological change, time and uncertainty in environmental decision making. In R. Costanza, and L. Wainger (eds.), *Ecological Economics: Science and Management of Sustainability*, Columbia: Columbia University Press.

Peterson, D.C. 2006. Precaution: principles and practice in Australian environmental and natural resource management. *Australian Journal of Agricultural and Resource Economics* 50:469-489.

Prato, T. 2005. Accounting for uncertainty in making species protection decisions. *Conservation Biology* 19:806-814.

Ready, R.C., and Bishop, R.C. 1991. Endangered species and the safe minimum standard. *America Journal of Agricultural Economics* 73:309-311.

Rogers, M.F., and Sinden, J. 1994. Safe minimum standard for environmental choices: old growth forest in New South Wales. *Journal of Environmental Management* 41:89-103.

UN doc. A/CONF.151/5. 1992. Rio Declaration on Environment and Development. The United Nations.

Wills, I. 1997. The environment, information and the precautionary principle. *Agenda* 4: 51-62.

Wright, P. 1993. The enduring values of wilderness in Wilderness - the future. Paper presented to the Fourth National Wilderness Conference, W. Barton (ed.). Sydney: Envirobook.

In: Handbook of Environmental Policy
Editors: Johannes Meijer and Arjan der Berg

ISBN 978-1-60741-635-7
© 2010 Nova Science Publishers, Inc.

Chapter 13

SOCIAL LEARNING PROCESSES OF ENVIRONMENTAL POLICY

Sanna Koskinen and Riikka Paloniemi

Environmental Science and Policy, University of Helsinki, Finland

ABSTRACT

Involving citizens in the design and provision of government policies and services has never been more in public demand. One remarkable area of progress in environmental policy during the last decade has been a shift from governing by national governments to multilateral participatory processes in which different stakeholders participate together in governance. Such an approach implies that different members of a community or organization can take part in planning and decision-making processes. However, not only who participates, but also in what, why and how they participate are relevant aspects in efforts to increase the legitimacy of environmental politics. In this chapter, we introduce a model of Environmental Policy Action as a Social Learning Process. We discuss participation in environmental policy action as a contextual and societal process and study social learning results produced in and by these processes. Such participatory processes are either driven by an actor's inner motivation or organized by society. In addition, the context of participation affects participants meanwhile they— through their action—drive the evolution of the context. We argue that if participants find environmental policy action processes both reasonable for themselves and their communities, and effective in regard to protecting the environment according to their personal perspectives, they experience self and social empowerment.

INTRODUCTION

There is a long tradition of citizen participation in environmental policy. Ordinary people, academics, students and civic organizations initiated the discussion on environmental policy in the 1960s and '70s. Since then, environmental policy has become institutionalized, but it is still shaped by social interaction among different stakeholders. Civic participation has a strong presence in both international and national environmental policy processes. The United

Nations Framework Convention on Climate Change and conferences on the environment are examples of internationally significant processes. In Finland, enthusiasm for participation reaches across most sectors of society: from environmental protection and land use planning to energy and transport policies and schools, to mention but a few examples. Despite this great eagerness, we find that participation has not yet been considered comprehensively enough. Participation is still perceived as something project-oriented rather than as an enduring course of action and practice. We are still lacking participative governance through which different members of a community could participate equally in various processes (Haikkola & Rissotto 2007).

Functional, multifaceted and effective citizen participation is both important and demanding. Citizens need new skills to be able to participate in planning processes and decision-making in society. These skills are constantly more in demand, since at the same time that problems in society increase in complexity, the community that should be able to solve those problems is also becoming more diverse and pluralistic. An active citizen is forced to act in a fragmented world marked by change and regeneration (Chawla & Heft 2002).

An active civil society is necessary for the realization of sustainable development. Sustainable development can be seen as a continuous collective learning process, which involves different actors in creating a shared vision of the future, taking action, and assessing change (Tillbury 2007). According to our view, it is precisely this processual perspective of collective learning that is needed in environmental policy. In this chapter we examine participation and involvement in environmental policy from the perspective of social learning.

The standpoint of environmental policy challenges us to consider learning as a socialization process. Learners need resources and social structures that enable them to participate in "communities of practice" from an early stage in their learning process. Initially, individuals are allowed to participate without being required to completely master actions or take full responsibility, but through the learning process they can gradually take on more responsibility and become more independently involved (Lave & Wenger 1991; Wenger 1998; Sfard 1998).

ENVIRONMENTAL POLICY ACTION AS A SOCIAL LEARNING PROCESS

In this chapter, we introduce a model of environmental policy action as a social learning process (figure 1). This model was first published in an article written by us for the Finnish geographical journal *Terra* in 2005 (Paloniemi and Koskinen 2005). At that time, the spiral model worked in two ways: as the model emerged, it steered the way in which we compared and interpreted our considerably divergent data side by side*, and eventually the model became the central outcome of our research.

---

* We both conduct research on participation and agency in two different contexts. Koskinen studies participation and environmental citizenship in schools, particularly through collaboration and participation with different community planning projects and instances. Paloniemi's research concentrates on the strengthening of the role of participation in everyday environmental protection by studying landowners' and nature conservation authorities' like-minded and discordant views on the conservation of private land in southern Finland.

In this chapter, however, we take a very different view from our earlier publication, as we now put our emphasis on social learning and its analysis. Our starting point this time is an image, which we will proceed to analyze in the text that follows. With the help of our model, we answer the following questions:

- What is environmentally responsible participation?
- What is the role of social learning in participation processes?
- What kind of learning takes place in participation processes?

The chapter proceeds according to the model's spiral. We begin with the context of participation. Then we distinguish the parallel curves of the spiral: the processes of participation and involvement. We define the concept of environmental responsibility and consider the environmental impacts of action. The model's concept of learning is that of social learning. We unfold the concept of reflection and its significance to learning. Then we present the outcomes of an environmentally responsible learning process: environmental action competence, self-empowerment and social empowerment, disempowerment and rejection. Finally, we elaborate on the model's central feature: the possibility of moving from curve to curve on the spiral.

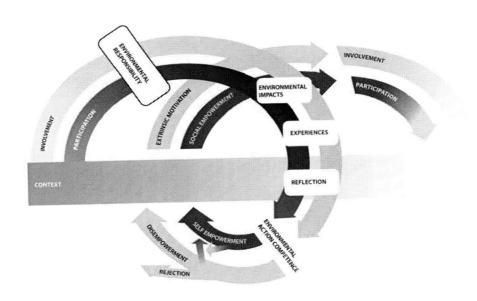

Figure 1. A model of environmental policy action as a social learning process.

THE CONTEXT CREATES THE FRAMEWORK FOR ACTION

The spiral starts from the context, which determines the prerequisites and limits for action. The context may support, limit or prevent environmentally responsible action.

Historical and societal contexts, as well as temporal and spatial aspects, have a crucial impact on an individual's scope for action. It is precisely these circumstances and the way they are interpreted that determines, whether intention turns into environmentally responsible action.

The various features of the context can be categorized in several different ways. On the one hand, the environment is divided, for instance, into the global, regional and local dimensions, which all have their own ecological, sociocultural and economic dimensions as well. On the other hand, the environment is perceived only partly: the boundary between the objective and subjectively experienced environment varies according to individuals, history, society and culture, and from one moment to the next. In our view, the third important feature of the context is its processual nature, as Anthony Giddens proposes (1998, p. 570): people constantly renew the structures of their environment through their everyday actions.

From the perspective of environmental policy action, it is important to note that contexts do not merely "exist", but that different individuals interpret circumstances differently. Some see many possibilities for environmentally responsible action in their everyday lives, when as to others the world appears more through various limitations (Thøgersen 1994; Ölander & Thøgersen 1995). Environmental psychologist Marketta Kyttä (2003) speaks of "potential and actualized affordances." The first are possibilities that an individual *could make use of*. The latter are the ones that he or she actually *does use*. Moreover, it is crucial to note that individuals are not fully aware of their contexts, but act on the basis of fragmentary, limited information in accordance with their own values (Raitio 2008).

We examine the regional and local dimensions of a context from the perspectives of different individuals and institutions. Different dimensions of the environment come to the forefront at different times, but eventually the whole picture is formed from all these various factors, rendering a sort of multifaceted both-and-situation. Even though the context provides the limits and framework for action, it is at the same time itself subject to change. The stakeholders of environmental policy processes constantly shape and renew the society's structures and practices.

AIMING FOR ENVIRONMENTAL RESPONSIBILITY

Our model demonstrates participation in environmentally responsible action. Environmental responsibility includes a conscious effort to promote environmental protection. An environmentally responsible individual or community aims to act in each context, according to their ability, in the best possible way for the environment (e.g. Hungerford & Volk 1990; Kollmus & Agyeman 2002; Osbaldiston & Sheldon 2003). Environmental responsibility includes an ethical dimension, in other words, *environmental values*. Values determine what individuals and communities perceive as the right ways of acting, and concepts of right and wrong are molded by communities' and society's values, which are reflected in different contexts.

The action can be either direct or indirect. In other words, an individual may aim to directly change a state of affairs (e.g. by organizing a reclaiming the streets event, thus momentarily reducing carbon dioxide emissions from road traffic), or try to promote changes in environmental policy through reforms in the political context (e.g. by voting in parliamentary elections for a candidate who proposes legislative changes that will benefit the public transport system).

Environmental responsibility is subjective and contextual. An individual or a community acts within the limits of their own knowledge, ability and capacity – by what they believe to be best for the environment. At the same time as they aim to influence prevailing conditions in society, those conditions also set the limits for their actions. The context of action determines what kind of prospects and obstacles there are for environmentally responsible choices (cf. Kollmus & Agyeman 2002, p. 257; Ölander & Thøgersen 1995).

Environmental responsibility can be described as a continuum along which individuals and communities move. It is not possible to specify a perfect, ideal situation – nor would it be ethically conceivable. Thus, citizens could be encouraged and supported to act with *increased* environmental responsibility; to move to a more responsible direction on the continuum. One possibility of doing so is to participate in environmental policy.

PARTICIPATION OR INVOLVEMENT?

In our model, we distinguish between participation and involvement in environmentally responsible action by placing one on the outer and the other on the inner curve of the spiral. *Participation* is individual citizens' or their interest groups' societal action which aims to influence planning, decision-making and public debate. In participation the social community through which people participate is crucial. An individual's acting alone does not amount to participation. Secondly, participation means active doing, being a part of a political process. And thirdly, it is ordinary people who are the participants, not Parliament members or government officials (Horelli 2002).

An important feature of self-determined participation is its voluntary nature. It begins with an individual's or a community's intrinsic need to act and take part (cf. Arnstein 1969). An environmentally responsible citizen, who is intrinsically motivated, acts out of his or her own will, is curious, spontaneous, and interested without extrinsic rewards (cf. Ryan & Deci 2000). Individuals can participate in various different ways in different contexts. They can, for instance, make a formal conservation agreement concerning the preservation of their own land, or act as lobbies in politics through active membership in an NGO, or formulate an environmental action plan for a school.

Involvement, then, takes place, when individuals or communities are encouraged or told to take part in certain projects or in a particular activity. In involvement the need to take part has been created elsewhere than in the participant's mind (cf. Arnstein 1969). The need may be created by the country's legislation, which obliges authorities to involve citizens in different kinds of planning and decision-making processes, or it may come from an official body or from municipal decision-makers, who urge schools to draw up their own environmental action plans. Yet, involvement may also be initiated in less formal circumstances "from bottom-up"; by, for instance, a teachers' initiative to carry out an environmental education project in school and involve students in environmental action through the project.

The assumptions behind participatory measures are that all stakeholders have the right to take part in defining the content of involvement, developing a common vision, mapping the possibilities for action, taking action, and evaluating the whole process (Percy-Smith 2006, p. 162).

Sometimes the difference between participation and involvement may, of course, be difficult to detect in practice. One person may view something as self-determined participation, when as another person interprets the same project as involvement initiated from the outside.

We see the potential of involvement in the fact that it enables even those individuals to participate meaningfully who would not otherwise take part in action. At the same time, it also offers new channels and forms of participating to those individuals, who act voluntarily but feel the need for support and partners for their action.

All participation always has an effect on the environment, and individuals experience both, the processes and their effects, in different ways. Environmental policy action also has both positive and negative ecological, societal, cultural and economic effects. Single environmentally responsible projects' effects often turn out to be rather minor. And yet, even those small steps are by no means insignificant, since environmental protection is regenerated on the societal level through small changes. If we consider these small changes from an environmental perspective, the essential thing is that environmental impacts diminish, and not, whether this happens through participation or involvement.

THE SOCIAL LEARNING PERSPECTIVE

We consider participation in environmental policy action as a social learning process, as Joanne Tippet et al. (2005) and Richard Bull et. al (2008) do. Social learning is the generation of new ideas and outcomes together, a collective learning process, which is connected to real life situations and practices. Social learning takes place in groups, networks, organizations and communities. According to Wenger, "learning is a matter of engagement: it depends on opportunities to contribute actively to the practices of communities that we value and that value us" (Wenger 1998, p. 227).

Danny Wildemeersch et al. (1998) consider social learning as a superior form of learning, when a particular group needs to solve extraordinary, controversial, and new kinds of problems. Thus, we regard this perspective as a particularly suitable one for the participation processes of sustainable development and environmental policy. However, Richard Bawden et al. (2007) prefer to use the concept of "societal learning", when they refer to learning that tackles the question of how to find new solutions and adjust to changes in society and the environment. According to them, it is crucial to get people actively involved in dialogue to make sense for action. We see features of both social and societal learning in participation in environmental policy, but for the sake of clarity, we shall refer to the more commonly used concept of social learning.

In their article, Wildemeersch et al. (1998) have created a model of social learning. Processes of action, reflection, communication and cooperation are central to the concept of social learning. Social learning is proactive in the sense that the learner takes an active role in his or her surroundings. Competencies are being acquired, restructured and developed in the interaction of actor and context. Reflection is the basis of all learning through which methods and patterns for action can be developed. Reflection should not only compare goals and achievements but also call into question the very goals and methods of action. Attention should be paid to a particular group's internal and external communication in social learning processes, particularly when there are both laymen *and* experts involved. In such cases, the language that experts use may be so complex that other stakeholders are not able to follow it. Similarly, experts may consider ordinary citizens' experiential, tacit knowledge as insignificant or they may not even notice it (cf. also Wenger 1998). Participatory processes that try to achieve particular goals on a cooperative basis are continually involved in processes of negotiation. Negotiations are efforts to reach agreements about the goals and means. They deal with differences in perception, interpretation and interest among the actors (Wildemeersch et al. 1998).

THE SIGNIFICANCE OF REFLECTION IN LEARNING

The spiral in our model proceeds towards reflection, i.e., self-assessment. In the reflection phase, an individual assesses, on the one hand, him/herself and the framework of his or her action, and on the other hand, the action as a process and the outcomes of this action (cf. Freire 1970; Bourdieu & Wacquant 1992). He or she reflects on these issues and turns to a mirror, where, in our view, a diverse collage of even up to four images may appear. In the first image, the individual sees him/herself as a thinking subject and concentrates on self-reflection: on his or her own thoughts and experiences and on him/herself as a conscious being who perceives things. Thus, the individual is able to locate him/herself and his or her social position. In the second image, the individual looks at him/herself in relationship and social networks, as a part of power structures. In the third image, he or she reflects upon his or her own agency in terms of the limits set by knowledge and action, and in the fourth image sees his or her action as a part of the context. With the help of this multiple image, the person looking at the mirror may consider what kind of environmental impacts his or her actions have had, and what is his or her own experience of action and its effects.

Reflection is a central part of learning (cf. Kolb 1984). Through it *metacognition* increases, the actor's self-awareness as an actor and learner. Awareness of one's own learning process changes the ways of knowing: individual issues form larger entities in the learner's mind, which makes it possible to utilize knowledge and skills in different ways in practice.

People should be encouraged to self-assessment. Critical reflection is demanding, since obvious goals and the modes of interpretation and patterns of acting that influence action are concealed. An individual needs support—a partner and methods—to gain awareness of his or her own rigid ways of acting and thinking and to be able to change them (Wildemeersch et al. 1998). Support is also needed for enduring the uncertainty brought on by critical thinking and raised awareness. In environmental policy processes, the problem is often that both the facilitator and participants lack communication skills, and there is also a lack of assessment. Furthermore, there is very little open discussion, despite the need for it.

Our model emphasizes the self-reflection of what is learnt and experienced, and awareness of the learning process. At the same time as individuals gain more environmental action competence, they assess their own learning process: the individual actor may either experience self-empowerment, social-empowerment or disempowerment, or reject environmental responsibility altogether. It is essential to understand that *something is always learnt* through involvement and participation. This very fact challenges the facilitator to see the learning possibilities of his or her project, but also to define the project's aims together with the participants involved.

ENVIRONMENTAL ACTION COMPETENCE

By *environmental action competence* we refer to an individual's capability, i.e., competence, to act in an environmentally responsible way, and the individual's confidence in his or her own capacities (cf. Bandura 1977; Jensen 2004). It is linked to the informational and skill-oriented aspects of environmental awareness. *Environmental awareness* is made up of the conceptions of environmental problems, their causes and solutions to them, and, at the same time, of one's own possibilities and willingness to influence problems (Jensen 2002; Kollmus & Agyeman 2002). Next, we analyze environmental action competence from the viewpoint of learning participation skills and developing into a competent participant.

On one hand, competence means knowledge, skills and experience, and on the other, it means connections, networks and sources. Sometimes competence appears as controlled explicit knowledge connected to a particular situation, at other times, as tacit knowledge, which is impossible to identify separately from the whole action but which still affects it. The environmental action competence of a layman and an expert are different in nature. Lay knowledge is based on everyday experiences, expertise on education and status. Environmental knowledge may well be based on everyday experiences, but it is important to be able to identify what environmental knowledge covers, and what kind of an entity this knowledge forms.

Bjarne Bruun Jensen (2002) categorizes dimensions of environmental knowledge that are essential from the viewpoint of environmental responsibility. Often the knowledge of environmental issues comes from the natural sciences' point of view, and the focus is on problems and their extent. Scientific information may invoke responsibility in individuals, but when it is out of context, it may just as easily cause anxiousness and inability to act. A second dimension of environmental knowledge is awareness of the social and cultural causes of environmental problems. The third dimension is knowledge of individual and community level aspects, which enable change for the better in terms of environmental protection. The fourth, and important aspect from the perspective of environmentally responsible action, is knowledge of alternative courses of action in the future and the creation and internalization of a vision of one's own, increasingly responsible agency.

Developing environmental action competence requires concrete possibilities for action and opportunities to shape the environment, and a chance to adopt active and meaningful roles as an environmental advocate (Horelli 2006). An individual's environmental action competence increases most obviously, if he or she is willing to learn new things. Yet, we think individuals cannot take part in environmentally responsible action without their participation having some effect on their environmental action competence. Learning takes place, even if the individual is not planning to act now or in the future according to what he or she has learnt. Different people learn in different ways.

Official goals of environmental policy rarely include increasing environmental action competence. If such processes were considered as learning processes, both citizens' and authorities' environmental action competence could be put to use more effectively than they are now. People's everyday tacit knowledge could be better integrated into political practice and into dialogue with authorities' expertise knowledge. Thus, there could be local expert knowledge, which would be more alive and more diverse than either side's skills alone (Heft & Chawla 2006, p. 200).

SELF-EMPOWERMENT

In the same way that we have distinguished between participation and involvement in our model, we have also divided the concept of empowerment into two brands. Self-empowerment represents the individual level, and social empowerment the community level. The two go hand in hand – without the one there is usually not the other.

Self-empowerment is what we call the process in which an individual's confidence in his or her own abilities increases. Self-empowerment is an individual's inner feeling of strength, a personal experience of capability and competence (Bandura 1977; Thomas & Velthouse 1990; Siitonen 1999). Self-empowered persons believe in their own degree of influence, consider their own actions as meaningful, and wish to take action.

An individual's self-empowerment is a personal process, and yet, it happens within a community context and surrounding circumstances may affect it. A person cannot be empowered from the outside and the empowering process cannot be given to another person. In the process of self-empowerment, a person finds his or her own strengths, learns self-determination, and detaches him/herself from outside constraints. Self-empowerment occurs differently in different people; in their actions, skills and beliefs. These qualities and their varying degree of intensity may also differ depending on the context and point in time (Zimmerman 1995; Siitonen 1999). Competence, the capacity to act, is tied to a particular context. It means readiness to act under the prevailing circumstances.

Self-empowerment requires positive experiences from involvement and participation processes, and a feeling that these processes do have an effect. It is difficult, however, for outsiders to assess self-empowerment. Sometimes participation in action, which does not reach its original goals, may be significant in terms of self-empowerment. This is because action itself produces important information to the actors about the structures and mechanisms of participation, and this gained knowledge helps them to act in even better ways in the future (Carr 2003).

POWER TO PARTICIPATE

Self-empowerment takes the actor towards the next curve on the spiral – possible self-determined participation in further environmentally responsible action. This move to the next curve will take place, if the individual has also experienced *social empowerment*, i.e., the individual feels that he or she has possibilities and power to act. Social empowerment means change in communities and in society. It is linked with contexts, and from the actor's viewpoint it means gaining power and opportunities to act, when as for those in power, it means sharing or giving away power. Social empowerment means the breaking down of oppressive structures, and most importantly, increased powers for individuals.

Sharing power is easier in principle than in practice, as Bent Flyvbjerg (1998) presents by comparing Jürgen Habermas's and Michel Foucault's viewpoints. *In principle*, Habermas's theory of Communicative Action helps to produce universal, widely accepted and justified solutions to questions, which participants may disagree on perhaps even strongly. According to Habermas's theory, power is shared so that people participate in debate with open minds and participants try to understand each other as best they can. Eventually, the most skillfully justified arguments win, and everyone accepts the solution. Yet, *in practice*, the nature of exercise of power makes the application of Habermas's theory difficult. In Foucault's view, participants are hopelessly late in participating in a process determined by the facilitator: they do not get the chance to formulate the questions for debate, nor do they get to define universal good, which is what Habermas's theory requires. Participants end up in the roles of fine-tuners and justifiers of the process. Perhaps there should be no attempt to share power broadly but instead, according to each situation, and with an attempt to be conscious of the hidden power structures in each particular situation.

From the perspective of environmentally responsible action, the essential thing is the initiation of social empowerment, gradual change through which individuals learn to see themselves as users of power, learn to search for opportunities to act, and find individually suitable ways of acting. It is hard to influence environmental issues and especially children and young people need adult support and companionship in the process – at least in the initial stages of action. Indeed, this need exists among all commencing civil society actors: the complexities of bureaucracy and administrative language require unfolding for actors of any age. Civic initiatives and the general public's views are accepted only, if they have been formulated in the formal language of bureaucracy. Yet, everyone should have the chance to take part in environmental policy processes from their own perspective and in their own way.

FROM DISEMPOWERMENT TO NEW ACTIVITY

If an actor estimates his or her experiences of action as mostly negative, he or she may end up on our spiral's outer curve and become disempowered or drop off the spiral entirely and reject action altogether. Negative experiences of involvement or participation leave a mark on an individual's conception of both the action and of oneself as an active agent. Individuals may experience a sense of disempowerment instead of feeling self-empowered to continue environmentally responsible action. Like its opposite, self-empowerment, disempowerment is also an individual's own, personal experience. Defining disempowerment from the outside would give the definer the authority to denounce the defined as a failure and "discard" them (Freire 1970). This kind of exercise of power is unnecessary and perhaps even harmful from the perspective of learning and the general increase in environmental responsibility. At its worst, outside definition only deepens the sense of disempowerment.

According to the educationalist Juha Siitonen (1999), a disempowered person experiences a participation and learning process and its environmental impacts negatively to a certain extent. Disempowerment may be caused by the individual's situation in life, limitations in one's own physical and mental abilities, the surrounding community's belittling attitude towards the individual, or because the action did not have the effects desired by the individual. These factors may cause a conflict situation, which diminishes the individual's capacity and willingness to participate in a particular context.

Sometimes an environmentally responsible actor may feel so frustrated and unexcited that he or she experiences a sense of being a complete outsider to the action and wishes to withdraw from it completely. This kind of *rejection*, which is stronger than disempowerment, may happen particularly, if the process lacks the community's support or the project in some way calls into question the actor's self-image and values. A situation like this may occur in environmental protection policy, for instance. A forest owner, devoted to utilizing forests for economic benefit, may well feel that forest conservation calls into question his or her own, as well as his or her community's internalized views of worthwhile forest use.

Disempowerment and rejection are unfortunate, though not permanent states. They have often been considered the end point of involvement and participation projects. This need not be, however. Therefore, we want to introduce to the discussion a possibility of continuing from disempowerment towards new, environmentally responsible action through extrinsic motivation. Motivation arouses an individual's interest, steers him or her towards action and prepares for continuing action. Extrinsic motivation is needed when a person's own motivation is not strong enough for some reason. The extrinsic motivating factors may be, for instance, a reward, merit or acknowledgement. Motivators for continuing action may, for instance, be a legislative change, a reward promised for participating in a project, international pressure to make local forestry more ecological, a strong initiative from authorities, or a sustainable development program in school. Extrinsic motivation is instrumental, when as intrinsic motivation means becoming motivated because of the cause or action itself. An intrinsically motivated actor does not need external rewards, but an extrinsically motivated actor may need much and frequent external support (Ryan & Deci 2000). In our view, only intrinsic motivation, the inner curve of our model, ensures that responsible action is firmly rooted.

FROM CURVE TO CURVE

The possibility of moving from curve to curve is a central feature of our spiral model. An individual, who participates in environmental policy action out of his or her own initiative, may through negative experiences feel *disempowered* and move from the inner curve to the outer one. Since self-empowerment is not a permanent condition, participation must be motivating and effective. Participating for the sake of participation should be avoided. In the same way, someone who takes part in action through involvement, might through positive experiences feel self-empowered and move from the outer curve to the inner one. The strength of such involvement projects is in this shift, since through them individuals may also become self-empowered, active and independently participating citizens.

Environmentally responsible action does not start from scratch, nor do its effects end after particular projects are over. Instead, it forms the basis for future processes and their success. This is what our model's spiral shape and processual nature represent. Although we have concentrated on describing only one round on the spiral – one environmental policy action process – we believe that at its best, environmental responsibility will develop into a process, which will continue its path in new participative projects. Thus, through different participation and involvement projects, as many and as diverse a group of people will get a chance to take part in environmentally responsible action.

CONCLUSION

In our view, involvement or participation as such is neither good nor bad. Both have their benefits, which can be brought to light through *awareness of the practices* of involvement and participation. It is crucial that people feel that their participation in environmental policy is both meaningful and effective. All stakeholders should have a say in defining the goals of the process and the rules and means of participation. Power struggles are demanding, however. Giving or sharing power and, at the same time, taking power are the biggest stumbling blocks of participation. Particularly, those who are only just getting involved in participation need a partner to support and defend them and keep them motivated. A partner and ally like this may also be a facilitator in the project—someone who is aware of the process and has good activating skills. An experienced facilitator's role should be considered more that of a professional who provides a favorable and safe setting for participation (Innes & Booher 2003; Heft & Chawla 2006; Tilbury 2007).

It is important to understand that, in addition to power and support, a full participant—*on both the inner and outer curves of our model*—needs sufficient capacities to participate. Considering participation as a continuous learning process creates a possibility for social learning. In such a process everyone is learning—regardless of age or profession. Taking part in a learning process requires a positive attitude, an open mind, and a willingness to communicate genuinely and enter into dialogue with other stakeholders (Innes & Booher 2003). Another challenge to participation is the creation of such shared spaces that have room for diversity and that promote the culturally varied and active use of public space. Barry Percy-Smith (2006) describes such spaces—which enable community encounters, interaction and social learning—as *"communicative action spaces"*. Barry Percy-Smith and Karen Malone (2001), on the other hand, speak of *"opportunity spaces"*, in which meaningful participation can take place and develop in an equal, dialogue-oriented and respectful atmosphere. In this case, space refers not only to a physical space but also to a safe and open atmosphere (Freire 1970).

Participation processes are rarely assessed. Yet, assessment is absolutely central for the sake of motivated learning and participation. Participation should be assessed and reflected upon together with the participants from at least two different viewpoints. First of all, participation should be evaluated from the viewpoint of its effectiveness. Did the process have any influence on anything? If so, on what? If not, then why not? Secondly, participation should be evaluated from the perspectives of experiencing and learning. What have we learned from this process? What kinds of experiences did we gain during the process? How did it feel if we managed to have an effect or if we didn't? What did we learn from this for the future?

ACKNOWLEDGMENTS

This research has been funded by the Maj and Tor Nessling Foundation and the Academy of Finland (Grant no. 118274).

REFERENCES

Arnstein, S. (1969). A ladder of citizen participation. *Journal of the American Institute of Planners* 35 (4), 216–224.

Bandura, A. (1977). *Self-efficacy: The Exercise of Control.* New York: Freeman.

Bawden, R., Guijt, I. & Woodhill, J. (2007). The Critical Role of Civil Society in Fostering Societal Learning for a Sustainable World. In A.E.J. Wals (ed.), *Social Learning Towards a Sustainable World. Principles, Perspectives, and Praxis* (pp. 133–148). Wageningen, The Netherlands: Wageningen Academic Publishers.

Bourdieu, P. & Wacquant, L. (1992). *An Invitation to Reflexive Sociology.* Chicago: University of Chicago Press.

Bull, R., Petts, J. & Evans, J. (2008). Social learning from public engagement: dreaming the impossible. *Journal of Environmental planning and Management* 51 (5), 701–716

Carr, E. S. (2003). Rethinking empowerment theory using a feminist lens: The importance of process. *Affilia* 18 (1), 8–20.

Chawla, L. & Heft, H. (2002). Children's Competence and the Ecology of Communities: A Functional Approach to the Evaluation of Participation. *Journal of Environmental Psychology* 22 (1/2), 201–216.

Flyvbjerg, B. (1998). Habermas and Foucault: Thinkers for a Civil Society? *British Journal of Sociology* 49 (2), 210–233.

Freire, P. (1970). *Pedagogy of the Oppressed.* New York: Continuum.

Giddens, A. (1998). *Sociology.* 3. Cambridge: Polity.

Haikkola, L. & Rissotto, A. (2007). Legislation, Policy and Participatory Structures as Opportunities for Children's Participation? A Comparison of Finland and Italy. *Children, Youth and Environments* 17 (4), 352–387.

Heft, H. & Chawla, L. (2006). Children as Agents in Sustainable Development: The Ecology of Competence. In C. Spencer & M. Blades (eds.), *Children and their Environments. Learning, Using and Designing Spaces* (pp. 199–216). Cambridge: University Press.

Horelli, L. (2002). A methodology of participatory planning. In R. Bechtel & A. Churchman (eds.), *Handbook of environmental psychology* (pp. 607–628). New York: John Wiley.

Horelli, L. (2006). A learning-based network approach to urban planning with young people. In C. Spencer & M. Blades (eds.), *Children and their environments. Learning, Using and Designing Spaces* (pp. 238-255). Cambridge: University Press.

Hungerford, H. & Volk, T. (1990). Changing learner behavior through environmental education. *The Journal of Environmental Education* 21 (3), 8–21.

Innes, J. & Booher, D. (2003). Collaborative policymaking: governance through dialogue. In M. Hajer & Wagenaar, H. (eds.), *Deliberative policy analysis. Understanding governance in the network society* (pp. 33–59). Cambridge: Cambridge university press.

Jensen, B. B. (2002). Knowledge, Action and Pro-Environmental Behaviour. *Environmental Education Research* 8 (3), 325–334.

Jensen, B. B. (2004). Environmental and Health Education Viewed from an Action-Oriented Perspective: a Case from Denmark. *Journal of Curriculum Studies* 36 (4), 405-425.

Kolb, D. (1984). *Experiential Learning: Experience as a Source of Learning and Development*. Englewood Cliffs. Prentice- Hall.

Kollmuss, A. & Agyeman, J. (2002). Mind the Gap: Why do people act environmentally and what are the barriers to pro-environmental behavior? *Environmental Education Research* 8 (3), 239–260

Kyttä, M. (2003). *Children in Outdoor Contexts: Affordance and Independent Mobility in the Assessment of Environmental Child Friendliness*. Espoo: Helsinki University of Technology, Centre for Urban and Regional Studies, Publication A 28.

Lave, J. & Wenger, E. (1991). *Situated Learning: Legitimate Peripheral Participation*. Cambridge: Cambridge University Press,.

Ölander, F. & Thøgersen, J. (1995). Understanding of Consumer Behaviour as a Prerequisite for Environmental Protection. *Journal of Consumer Policy* 18 (4), 345–348.

Osbaldiston, R. & Sheldon, K. (2003). Promoting Internalized Motivation for Environmentally Responsible Behavior: A Prospective Study of Environmental Goals. *Journal of Environmental Psychology* 23 (4), 349–357.

Paloniemi, R. & Koskinen, S. (2005). Ympäristövastuullinen osallistuminen oppimisprosessina. *Terra* 117 (1), 17–32. (Participation as a Learning Process. In Finnish, Abstract in English.)

Percy-Smith, B. (2006). From Consultation to Social Learning in Community Participation with Young People. *Children, Youth and Environments* 16 (2), 153–179.

Percy-Smith, B. & Malone, K. (2001). Making Children's Participation in Neigbourhood Settings Relevant to the Everyday Lives of Young People. *Participatory Learning and Action Notes* 42, 18–22.

Raitio, K. (2008). "You Can't Please Everyone" – Conflict Management Practices, Frames and Institutions in Finnish State Forests. Joensuu: University of Joensuu, Publications in Social Sciences N:o 86.

Ryan, R. & Deci, E. (2000). Self-Determination Theory and the Facilitation of Intrinsic Motivation, Social Development, and Well-Being. *American Psychologist* 55 (1), 68–78.

Sfard, A. (1998). On Two Metaphors for Learning and the Dangers of Choosing Just One. *Educational Researcher* 27 (2), 4–13.

Siitonen, J. (1999). *Voimaantumisteorian perusteiden hahmottelua*. Oulu: University of Oulu, Department of Teacher Education. (Conceptualisation of Empowerment Fundamentals, In Finnish, Abstract and Summary in English.)

Tilbury, D. (2007). Learning Based Change for Sustainability: Perspectives and Pathways. In A.E.J. Wals (ed.), *Social Learning Towards a Sustainable World. Principles, Perspectives, and Praxis* (pp. 117–131). Wageningen, The Netherlands: Wageningen Academic Publishers.

Tippet, J., Searle, B., Pahl-Wostl, C. & Rees, Y. (2005). Social Learning in Public Participation in River Basin Management – Early Findings from HarmoniCOP European case studies. *Environmental Science & Policy* 8 (3), 287–299.

Thomas, K. & Velthouse, B. (1990). Cognitive elements of empowerment: An "interpretive" model of intrinsic task motivation. *Academy of Management Review* 15 (4), 666–681.

Thøgersen, J. (1994). A Model of Recycling Behaviour. With Evidence from Danish Source Separation Programmes. *International Journal of Research in Marketing* 11 (2), 145–163.

Wenger, E. (1998). *Communities of Practice. Learning, Meaning, and Identity.* Cambridge: Cambridge University Press,.

Wildemeersch, D., Jansen, T., Vandenabeele, J. & Jans, M. (1998). Social Learning: a New Perspective on Learning in Participatory Systems. *Studies in Continuing Education* 20 (2), 251–265.

Zimmerman, M. (1995). Psychological Empowerment: Issues and Illustrations. *American Journal of Community Psychology* 23 (5), 581–599.

In: Handbook of Environmental Policy
Editors: Johannes Meijer and Arjan der Berg

ISBN 978-1-60741-635-7
© 2010 Nova Science Publishers, Inc.

Chapter 14

INCENTIVE BASED ENVIRONMENTAL POLICIES AND COLLECTIVE RESPONSE TRENDS; SPATIO-TEMPORAL PATTERNS OF LAND MANAGERS' ADOPTION OF AGRI-ENVIRONMENTAL MEASURES

Dan van der Horst
University of Birmingham, UK

ABSTRACT

A comprehensive protection and enhancement of ecosystem services cannot be limited to the management of designated protected areas. Much of the earth is privately owned and many governments are now offering incentive payments to these owners (farmers or other land managers) to adapt their management of the land in such a way as to enhance ecosystem service provision. These incentive payments are consistent with the currently dominant model of market-led approaches to environmental management. But while these payments are offered to individual land owners, their decision making is not always the simple product of economic rationality. In the European Union, various agri-environmental schemes (AES) have been implemented and studied since the early 1990s. AES are incentive-based approaches to ecosystem services provision in agricultural landscapes that are rich in semi-natural and cultural features. Farmers' decisions to enter voluntary agri-environmental schemes have been explained in the literature in a number of ways, including the availability of information. Some authors suggest that informal neighbourhood networks impact on the penetration of information through some farming communities, and indicate that certain aspects of community cohesion and social capital are key factors influencing collective attitudes with regards to farm management, especially where management relates to the provision of novelty 'products' such as non-market goods and services. To date little empirical work has been carried out to estimate the extent and relative importance of farmer networks on entry into an AES. This study sets out to detect possible relationships between farm locations and farm entry time for the entrants into a specific agri-environmental scheme; the Environmentally Sensitive Area (ESA) Scheme in Scotland. Using quantitative measures based on Hagerstrand's model of innovation-diffusion as a spatio-temporal model, and GIS as a visualisation tool, clear spatio-temporal uptake patterns are found at different spatial scales and in different types of rural spaces. These findings are critically discussed.

1. INTRODUCTION

Financial incentives are currently the most widely used policy instrument for conserving biodiversity in traditional agricultural landscapes in Europe. The design of such incentive policies (also known as agri-environmental schemes, or AES) requires (a) knowledge of potential biodiversity conservation gains from different land management decisions and (b) understanding of what it takes to motivate farmers (and other land managers) to participate in these policies. In recent years there has been a growing body of literature about the conservation gains from agri-environmental schemes, including both empirical studies to asses the impacts on the ground of schemes that have been in place for a number of years (e.g. Klein and Sutherland, 2003) and simulation or modelling studies to asses the potential impacts of proposed or ongoing schemes (e.g. van der Horst and Gimona, 2005; van der Horst, 2006; 2007). Also on farmer behaviour there has been a growing body of literature, focussing strongly on the factors influencing the choices made by *individual* farmers (discussed in more detail below). These studies employ either survey methods to assess behavioural patterns (of representative samples) of the farmer population, or qualitative approaches to develop a more in-depth understanding of values and motivations. This paper sets out to explore one aggregate aspect of farmer behaviour in relation to these policies, namely the pattern of response over space and time. However first we will provide an overview of the literature on farm adoption of agri-environmental schemes and we will introduce our case study.

2. WHY FARMERS JOIN AES

An extensive literature exists on farmers' motivation and factors influencing farmers' decision to join such agri-environmental schemes. Entry decisions have been found to be highly influenced by the consequences for farm income and farmers with fewer financial constraints are much more likely to be influenced by potential conservation considerations (Morris et al., 2000). The degree to which the scheme prescriptions fit the existing farm system, determines its 'popularity' with farmers, i.e. farmers are more likely to enter the scheme when only minimum effort would be required (Wilson, 1997a). Wynn et al (2001) observed that the ESA Scheme in Scotland (discussed in more detail below) favoured extensive farms as the prescriptions with minimal opportunity costs (eg stock management, herb-rich grassland, woodland, and wetland) applied to these extensive farms; in addition some of these prescriptions included improved fencing, reducing the input costs to agriculture.

Nevertheless, conservation interest can be an important motivation to join a scheme for some farmers. Morris and Potter (1995) subdivided adopters to the ESA scheme into passive and active, passive farmers joining for business reasons and active ones more motivated by the environmental objectives of the scheme. A number of other farmer and situational factors have been shown to influence entry decisions. Potter and Lobley (1992) found that farmers without successors were most likely to be disengaging from full-time agriculture and extensifying. A CEAS (1997) analysis of ESAs found the main factors determining entry to

ESAs were the level of payments and the changes required to the farm; they also found entrants to be marginally older and more environmentally concerned than non-entrants.

2.1. The Role of Information

The importance of the source of information on farmers decision to enter is stressed by a number of authors. McHenry (1995) found that farmers tend to suffer from a lack of information and make decisions based upon rumour. In a study of the Mourne Mountains and Slieve Croob ESA (Ireland), Moss (1994) found that channels of information differed between non-participants and participants with the latter most likely to have heard about the scheme through County Advisory Staff and public meetings. Burton *et al.* (1997) however, found that probability of adoption of organic farming increased if the farmer obtained information primarily from other farmers. Skerratt's (1994) analysis of uptake in Breadalbane ESA confirms that both the farm advisors and the neighbourhood network play an important role in the adoption decision. In a study of the uptake of the ESA scheme in the Cambrian Mountains in Wales, Wilson (1997b) has found that some farmers were perceived as 'leaders' by their neighbours and their participation can accelerate the uptake by other farmers in the area. Wynn *et al.* (2001) found that some farmers have a 'wait and see' approach, postponing their decision until they learn of other farmers' experiences with the scheme.

It is important to note that *adoption or non-adoption of an ESA cannot be reduced to a 'single dichotomous decision taken at one point in time'* (Skerratt, 1994). Some ESA studies (e.g. Morris and Potter, 1995) have been carried out from the viewpoint that adoption is a process. This approach is not new as Taylor and Miller (1979) found that formal communication and informal communication both play an important role at different stages in the adoption process. Black and Reeve (1993) provide an in-depth literature review as to whether adoption/diffusion theory applies to the introduction of conservation practices. This study however is not so much concerned with theoretical considerations but with the empirical observations and practical explanation of spatio-temporal patterns of uptake.

2.2. Temporal and Spatial Trends of Uptake

Temporal trends of uptake have received attention in a number of studies. Fischer *et al.* (1996) concluded that the rate at which information becomes available is a critical factor in the adoption decision and the speed of that decision. Moss (1994) found that 40% of the farmers in the Mourne Mountains and Slieve Croob applied to the scheme as soon as it was brought to their attention; for the remainder there was a 6 months time lag. Evans (1997) examined the uptake levels for the ESA scheme, the Countryside Stewardship scheme (CSS) and the Farm and Conservation Grant Scheme (FCGS). With the CSS he found the uptake rates declining since the scheme's inception, and he suggests this is partly explained by selective targeting of large farms and active early promotion. Skerratt (1994) suggest that farmers in Breadalbane ESA would not have joined (or taken much longer to join) had the agricultural advisor not possessed his negotiating skills and the ability to appreciate both the farming and the conservation objectives of the ESA. The work of Morris and Potter (1995) leads one to expect differences in the motivation of early and late adopters. Fischer *et al.*

(1996) suggest that a Beyesian learning model would be well suited to account for the time lag between when farmers first hear about an innovation and the time when they adopt it, and the model can be used to observe laggards and partial adopters.

Empirical work on the spatial patterns of uptake of agri-environmental schemes is less common. Battershill and Gilg (1996) have studied the role of geographical location on the uptake of a number of voluntary conservation schemes (including ESAs) in south-west England. Location and attitudes were found to be far more important as determinants of adoption than socio-economic characteristics. Wilson (1997a) found that farmers located in the centre of ESAs were more likely to sign an agreement than those farmers located on the boundaries.

The above literature presents a variety of factors which influence the entry decision and the speed with which that decision is taken. Many of these factors differ from place to place and also over time. It is therefore expected that if entry is influenced by such factors, this may manifest itself in spatial and/or temporal patterns of uptake. So far, little empirical research has been carried out to investigate such patterns. The aim of this study is to explore and (where possible) to explain spatio-temporal patterns of uptake.

3. Case Study: The ESA Scheme

The Environmentally Sensitive Area (ESA) Scheme is an agri-environmental scheme which aims to protect flora and fauna, geological and physio-graphical features, buildings and other objects of archaeological, architectural or historic interest in an area or protect and enhance the natural beauty of that area through payments made to farmers to maintain or alter their current practices. In 1997, the Scottish Office Agriculture, Environment and Fisheries Department (SOAEFD) commissioned a study on the agricultural and socio-economic impacts of the ESA Scheme in Scotland (Crabtree et al 2000). One aim of this project was to investigate the delivery mechanisms of the policy and to identify the factors determining participation in the scheme. As part of this project, Wynn et al. (2001) investigated the factors which distinguish entrants from non-entrants and early entrants from late or potential entrants. Amongst others, they found that information was a significant explanatory variable influencing both the probability and rate of uptake. This study investigates further the importance of information in determining uptake to the ESA scheme, focussing on networking between farmers and the spatio-temporal clustering of uptake.

In the next section, the ESA scheme is analysed especially with respect to how features of the scheme may influence neighbour networks and clustering of uptake.

The ESA Scheme applies only to designated Environmentally Sensitive Areas (ESAs). These are often areas characterised by traditional agricultural systems and especially areas where this traditional agriculture is threatened by rural depopulation or agricultural intensification (Wilson, 1997b). The ESA Scheme is voluntary, offering incentive payments to entice farmers to join. The ESA Scheme is a European scheme, first implemented in the UK in 1986. In 1996, 43 ESAs had been designated in the UK, 10 of which were in Scotland.

The ESA scheme imposes certain activities ('prescriptions') on the farmer in exchange for financial compensation. These prescriptions are divided into two categories: tier 1 and tier 2. Tier 1 prescriptions are a formalisation of what may be considered good farming practice; tier 2 payments are made at higher rate and require farmers either to continue farming in a particular way or to modify their activities in some way. There are three local agencies involved in delivery of the ESA scheme: the Scottish Agricultural College (SAC), the Farming and Wildlife and Advisory Group (FWAG) and the Scottish Executive Rural Affairs Department (SERAD). Although there are local idiosyncrasies in different ESAs, FWAG input a conservation audit of a farm, SAC are principally involved in promotion and drawing up a farmer's application, and SERAD evaluate applicants and monitor entrants.

Having agreed in principle to join the scheme, a farmer will submit an application to the local SERAD office. If the application is acceptable, SERAD will return the application to the farmer, who must return a signed copy to SERAD before s/he is officially an entrant.

3. METHODOLOGY

The methodology consists of three stages. Firstly, some quantitative indicators of spatio-temporal clustering are applied to each ESA. Secondly, the ESAs for which these indicators suggest a possible occurrence of neighbour networks, are subjected to a closer, visual study. A Geographical Information System (ArcView 3.1) is used to display the farm locations on a map (a farmers's postal address is translated into x,y coordinates using Ordnance Survey's Address-Point database) with the time of entry printed at each farm's location.

Observations of spatial clustering of entrants can be an indication of neighbour network effect, but only once other explanatory factors have been accounted for. First of all, some entrants may be spatially clustered simply because their farms happen to be close together. Only when entrants are spatially clustered relative to the spatial distribution of non-entrants can a localised decision to enter be assumed. These localised entry decisions do not necessarily have to be due to a neighbour network effect. It is possible that these farmers do not communicate with each other but have common characteristics which increase the likelihood of uptake, such as a relatively high amount of rough grazing. Only when a local cluster of entrants have similar characteristics as many non-entrants elsewhere can a neighbour network effect may be assumed. Unfortunately data on the locations and characteristics of all non-entrants were not available for this study so spatial clustering of entrants could not be used to identify neighbour networks. Temporal clusters of uptake can be much better indicators of neighbour network effects. There are two reasons for this. The farmer can freely choose the time of entry, whereas he/she cannot chose the farm location. This makes entry time a stronger indication of the decision making process a farmer goes through. And unlike spatial distance, temporal distance has a set direction, suggesting who may influence who. The latter can aid in the identification of a network as it suggests who may be a leader and who may be followers.

A simple measure of temporal clustering is the coefficient of variation of the entry time, i.e. the average entry time, divided by the standard deviation. More close-knit communities are expected to have a more temporally clustered uptake. Another indication is the uptake trend over time. If all farmers are perfectly informed from the onset and are completely independent in their response to the scheme, a linear cumulative entry trend can be envisaged. In reality information provision is more likely to be patchy, for example in the case of geographical promotional targeting. With a patchy information regime where farmers are unaffected by each other's decisions, a linear cumulative trend can still be expected but it will have stepwise interruptions. If farmers do influence each other, a more curved cumulative trend of uptake can be expected with a few early entrants to begin with, then a rapid increase as a lot of farmers follow suit then tailing off again with just some laggards still entering. This is the typical 'elongated S-shape commonly associated with innovations' (Hägerstrand, 1967, p 57. see also Rogers and Shoemaker, 1971). If more than one network or group of farmers are captured within the same data set and the response of these groups ('adoption wave') is sufficiently temporally distinct, the curved trend may repeat itself. In reality we can expect a lot more 'noise' in the uptake trends over an entire ESA. Several adoption waves can overlap each other temporally, each with a different size s-shaped curve. Also we can anticipate horizontal 'breaks' in the uptake curve for those ESAs which have an application window (see above), possibly followed by a relatively steep increase when the backlog of applications (submitted during the closed period) is processed.

The best indicators for a neighbour network effect are measures of spatio-temporal clustering. The choice of method for the measurement of spatio-temporal patterns of entry depends on the characteristics of the dataset. The most conventional methods to measure spatio-temporal clustering might be some type of cluster analysis, possibly preceded by a semi-variate approach. These methods are deemed unsuitable in this case because both the distance over which farmers may influence each other and the 'following time' (between adoption by leader and follower) are expected to vary strongly and unpredictably, with temporally or geographically out-lying farms creating a 'noisy' data set. An additional problem is that single observations in each dataset cannot be assumed to be independent of each other (for example promotional visits can be clustered in time). A simulation approach is therefore needed to test for the significance of association between temporal and spatial pattern of uptake. A simulation method appropriate for this type of problem is the Mantel test (see Sokal and Rohlf, 1995). The Mantel test is used to estimate the association between two independent dissimilarity matrices and test whether the association is stronger than one would expect from chance. This method permits to test for association between a 'euclidean distance matrix' (X_{ij}) and a 'difference in entry time matrix' (Y_{ij}) for any set of farms ($N>7$) under the null hypothesis that no association exists. The observed Z statistics, for the test is computed by multiplying two matrices with each other, and summing the results:

$$Z_o = \sum_{i=1}^{n-1} \sum_{j=i+1}^{n} X_{ij} Y_{ij} \tag{1}$$

After a random permutation of the elements of one matrix, the Z value is re-calculated. By repeating this procedure 1000 times, a reference distribution of simulated values, Z_s, is obtained. Z_o (Equation 1) is then compared to the distribution of Z_s to assess if the observed level spatio-temporal clustering is stronger ($Z_o > Z_s$) or weaker ($Z_o < Z_s$) than the level of spatio-temporal clustering that is likely to occur by chance.

4. RESULTS

4.1. Statistical Indicators of Clustering, A Comparison between ESAs

Table 1 lists the results of the statistical analysis for each ESA. The coefficient of variation records the degree of temporal clustering about the average entry time. The higher the values for the coefficient of variation, the stronger the temporal clustering is. Mantel p measures the level of spatio-temporal clustering in comparison to a random distribution, with $p>0.5$ indicating stronger clustering and $p<0.5$ indicating weaker clustering than what could be expected by chance. Table 1 shows that the Western Isles have the fastest average entry time from the start of the scheme, and the Shetlands the slowest. With the exception of the Shetlands, temporal clustering is or exceeds 0.5 – tough it never exceeds 0.74. Spatio-temporal clustering can be much stronger, reaching .098 or 0.99 for four out of ten ESAs. However for four other ESAs it is significantly below what may be expected from chance (0.23-0.31).

Table 1. Comparison between the 10 Scottish ESAs.

Environmentally Sensitive Areas (ESA) in Scotland	date first entrant	Number of entrants	Average time of entry since start of the scheme (days)	Coeff. of Var.	Mantel P**
Western Isles Machair	14/06/90	172	396	0.58	0.98
Argyll Islands	21/12/94	168	458	0.73	0.99
Breadalbane	27/08/88	83	499	0.71	0.99
Central Southern Uplands	25/10/93	155	591	0.62	0.28
Central Borders	27/07/94	72	602	0.50	0.86
Stewartry	05/05/89	151	609	0.59	0.28
Cairngorm Straths	30/09/94	76	609	0.55	0.99
Western Southern Uplands	16/12/93	100	624	0.65	0.31
Loch Lomond	07/05/88	34	632	0.74	0.76
Shetland Islands	23/06/94	172	776	0.38	0.23

The uptake trends can be easily examined graphically by plotting the cumulative number of farmers joining over time. Breadalbane seems to have one major continuous S-shaped curve. By comparison, the four southern ESAs seem more linear and continuous, albeit with one or two stepwise breaks. Multiple curves can be identified in the irregular uptake patterns of Western Isles Machair, Cairngorm Straths, Argyll and especially the Shetlands. For the Shetlands, which have by far the lowest spatial and spatio-temporal clustering scores in Table 1, two clearly separate S-shapes can be identified. The graph also shows a horizontal stretch before and after the second S-shape, which represents an application window. In other words Shetland's scores are low because they are the aggregation of two separate uptake events over time.

4.2. Visual Identification of Mini Clusters within an ESA

Following on from the statistical indicators of clustering calculated across all entrants of each individual ESA, four ESAs are selected for closer, 'eye-ball' examination of the patterns of uptake by zooming in on the GIS maps on screen: Western Isles Machair, Cairngorm Straths, Argyll and the Shetlands. The Shetlands are selected because of the outlying and conflicting clustering indicators. The others three ESAs are selected on the basis of their high p score (significant clustering). Breadalbane was at first also selected on that basis but a visual analysis revealed only one, relatively minor cluster. Table 2 contains an overview of the main clusters identified in these ESAs. A number of clear clusters are identified on a number of the smaller Argyll islands (Coll and Tyree, Colonsay, Lismore, see figure 1), parts of the Western Islands (northern Barra) and valleys of the high Cairngorm mountains and to a lesser extent Breadalbane. Localised 'micro-clusters' can be identified in many cases (Upper Strath Spey, Loch Rannoch). The selected ESAs are discussed in turn below.

Table 2. Examples of small/local clusters found within ESAs by visually analysing the GIS maps.

ESA	Area of cluster	Farmers in entry cluster / total no. of farmers in area	Entry days
Cairngorm Straths	Glen Avon	10/10	941-1113
	Dee valley	5/7	336-465
Argyll Islands	Coll	7/8	46-175
	Tyree	15/16	769-1029
	Colonsay	5/8	62-86
Breadalbane	Strath Ardle	9/14	208-364
Western Isles Machair	Barra/Eolaigearraidh	4/5	883-878
	Barra/Aird Mhor	11/11	105-457
	South Uist/...	6/13	569-650
Shetland Islands	Fair Isle	8/9	523-564
	Mainland south of Stove	8/10	1103-1208

The Shetlands

The uptake pattern on the Shetlands is quite irregular and is subdivided into three subsets. The first subset (days 0-707) is actually significantly unclustered (p = 0.09). The exceptionally slow rate of uptake in the first year (13 entrants) is blamed by the local SERAD advisor on negative local media publicity at the scheme launch.

The large time lag at the beginning of the scheme and the application window time lags make the dataset for the Shetlands as a whole seem rather unclustered.

Farmers belonging to subset 1 are scattered all over the Shetland Islands with the exception of the southern part of Mainland (south of Fladdabister). One notable spatio-temporal cluster is identified within subset 1: this is a cluster of 8 farms (5% of all entrants at 1997) on Fair Isle, a remote island lying halfway between Shetland and Orkney. This cluster clearly suggests a collective approach or response by the farmers on this very remote island, who are responding in a manner that is quite contradictory to the pattern found at the same time in the rest of the Shetland islands.

Subset 2 (days 708-1013) is significantly clustered at p = 0.05. This subset contains a third of all entrants in Shetland. However more than half the entrants from this subset live on Unst or on the western part of Mainland (west of Aith) where they make up half the total number of entrants. That this subset represents a cluster could be explained by a rush of entrants after the initial unfoundedly negative publicity subsided – yet their peculiar spatial pattern (Unst and western Mainland) suggests the possibility of neighbourhood networks. Subset 3 (day 1082-1226) is randomly distributed (p = 0.5). Of the 37 entrants in this subset, 15 are concentrated on the southern part of Mainland (south of Lerwick) where they make up about 50% of all entrants. The remaining 60% of this subset are found scattered over the Shetland islands with the exception of Yell and the north-eastern part of Mainland.

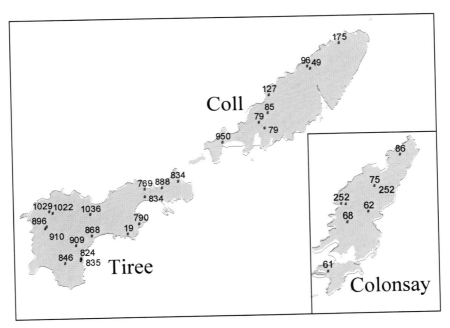

Figure 1. Entrants to the ESA scheme, showing number of days in the scheme, on some of the smaller Argyll islands.

Argyll Islands

On the Argyll islands uptake patterns are affected both by an application window and spatial targeting of promotion, especially on the smaller islands. The latter especially gives rise to spatio-temporal clusters, accentuated by local idiosyncrasies of the island community networks eg on Coll and Tiree. Isolated from other islands and only 4 km apart, the two island communities respond quite differently to the scheme (see Table 2 and Figure 1). According to the farm advisor, the difference in entry time of almost three years was a result of the more sceptical attitude of the crofters on Tiree who only joined after the prescribed time of haycutting had been adjusted in their favour. On Coll the farmers are not traditional crofters and were much quicker to join. The farming communities on both islands responded as one group (clustered in time) with only one outlier. On Coll this outlier (entry day 950) is a farm run by RSPB (the Royal Society for the Protection of Birds – a national conservation NGO), indeed an outsider to the community. There was no information on the outlier (entry day 19) on Tiree.

There is also some evidence to suggest that farm size can be a predictor to the time of entry. The first five entrants (day 62-86) on Colonsay have relatively big farms (400-1300 ha), the two entrants who joined at day 252 have around 25 ha while the last joiner (day 658) has only 7 ha. Clearly the financial benefits of joining are higher for larger farms, whilst the transaction costs (i.e. the hassle of form filling) do not differ so much between large and small farms.

Western Isles

For the Western Isles, larger clusters are only identified at the nothern tip of Barra where two crofting communities only 3 km apart join the scheme more than 400 days apart. There was no information from the local advisor to explain this pattern. Many smaller clusters can also be identified. One example is the six entrants living at the southern most tip of South Uist, three of which entered between day 14 and 36 (subset 1) and the other three entered between day 636 and 658 (subset 4). For subset 2 and 3 –relatively unclustered according to the Mantel test- it is indeed more difficult to find such small patterns. Visual analysis makes it clear why subset 4 is significantly clustered ($p = 0.01$). Almost 60% of the entrants of this subset (days 548-715) are found on South Uist.

Cairngorm Straths

The significant clustering ($p = 0.99$) in the entrants dataset for the Cairngorm Straths ESA can be explained by both large and small clusters. Surrounded by Britain's highest mountain range, these 3 valleys can be expected to represent separate farming communities. However the explanation of the difference in uptake time may also lie in the mode of operation of the regional advisors. The three valleys belong to three different regional offices of farm advisors. The advisor for Strath Avon explained that he made use of existing farm networks to persuade farmers to join. The highly concentrated time of entry suggests he was successful.

5. CONCLUDING THOUGHTS

The analysis presented in this chapter has yielded some strong and diverse evidence of spatio-temporal clustering of farmer decision to join a voluntary agri-environmental scheme. However drawing generic conclusions from these findings is not easy – for two different reasons of (spatial) scale, namely inter and intra ESA. With regards to the former, similar trends or their absence may be found in different local circumstances as all ESAs differ from one-another and all have their own unique circumstances and local histories. With regards to the latter, the trends found at the population level (all entrants of each separate ESA) are not always recognisable on the ground, where micro-clusters may be encountered, or larger trends may be harder to spot. This problem is generally known as the ecological fallacy, a term used to highlight the inappropriateness of extrapolating findings from a population-level study to a more individual-level study or vice-versa. This problem is of course associated with the Modifiable Areal Unit Problem (MAUP) which is encountered when an analysis is rerun at different spatial levels/scales; it tends to yield different results.

However there are some tentative conclusions we can draw from the analysis.

It is noted with respect to Table 1 that uptake in the six later ESA schemes (designated 1993-1994) is neither quicker nor more clustered than that of the four earlier ESA schemes (designated 1988-1990), suggesting favourable opinions and information do not easily cross separate (spatial) ESA designations over time .

Cairngorm Straths, Argyll, Breadalbane and Western Isles all have a positive association between location of farms and time of entry significant at the 95% level of confidence. By contrast, three of the four southern ESAs score very similarly with an observed association which is less positive than what would be expected from random ($p = 0.28$-0.31). It is tempting to hypothesise that this may have something to do with the very different socio-economic and geographical nature of the farming communities in the south which is (by comparison) highly accessible and close to major cities. These communities are much larger, and much more diverse, with more new-comers, more farmers with opportunity for off-farm earnings, more farm-diversification options and more/diverse sources of information to individual farmers. Collective or identical decision making is much less likely to take place here and spatio-temporal patterns of uptake are much less likely to be found.

The frequent occurrence of micro-clusters in the more remote ESAs suggest strongly that neighbourhood effects can be very prominent in small and remote communities. Statistical or map analysis does not yield information about the personal motivations and group/community dynamics on the ground, but two related explanations may be proffered. The first explanation is that group deliberation and collective decision making is taking place in these tight-knit communities which are remote from other communities and small enough to make people more collaborative with, reliant on and trusting towards their neighbours. Whether on small remote islands or isolated mountain valleys, these people form island communities in a physical, socio-economic and cultural sense). The second explanation relates to the role of the farm advisor in information provision and 'selling' the scheme. This too could be a major explanatory variable for the observed pattern of uptake in these remote areas where a trusted farm advisor may act as a (surrogate) 'leader'. This would be consistent with Skerratt's (1994) suggestion that some farmers would not have joined (or would have taken much longer to join) had the agricultural advisor not possessed the necessary negotiating skills. It could be hypothesised that promotion is more critical in remoter areas, because of the opportunity for exploiting neighbour networks and because of fewer opportunities for farmers to access information through a range of different channels. The impact of negative local publicity on the uptake in the Shetlands is a strong indication of the high impact of localised information provision, resulting in collective inaction by the farmers.

In the analysis presented here, the data was missing to fully account for farm type in observed spatial clusters of uptake. It would be logical to expect that neighbours are more likely to reach similar conclusions and influence each other's decision making if their farms are more similar. Wynn et al (2001) reported that farmers with more rough grazing had the tendency to join *earlier*. Since extensive farms, e.g. those associated with more rough grazing, are found in similar biophysical environments and are thus likely to be spatially related, observed *spatial* clusters of uptake are not necessarily evidence of neighbourhood networking. In short, the analysis of spatio-temporal clustering should ideally account for both biophysical farm type and the level of economic benefit offered by the agri-environmental scheme. However whilst high economic incentives may result in neighbouring farmers making identical *individual* decisions, it can also be hypothesised that in relatively close-knit communities lower economic incentives result in less individualistic responses and the stronger the role of neighbour networks become. It would require a combination of the quantitative and top-down analysis presented here (extended with data on farm typology), and local qualitative and in-depth fieldwork to really unpack the patterns of adoption of agri-environmental measures and to account more realistically and accurately for the role of neighbourhood networks. However this clustering analysis alone is a quick and useful step to identify clustering of uptake and to guide more extensive and expensive local qualitative fieldwork.

The analytical approach in this chapter is based on the classical work of Hägerstrand (1967) who explicitly presented his Innovation Diffusion Theory as a spatial process. Observing a clear occurrence of a neighbourhood or proximity effect, he concluded that for the adoption of subsidised grazing-improvement (and other cases of innovation), private communication is a far more powerful agent of diffusion than public announcement. This chapter has shown that his method is still very useful today, but that drawing hard conclusions from the findings is rather less straight-forwards.

ACKNOWLEDGMENTS

The author would like to thank Dr Alessandro Gimona for his invaluable advice with the Mantel test, Dr Gerard Wynn for his feedback on an earlier version of this chapter and the UK Economics & Social Sciences Research Council (ESRC) for fellowship funding (Res-152-27-0004).

REFERENCES

Battershill M.R.J. and Gilg A.W. (1996) New approaches to creative conservation on farms in South-west England. *Journal of Environmental Management*, 48, 321-340.

Burton M., Rigby D. and Young T. (1997). *Sustainable Agricultural Technologies: A Quantitative Analysis of the Adoption Process*. Paper presented at the Agricultural Economics Society Annual Conference, Edinburgh, 21-24 March, 1997.

CEAS (1997). *Economic Evaluation of Stage II and III ESAs*. Final report for MAFF, Wye College, Ashford.

Crabtree, J. R., Thorburn, A., Chalmers, N., Roberts, D., Wynn, G, Barron, N., Macmillan, D. and Barraclough, F. (2000). Socio-economic and agricultural impacts of the

Environmentally Sensitive Areas (ESA) Scheme in Scotland. SOAEFD Contract Report. Macaulay Land Use Research Institute.

Evans N.J. (1997). Something Old, New, Borrowed and Blue: The Marriage of Agriculture and Conservation in England. In: Ilbery, B., Chlotti, Q. and Rickard T. (Eds.). Agricultural Restructuring and Sustainability: A Geographical Perspective. Sustainable Rural Development Series No. 3, CAB International.

Fisher A.J., Arnold A.J. and Gibbs M. (1996). Information and the Speed of Innovation Adoption. American Journal of Agricultural Economics, 78, 1073-1081.

Gimona A., and van der Horst, D. (2007), Mapping hotspots af multiple landscape functions:a case study on farmland afforestation in Scotland, Landscape Ecology, 22:1255-1264.

Hägerstrand T. (1967) Innovation Diffusion as a Spatial Process. University of Chicago Press. (a translation of Hägerstrand T. (1953) Innovationsförloppet ur korologisk synpunkt, Gleerup, Lund, Sweden).

Kleijn D., Sutherland W.J. (2003) How effective are European agri-environmental schemes in conserving and promoting biodiversity? Journal of Applied Ecology 40, 947-969.

McHenry H.L. (1995). Farmers' Interpretation of their Situation: Some Implications for Environmental Schemes. In: Corpus A.K. and Marr P.J. (Eds.). Proceedings of the 35th EAAE Seminar.

Morris C. and Potter C. (1995). Recruiting the New Conservationists: Farmers' Adoption of Agri-environmental Schemes in the UK. Journal of Rural Studies, 11, 51-63.

Morris, J Mills, J, and Crawford, I, M (2000), Promoting farmer uptake of agri-environment schemes: the Countryside Stewardship Arable Options Scheme, Land Use Policy 17:24-254.

Moss J.(1994) A baseline assessment for a new ESA: the case of the Mourne Mountains and Slieve Croob. In: Whitby, M.C. (Ed.) Incentives for Countryside Management. CAB International, Wallingford, UK.

Pampel F. and van Es J.C. (1977). Environmental Quality and Issues of Adoption Research. Rural Sociology, 42, 57-71.

Potter C. and Lobley M. (1992). The conservation status and potential of elderly farmers: Results from a survey in England and Wales. Journal of Rural Studies, Volume 8, Issue 2, Pages 133-143.

Rogers E.M. and Shoemaker F.F (1971). Communication of Innovations. New York: Free Press.

Skerratt, S.J. (1994). Farmers's Adoption and Non-adoption of Agri-Environmental Incentives: The Case of Breadalbane ESA, Scotland. In: Jacobson B.H., Pederson D.E., Christensen J. and Rasmussen S. (Eds.). Farmer Decision Making – A Descriptive Approach. Proceedings from the 38th EAAAE Seminar, Copenhagen.

Sokal R.R and Rohlf F.J. (1995). Biometry (3rd Ed) W.H.Freeman and Company, New York. p 815.

Taylor D.L. and Miller W.L. (1979). The Adoption Process and Environmental Innovations: A Case Study of a Government Project. Rural Sociology, 43, 634-648.

Van der Horst D. and Gimona A. (2005). Where new farm woodlands support Biodiversity Action Plans: a spatial multi-criteria analysis. Biological Conservation 123, 421-432.

Van der Horst D. (2006). A prototype method to map the potential visual amenity benefits of new farm woodlands. Environment and Planning B 33, 221-238.

Van der Horst D. (2007). Assessing the efficiency gains of improved spatial targeting of policy interventions; the example of an agri-environmental scheme. *Journal of Environmental Management* 85, 1076-1087.

Wilson G.A. (1997a). Selective Tartgeting in Environmentally sensitive Areas: Implications for farmers and the Environment. Journal of Environmental Planning and Management, 40, 199-215.

Wilson G.A. (1997b). Factors Influencing Farmer Participation in the Environmentally Sensitive Area Scheme. *Journal of Environmental Management* 50, 67-93.

Wilson, G A, and Hart, K, (2000), Financial imperative or conservation concern? EU farmers' motivations for participation in voluntary agri-environmental schemes. *Environment and Planning A* 32(12):2161–2185

Wynn, G. Crabtree, R. and Potts, J. (2001) Modelling farmer entry into the ESA scheme in Scotland. *Journal of Agricultural Economics* 52 (1): 65-82.

In: Handbook of Environmental Policy
Editors: Johannes Meijer and Arjan der Berg

ISBN 978-1-60741-635-7
© 2010 Nova Science Publishers, Inc.

Chapter 15

CAPITALISM, STATE, AND ENVIRONMENTAL MOVEMENTS: AN ANALYSIS FROM POLITICAL ECONOMY PERSPECTIVE

Md Saidul Islam[*]

Division of Sociology, Nanyang Technological University
Singapore

INTRODUCTION

Scholars of political economy view that environmental problems are deeply embedded in the reified nature of capitalism. The inherent nature of capitalism is exploitation of labour and nature with a view to maximizing profits. As capitalism expands, different legitimizing agencies for the capitalist class also emerge. State is one of such agencies that, to Marxist sense, serves as *accumulation* and *legitimation[+]*. In the discourse of development, it has been made widely accepted that development activities must go on inspite of its severe environmental costs. "All natural resources now became strategic geo-power asset to be mobilized, not only for growth and wealth production, but also for market domination and power creation. To resist growth is not only to oppose economic prosperity, it is to subvert the political future, national interest, and collective security for the nation state!" (Luke 1999, p. 125).

In this context, many believe that ecological consideration can be ignored, or at best, given only meaningless symbolic responses, in the quest to mobilize as many of earth's material resources as possible. Having more material wealth or economic growth in one

---

[*] Email: msaidul@ntu.edu.sg
[+] This may not be applicable to all the nation states, especially so-called third world countries. State can be explained as much more than an agency of legitimation. At the same time, it needs/seeks legitimation within hegemonic ideology [which it produces and protects, and perpetuates]. Marx explicitly talks about America and its capitalist expansion. See, Marx, Karl, *The Communist Manifesto*. In case of Canada, see Bannerji, Himani (1996). 'On the Dark Side of the Nation: Politics of Multiculturalism and the State of Canada' in *the Journal of Canadian Studies*, 31(3): 103-128; Whitaker, Reg (1977). "Images of State in Canada", in Leo Panitch (ed.). The *Canadian State: Political Economy and Political Power*. Toronto: University of Toronto Press.

place, like a particular nation state, means not having it in another places—namely rival foreign nations. It also assumes that material scarcity is a continual constraint; hence, all resources, everywhere and at any time, must be subject to exploitation (Luke 1999). Roy Rappaport (1993) calls it "subordination of the fundamental to the contingent and instrumental" (P. 298).

Ecological problems are deeply rooted in this ideological underpinning of capitalism. Environment movements emerged out of this environmental problems initiated by capitalist economy. Here we are not suggesting that all environmental movements are anti-capitalist and anti-accumulation. There are many environmentalists who would hesitate to define their position clearly as anti-capitalist. The chapter examines the root cause of environmental problems by showing the interconnectedness between capitalist mode of production and environmental problems. Using political economy perspective, we will also examine to what extent the environmental movements/NGOs address those issues, and what kind of relationship the environmental movements present today with the capitalist enterprises marked by globalization.

CAPITALISM, STATE AND THE ENVIRONMENTAL PROBLEMS

A careful reading of Marx and Engels' work leads to the realization that their political economy, firmly grounded on materialist premises, contains important theoretical categories and methodological guidelines for the theoretical analysis of the determinants of the current ecological/environmental predicament, and for the development of a Marxist ecology based on ecological principles central to Marxist theory (See, Gimenez 2001; Bookchin 1980).

In *Communist Manifesto*, Marx explicates the rise and social effects of capitalism. Before the bourgeoisie rose to prominence, society was organized according to a feudal order run by aristocratic landowners and corporate guilds. With the discovery of America and the subsequent expansion of economic markets, a new class arose, a manufacturing class, which took control of international and domestic trade by producing goods more efficiently than the closed guilds. With their growing economic powers, this class began to gain political power, destroying the vestiges of the old feudal society, which sought to restrict their ambition. According to Marx, the French Revolution was the most decisive instance of this form of bourgeois self-determination. Indeed, Marx thought bourgeois control so pervasive that he claimed, "*The executive of the modern State is but a committee for managing the common affairs of the whole bourgeoisie*" (p. 16).

An interpretation of the Marxist theory of state as claiming that the state merely acts on the direct instructions of the bourgeoisie is a crude caricature of the concept of the modern state as 'a committee for managing the common affairs of the whole bourgeoisie,' a caricature which fails to distinguish between the state acting on *behalf* of the bourgeoisie and its acting on their *behest*. As Ralph Miliband (1973) has put it:

> The notion of common affairs assumes the existence of particular ones; and the notion of whole bourgeoisie implies the existence of separate elements, which make up that whole. This being the case, there is an obvious need for an institution of the kind they [Marx and Engels] refer to, namely the state; and the state *cannot* meet this need without enjoying a certain degree of autonomy. In

other words, the notion of autonomy is embedded in the definition itself, is an intrinsic part of it (p. 85).

For the state to act only at the behest of particular segments of the bourgeoisie would be dysfunctional to it managing the common affairs of that class, a 'relative autonomy'. A crude economistic interpretation of the state makes it in fact impossible to understand the real functions of the state performs for the capitalist class. And within this broad explanation and operation of the state, the basic functions of the capitalist state are *accumulations* and *legitimation* (Whitaker 1977).

There are debates whether all nation-states fall into this trap of Marxist explanation. However, American culture, and Western culture in general, may be characterized as the culture of capitalism, or more specifically corporate capitalism. The core premise of the culture of corporate capitalism is that commodity consumption is the source of well-being. Part of this corporate capitalism is the creation of so-called "modernity", which is a complex set of beliefs that came to dominate European culture in the modern era and subsequently spread to the Americas, Africa, and Asia via colonialism and other manifestations of European power. The worldview of modernity includes beliefs about autonomy of individuals, the power of science and technology, the desirability of increased consumption, and the inevitability of progress (Said 1993; Conca and Geoffrey 1998).

Dennis Pirages (1998) describes modernity as "dominant social paradigm"- a core set of ideas that shapes our understanding in ways most people never question. He sees underlying technological and economic forces as creating the dominant social paradigm of any given era. The paradigm in turn legitimizes the prevailing technological, economic, political and social practices, which started from Industrial Revolution. To him, the paradigm is unsuited to a world of resource scarcity and environmental vulnerability.

The central players in the culture of capitalism are the capitalist, the laborer, and the consumer, each operating according to rules largely orchestrated by the capitalist and enforced by the nation-state. The culture of corporate capitalism requires perpetual growth of material consumption, and hence ever-expanding exploitation of the world's resources and peoples. 'Capitalism has within its inherent logic always contained a motivation to ruthlessly expand…Since its inception the logic guiding capital has been to grow or die. This mentality and material manifestation has permeated throughout its history' (Black Star North Zine 2001, p. 1).

Successful operation of the culture of capitalism compels that consumers be segregated or masked from the consequences of their lifestyles on the laborer, on the environment, and on the ways of life of those whose degradation makes such consumption possible. Profit in a capitalist culture comes largely from the capitalist's control of both the surplus value of labour and the exploitation of nature. There is an inherent tendency of laborers to resist the discipline imposed on them by capitalists. Operation of the capitalist system results directly in a growing concentration of wealth in fewer and fewer hands (and hence in increasing polarization of wealth in society). As V.I. Lenin (1975) states:

> By destroying small-scale production, capital leads to an increase in productivity of labour and to the creation of a monopoly position for the association of the big capitalists. Production itself becomes more and more social-hundreds and thousands and millions of workers become bound together in a

regular economic organism, but the product of this collective labour is appropriated by a '*handful of capitalists*' (emphasis added) (p. 455).

As corporate capitalism has developed, the organization of capital and how it is controlled have evolved where just a few corporations now control vast wealth. The capitalist class exerts its growing power to direct social, political, economic, and cultural relations, in ways that will serve its class interests (Marx 1998).

A dominant historical trend has been the growing integration of the global economy (*globalization*), to the extent that events in one area of the world have repercussions in others. The role of the nation-state is being replaced by new institutions, most importantly the transnational corporation (or TNC) (Strange 1996). Capitalists have created international organizations (e.g., the World Bank and the IMF) and pacts (e.g., WTO and NAFTA) ostensibly to aid the "development" of impoverished peoples and countries, especially through "free trade"; but these mechanisms essentially serve to accumulate more capital in the hands of the wealthy, while imposing trouble on peripheral peoples and environments (Black Star North Zine 2001).

Democracy, as a system of government, has been largely superseded by the operation of corporate capitalism; the principle of one person, one vote, has largely been replaced by a system where money holds sway. In order to maintain corporate capitalism, the modern state must convince its populace that they and the capitalist class share a common destiny. Most of the major problems faced by countries in the periphery, such as poverty, hunger, and environmental destruction, are exacerbated by population growth. The spectre of population growth is a scheme used in the culture of capitalism to shift the blame for global problems to their victims, and to obscure a greater cause, capitalism's perpetual and uneven economic growth.

The evolution of agriculture in the culture of capitalism is characterized by the steadily increasing concentration of agricultural wealth (land and factors of production), and the growing dependency of the many on the few. Programs of so-called "food aid" are ways that the state funnels tax dollars to agribusiness, increases the influence of food aid organizations, and promotes the ruin of small, local food growers. The fact that people are starving to death because they lack the resources to grow their food, or the money to buy it, is obscured by calling starvation "malnutrition," and treating it as a medical problem. Consumption patterns, and even eating habits, are moulded largely to fuel economic expansion and maintain the society of perpetual growth. Every culture or age has its characteristic illnesses; for the culture of capitalism, characteristic diseases are those linked to environmental degradation, as well as to poverty and hunger on the one hand and to over-consumption on the other. The culture of capitalism has fostered the spread of alien organisms, including infectious diseases.

The cultures of indigenous peoples are vulnerable to destruction from capitalist expansion, in part because their way of life differs greatly from that in the culture of capitalism. Capitalism is revolutionary in the sense that to foster perpetual growth, it must constantly revolutionize the factors of production, promote ever-increasing consumption, and, consequently, regularly modify patterns of social, political, economic and environmental relations. The various forms of social protest such as workers organizations and strikes, national liberation, civil rights, feminist, militia, environmental, and fundamentalist religious movements can all be understood primarily as reactions to the expansion of the culture of capitalism. There exists a global environmental crisis, and corporate capitalism is the major cause. It is impossible to sustain the culture of capitalism at its present rate of consumption; the expansion of that culture to other areas of the globe will accelerate environmental collapse. Given the nature of the culture of capitalism, it is impossible to halt the destruction of the environment. According to Martha E. Gimenez (2001):

> As the world systemic nature of capitalism becomes increasingly visible, the accelerated nature of the circulation of capital and labour are creating the conditions for the emergence of regional transnational working-class organizations and movements. At the same time, the exploitation of nature and the circulation of waste, pollutants, viruses, infectious diseases, pests, plant diseases, and healthy animals and plants deliberately or unwittingly taken from their natural habitat intensify and highlight the global nature of most ecological problems. As the situation worsens at the local, regional, national, and world levels of analysis, it will call for the Marxist historical analysis of its conditions of existence and reproduction through time and will also call for the development of regulatory agencies and planning. Marxist contributions to ecology that, despite their importance and timeliness, are today largely the concern of academics will at that time become even more relevant (pp. 1-2).

Thus capitalism causes ecological destruction because it is based upon domination (of human over human and so humanity over nature) and continual, endless growth (for without growth, capitalism would die). According to Marx's prediction, since the culture of capitalism must continually destroy the environment, expand economic hardship, and create continual conflict and resistance, it must inevitably collapse (and be replaced by a socialist world government or highly localized, independent, and self-sufficient cultures).

ENVIRONMENTAL MOVEMENTS IN ADDRESSING THE ISSUES

Environmental damage has reached alarming proportions. Almost daily there are new upwardly revised estimates of the severity of global warming, ozone destruction, topsoil loss, oxygen depletion from the clearing of rain forests, acid rain, toxic wastes and pesticide residues in food and water, the accelerating extinction rate of natural species, etc., etc. 'Some scientists now believe that there may be as little as 35 years to act before vital ecosystems are irreparably damaged and massive human die-offs begin' (Meadows et al 1992).

As we have seen in the earlier section that, the root of environmental problems is the capitalist tendency of accumulation, which is inherently exploitative of human and natural resources. Wastefulness is inherent in capitalist tendency of private accumulation. Many anarchists and environmentalists see the ecological crisis or environmental problems as rooted in the psychology of domination, which emerged with the rise of patriarchy, slavery, and the first primitive states during the Late Neolithic. Murray Bookchin (1980), one of the pioneers of eco-anarchism, points out, "The hierarchies, classes, propertied forms, and statist institutions that emerged with social domination were carried over conceptually into humanity's relationship with nature. Nature too became increasingly regarded as a mere resource, an object, a raw material to be exploited as ruthlessly as slaves on a *latifundium*" (p. 40). In his view, without uprooting the psychology of domination, all attempts to stave off ecological catastrophe are likely to be mere palliatives and so doomed to failure. Bookchin adds:

> The conflict between humanity and nature is an extension of the conflict between human and human. Unless the ecology movement encompasses the problem of domination in all of its aspects, it will contribute *nothing* toward eliminating the root causes of the ecological crisis of our time. If the ecology movement stops at mere reformism in pollution and conservation control - at mere 'environmentalism' - without dealing radically with the need for an expanded concept of revolution, it will merely serve as a safety value for the existing system of natural and human exploitation (p. 43).

Since capitalism is the vehicle through which the psychology of domination finds its most ecologically destructive outlet, most eco-anarchists give the highest priority to dismantling capitalism. It's important to stress that capitalism must be eliminated because it cannot reform itself so as to become "environment friendly," contrary to the claims of so-called "green" or "natural" capitalists. This is because, as Bookchin sees,

> [C]apitalism not only validates pre-capitalist notions of the domination of nature, it turns the plunder of nature into society's law of life. To quibble with this kind of system about its values, to try to frighten it with visions about the consequences of growth is to quarrel with its very metabolism. One might more easily persuade a green plant to desist from photosynthesis than to ask the bourgeois economy to desist from capital accumulation (p. 66).

Recent years have witnessed the rapid proliferation and growth of numerous local, national, and transnational environmental movements to address the environmental issues. Most often today, they appear in the guise of what have become known as Non-Governmental Organizations (NGOs). 'These movements, representing new forms of political agency, stand at the forefront of a fundamental shift in the distribution of power- or at least they *appear* to do so' (Brosius 1999, P. 36). The questions arise, do environmental movements and NGOs* address and subvert the capitalist mode of production that is inherently exploitative and is regarded as the root cause of environmental problems? Can environmental movements avoid/evade confronting capitalist mode of production driven by maximization of accumulation? Are they used as an 'ideological device' by the capitalist as a tool of legitimation and progression/ accumulation? We will explore all these below.

ENVIRONMENTAL MOVEMENTS IN THE ERA OF GLOBALIZATION

Environmental movements are working worldwide to address the environmental predicaments. Some movements in the form of environmental NGOs are working in the international institutions with an intention to 'work within the system to change the system'. Sally Morphet (1996) talks about the detailed roles of environmental NGOs on addressing environmental issues, their impacts on the member countries of UN after World War II, and

---

* There may be differences between "environmental movements' and 'environmental NGOs'; however, in this chapter they have been used interchangeably.

on UN system itself. She explains that there was no environmental agenda at the beginning of UN charter, however, later it contained the Environmental programme and commission on sustainable development. International NGOs were primarily responsible for that. The author looks at NGOs from optimistic and comparative perspectives. She highlights on mainly two hybrid NGOs- ICSU (International Council of Scientific Unions) and IUCN (the International Union for the Conservation of Nature) that have tremendous impacts on UN system regarding environmental agendas. These two NGOs have continued to play a major role in developing ideas on environmental questions, for example:

IUCN	ICSU
• Nature conservation • Setting up new national parks in Latin America and Africa. • Helping ecological research worldwide. • Cooperation with conservation bodies in USA, Germany and else where.	• Gave UNESCO valuable advice on peaceful uses of atomic energy. • Regional centres for scientific cooperation and exchange of knowledge. • Calling of international scientific congress. • Liaison with other international agencies concerned with science, such as FAO for agricultural science and applied ecology, and WTO for medicine, physiology, and social well-being.

The NGOs interaction with certain specialized agencies, as well as main UN system, as she sees, resulted in tremendous progress in environmental issues in UN system, North America, and other member countries of UN. Some examples of the result of this wide range of interaction are:

1. In 1952 ICSU formed committee for International Geographical Year (IGY). The successful IGY project was followed by the setting up of 9 similar scientific committees within ICSU.
2. UN scientific conference on the Conservation and Utilization Of Resources (1949)
3. Conference on the Application of Science and Technology for the Benefit of Less-developed Countries (1963).
4. Establishment of World Fund for Nature (WFN).
5. The Paris Biosphere Conference (1968). It was organized by UNESCO, but influenced by ICSU and IUCN. It led to the formation MAB (Man and Biosphere), UNESCO's most influential interdisciplinary research programme (1971).
6. ICSU formed further committee- the Scientific Committee on the Problems of Environment (SCOPE). The first meeting was in Madrid (1971). It was a road to the Stockholm Conference in 1972. The committee's three completed projects were made available in the conference. (Observers from 52 international NGOs attended in the preparatory committee sessions of the conference).
7. After Stockholm concerned NGOs held conference in New York and later in Geneva to discuss their own interrelationship and their future relationship with new

environmental secretariat. They set up Environmental Liaison Centre (ELC) in Nairobi in 1974 as a communication link with UNEP and HABITAT.

8. NGOs maintained north-south relation. The result is the creation of World Charter for Nature (WCN) by Zaire to protect animals, planet and the environment. In 1994 UN general assembly members decided to issue an annual list of banned hazardous chemicals and unsafe pharmaceutical products. It was also backed by NGOs.

9. Brundtland Report, Our Common Future (1987)- also product and influence of NGOs. We find the increasing influence of NGOs on UN in 1980s. Other important NGOs including Greenpeace emerged in US. It has impact on World Bank to initiate environmental agendas.

10. Conference at Rio (1992) was attended by 178 countries and over 650 NGOs. NGOs had critical role in follow-up activities. The conference set 'a set of principles' and Agenda 21- a kind of road map pointing the direction to the sustainable development.

11. Event after Rio, NGOs continue to participate in UN body.

Her article shows a kind of optimism with regard to the role and influence of NGOs. However, the tension remains especially in the leadership role between NGOs and UN. It appears clear from her article that the international environmental NGOs are powerful, and many of them have close link with the powerful bodies of the world like World Bank, IFM, UN, WTO, and other visible institutions of the global capitalism. The question becomes more intense and crucial: can they evade themselves from the trap of global capitalism?

Chartier, Denis and Jean-Paul Deleage in their article '*The International Environmental NGOs: From Revolutionary Alternative to The Pragmatism of Reform*'(1996) address this crucial question in different ways. The authors in fact justify the changing nature of environmental NGOs in the modern globalized world without really problematizing 'capitalism' as the core of environmental degradation. The authors think, the environmental NGOs are of great need, as the nation-states cannot address all of their environmental problems. 'NGOs can offer a new mode of regulation through crystallizing the aims of an international civil society' (p. 26). They establish a link between local and international levels of politics and are at the core of conflicts and power struggles that determine the outcome of human and environmental disasters. But the definition of situation, as the authors think, though now accepted, raises problems, as major international environmental NGOs are pushed into making arguable compromises. An analysis of the link that the major international environmental NGOs have with certain political institutions, together with their chief modes of action and their discourse, reveals that '*they do not escape the dominant logic of capitalism*' (p. 26). Here is how the authors explicate the ambivalent and ambiguous roles of environmental NGOs:

> We must also emphasize that in a world dominated by a few superpowers, subjected to the political, military, techno-scientific and cultural hegemony of the USA, international institutional simply act as an ambiguous backdrop to the NGOs which play both sides of the coin, solicitude for the most destitute and cooption by the most powerful... the NGOs are certainly recognized as being potential enemies, but their links to the state and international organizations, and their institutionalization, render them vulnerable to the tactics of circumvention and to the means for corruption employed by the industrialist (p. 40).

In the similar tone, Heike Fabig and Richard Boele (1999), in *"The Changing Nature of NGO Activity in a Globalizing World"* talk about the dynamic nature and role of NGOs especially in the context of present globalization initiated by transnational corporations (TNCs). The article gives an overview of globalization's effect on both TNCs and NGOs. It's interesting to note that they view 'globalization' as a separate entity from TNCs, and *not* intertwined. In relating to globalization and targeting TNCs, NGOs have diversified their strategies. The authors have identified three main strategies of the NGOs: (a) Forging new alliance to each other (for example, for Ogoni people in Nigeria who suffered for the multinational oil company Shell), (b) Creating new types of NGOs which take an integrated approach (for example, environment and human right) by examining both environmental and social aspects of globalization, and (c) Establishing constructive business/NGOs relationship. Here they adopt two strategies- confronting, and engaging.

Confrontational NGOs position themselves as diametrically opposed to the corporations they campaign against. They have an intuitive distrust of business and to a large extent see themselves as "outsiders" to the current neo-liberal economic and political system, which they reject. Examples of such NGOs are the member-organizations of the People's Global Action, a transnational alliance of people's movements working under the motto that resistance will be as transnational as capital (Fabig, Heike and Boele 1999).

The engaging NGOs believe in changing the system and its effects by working with business. Using the tools of the system such as management processes and public relations, they aim to reward good business practice with cooperation and endorsement. Building on the free-market concept of consumer sovereignty, they enlist the idea of consumer's 'market vote' to encourage the corporate world to be more socially and environmentally responsible. Possibly the first example of such cooperation was when Greenpeace endorsed and ran an advertising campaign for a propane-butane refrigerator designed by Foron (Fabig, Heike and Richard Boele 1999; Porter and van der Linde 1996, p. 74).

With these strategies, to what extent the environmental NGOs are successful in reducing environmental hazards? If the answer is negative, then, what are the impediments? Robert Taylor (2001) looked at the global response to environmentalism in a very critical way. He is critical on the role of organizations, agreements, and particularly super-power USA. According to him:

> Since 1972, the nations of the world have convened thousands of meetings, unleashed torrents of rhetoric, and signed hundreds of agreements on shared environmental concerns. Yet today, most of these agreements remain ineffective and incomplete, and dismay at their failings is widespread... with some honourable exceptions,...responses are too few, too little, and too late (p.1).

Many agreements, plans, and proposals regarding environment remained ineffective. Ineffectiveness stems partly from the fragmentation of environmental policy making. The Stockholm Conference established a new organization to pursue several goals, but did so in an ambiguous way. At the conference's direction, the UN founded UNEP, but gave it a relatively small budget and general authority to catalyze international environmental agreements. The next two decades were busy times for environmental lawmakers and diplomats. The U.S. and most other industrialized economies enacted statutes to protect clean air, clean water, safe drinking water, endangered species, and the like. At the international level, with help from UNEP, nations began to negotiate agreements to respond to environmental problems. The top concerns were saving endangered species of animals and plants and their habitats and industrial pollution across borders (Taylor 2001).

In the 1980s and 1990s, agreements broadened to deal with global issues, such as climate change and biodiversity, which proved much more complicated and tougher to fix. The players expanded too. Civil society groups such as Greenpeace, multinational corporations, and regional and global trade organizations acquired influential voices in international environmental discussions. But even though environmental concerns have flowed into the mainstream of international policy dialogues, the results have been disappointing to almost everyone. Though some international environmental agreements have made strong progress, many remain empty shells, lacking agreement on the problem, the solution, or the means to achieve it. They have piled up without a blueprint or guiding architect. And beneath a veneer of rhetoric, environmental considerations tend to remain secondary, at best, in multinational finance and trade policy (Taylor 2001).

Globalization of commerce has fuelled the economic engine that is exacerbating many of these problems and highlighted the need for global solutions. Today's economies are dominated by multinational corporate behemoths and global markets. With global environmental problems heavily influenced by capital markets, environmental concerns could no longer be separated from international lending and trade, but multilateral banking and trade organizations were not set up to deal with environmental issues. IMF, World Bank, AID (US agency for International Development) pay a little heed to environmental issues. For example, more than half of WB's projects are unsustainable. IFM is worse than that (Taylor 2001).

In part, the failure to negotiate and implement effective solutions to global problems stems from an inability of nations to agree on the nature of a problem or the appropriate cure. The Biodiversity Convention written in Rio de Janeiro in 1992, for instance, encourages nations to conserve biodiversity, but contains no clear mandate to preserve the globe's richest and most endangered storehouse of species, its tropical forests. Scientists, the timber industry, and its critics have failed to agree on a prescription for sustainable forestry. Negotiation of international agreements has been complicated by deep distrust between the developed countries of the North and the less developed South.

USA appears as a big problem and great obstacles here. USA, for example,

(a) Fought to weaken biodiversity conservation at Rio
(b) Withdrew from Law of Sea Agreement
(c) Abstained from Ottawa Convention on Antipersonnel Mines and Rome Statue on International Criminal Court.
(d) Fought against conservation on Persistent Organic Pollutants
(e) Fails to meet its year 2000 goal for reducing carbon
(f) Kyoto protocol enters into force without USA (Taylor 2001).

Despite frustration over the lack of progress, the players in the policymaking arena on environmental issues have been multiplying. Private sector organizations, from environmental groups to corporations, are playing increasingly important roles in preparing the groundwork for, writing, and enforcing multilateral agreements or other environmental safeguards. Taylor (2001) suggests some important functions for environmental governance:

(a) Priority Setting/Strategic Planning to work toward goals that go beyond narrow national interests to pursue larger, international or global benefits.

(b) Coordination/Integration of Policies and Activities to bring together governments, international organizations, and interested and affected parties in the private sector to ensure that strategies, policies, and actions are coherent and effective.

(c) Data Collection to identify gaps in knowledge, build consensus, and choose appropriate policy responses to problems.

(d) Rule Making by entering into international environmental agreements, treaties, and protocols for actions at home and overseas.

(e) Standard Setting to facilitate assessment of progress toward environmentally sound practices.

(f) Compliance and Assistance depending largely on the voluntary actions of national governments.

(g) Dispute Resolution among the parties.

CONCLUSION

According to political economy perspective, the inherent nature of capitalism is greed, exploitation, extraction of resources, rather than cooperation, responsibility and caring for others. According to Mahatma Ghandi, "The world is sufficient for every body's need, but it is not sufficient for one person's greed" (Guha 2000). The present capitalism has created millions of greedy persons, nations, and even organizations, and hence, their solutions become the real problems. We need to keep in mind that exploitation of human and natural resources for the satisfaction of collective "needs" is different from their exploitation for private profit. The international cooperation is needed, no doubt, but that cannot be effective unless the nature of capitalism is changed.

We cannot deny the tremendous role of environmental NGOs and movements in addressing the issue of environmental problems. Their success lies in creating awareness among the people, in pursuing the governments to enact environmental laws, and hence we do not afford deny the credit they are supposed to get. However, in the era of globalization, their ambivalent role poses more questions than answers. For funding purposes, NGOs have to depend on the International Organizations/ institutions like World Bank, which are constantly blamed for perpetuating neo-colonization, or aggressive capitalism. (e.g., WB invests it 41% money to NGOs). In such a context, given that the NGOs sometimes receive funding from business or national or international organism, how great can their independence be? Are they not advocating the capitalists' accumulation and expansion? The NGOs clearly do not escape from the dominant logic of the capitalism. Then, are NGOs the 'legitimate guise' or 'ideological device' for the dominant capitalist institutions to perpetuate or normalize their actions, interventions, and accumulations over the globe? Are environmental NGOs, in collaboration with, or as an ideological device, trying to commoditize the environment that help to further expand capitalism? Can environmental movements avoid/evade confronting capitalist mode of production driven by maximization of accumulation?

REFERENCES

Bannerji, Himani (1996). 'On the Dark Side of the Nation: Politics of Multiculturalism and the State of Canada' in *the Journal of Canadian Studies*, 31(3): 103-128

Black Star North Zine (2001). 'Beyond Anti-Globalization: Towards a Deeper Understanding of Capital and State', in *Infoshop* (www.infoshop.org), March 28.

Bookchin, Murray (1980). *Towards an Ecological Society*. Montreal: Black Rose Books

Brosius, Peter J. (1999). Green Dots, Pink Hearts: Displacing Politics from the Malaysian Rain Forest. In *American Anthropologist* 101 (1):36-57

Chartier, Denis and Jean-Paul Deleage (1996). 'The International Environmental NGOs: From Revolutionary Alternative to the Pragmatism of Reform', in Peter Willetts (ed.). *The Conscience of the World: The Influence of the Non-Governmental Organization in the UN System*. Washington: The Brookings Institution.

Conca, Ken and Geoffrey D. Dabelko (eds.) (1998). *Green Planet Blues: Environmental Politics from Stockholm to Kyoto*. Bounder, Colorado: Westview Press.

Fabig, Heike and Richard Boele (1999), in "The Changing Nature of NGO Activity in a Globalizing World" in *IDS- Bulletin*, 30, 3, July 58-67.

Guha, Ramachandra (2000). *Environmentalism: A Global History*. Longman.

Luke, Timothy W. (1999). Environmentality as Green Governmentality. *Discourses of the Environment*. Eric Darier (eds.) *Discourses of Environment*. Blackwell Publisher.

Martha, E. Gimenez (2001). Does Ecology Need Marx? *Monthly Review*. Volume 52, Number 8

Marx, Karl (1998). *The Communist Manifesto* (edited by Mark Cowling). New York: New York University Press.

Marx, Karl, Frederick Engels, Valdimir Lenin. 1975. Dialectical and Historical Materialism. Moscow: Progress Publishers

Marx, Karl, Frederick Engels, Valdimir Lenin (1975). *Dialectical And Historical Materialism*. Moscow: Progress Publishers

Meadows, Donella M., Dennis L. Meadows, and Jorgen Randers (1992). *Beyond the Limits: Confronting Global Collapse, Envisioning a Sustainable Future*. Chelsea Green Publishing Company.

Miliband, Ralph (1973). 'Poulantzas and the Capitalist State', in *New Left Review*, 82, (November-December).

Morphet, Sally (1996). 'NGOs And Environment' in Peter Willetts (ed.). *The Conscience of the World: The Influence of the Non-Governmental Organization in the UN System*. Washington: The Brookings Institution.

Pirages, Dennis (1998). 'Global Technopolitics' in Conca, Ken and Geoffrey D. Dabelko (eds.) (1998). *Green Planet Blues: Environmental Politics from Stockholm to Kyoto*. Bounder, Col.: Westview Press

Porter, M., and C. van der Linde (1996). 'Green and Competitive: Ending the Stalemate' in R. Welford and R. Starkey (eds.) *The Earthscan Reader in Business and the Environment*. London: Earthscan

Rappaport , Roy A. (1993). Distinguished Lecture in General Anthropology: The Anthropology of Trouble. *American Anthropologist*, 95: 295-303.

Said, Edward (1993). *Culture and Imperialism*. New York: Random House.

Strange, Susan (1996). *The Retreat of the State: The Diffusion of Power in the World Economy*. New York: Cambridge University Press.

Tylor, Robert (2001). 'New World Old Order' in *Our Future, Our Environment*. Rand Corp. http://www.rand.org/scitech/stpi/ourfuture/

Whitaker, Reg (1977). "Images of State in Canada", in Leo Panitch (ed.). The *Canadian State: Political Economy and Political Power*. Toronto: University of Toronto Press.

In: Handbook of Environmental Policy
Editors: Johannes Meijer and Arjan der Berg

ISBN 978-1-60741-635-7
© 2010 Nova Science Publishers, Inc.

Chapter 16

Incentive Mechanism Design for Nonpoint Source Pollution in China: Group or Individual?

Han Hongyun[1] and *Zhao Liange[2]*

[1]China Academy for Rural Development, Zhejiang University,
Hangzhou Zhejiang, China
[2]College of Economics, Zhejiang Gongshang University, Hangzhou Zhejiang, China

Abstract

With the effective abatement of point source pollution, nonpoint source pollution (NSP) has become a major concern of environmental management in China. Agricultural pollution is predominantly nonpoint due to fertilizer runoff, pesticide runoff, and discharges from intensive animal production enterprises. Environmental quality is a pure public good. While markets are ideal for maintaining incentives to reduce costs and adapt autonomously, the incentives through the market weaken as adaptations require a coordinated response from involved parties. Neither the state nor the market is uniformly successful in enabling individual small farmer to sustain long-term productive use of natural resource systems. It is critical to develop a voluntary incentive program to induce a reduction in nitrogen fertilization levels that also avoids moral hazard and is politically acceptable to the farm community and legally enforceable. Rational economic agents choose a noncooperative strategy to maximize their own well-being. Limitation of group number, establishment of internal rules, group heterogeneity, fairness of norms, expectations of individual efficacy and maintenance of mutuality are possible factors influencing cooperative incentives to conserve environmental public assets.

Keywords: nonpoint source pollution, incentive mechanisms, collective action.

∗ Corresponding author: mailing address: 268 Kaixuan Road, Hangzhou, Zhejiang 310029, China. Tel.: 86 01 0571 88210030; Fax: 86 01 0571 86971645; E-mail: hhyzlg@yahoo.com.cn; hongyunhan@zju.edu.cn.

INTRODUCTION

With the effective abatement of point source pollution, agricultural nonpoint sources have contributed greatly to environmental degradation in China, especially degrading water quality. The growing recognition of the environmental impact of agricultural activities has lead to increasing public concern regarding NSP from agricultural production, and the control of agricultural nonpoint source pollution is emerging as a major priority of state and national pollution control programs. Best management practices (BMPs) are often proposed as an effective instrument of NSP control; however, they are perceived by farmers as having economic disadvantages when compared to conventional management systems. In the absence of tougher environmental restrictions on farmer behavior and complete observability of individual farmer actions, it is necessary to provide economic incentives to encourage farmers' adoption of BMPs. What is unclear is how incentive mechanisms to induce farmers' environmental conservative activities should be designed, and what factors influence farmers' cooperative behavior with environmental conservation.

Although NSP has attracted much attention from researchers in different fields, there are few reports on the management of NSP in China. "Pollution from agricultural areas and nonpoint sources is largely uncontrolled", and "there is little evidence that Asian policy makers take into account the benefits that arise from nonconsumptive or nonextractive uses of the environment" (Dudgeon, 2000, p. 795). This chapter aims to analyze incentive mechanism of agricultural NSP in China, and should be helpful to decision-making with environmental management not only in China but also throughout the world. This paper proceeds as follows: after the introduction, in section 1 the present situation of over-application in China is introduced; section 2 is an analysis of the nature of NSP with agricultural activities; section 3 is a comparison of group and individual incentives; finally, a brief conclusion and its implications for NSP policy options are discussed.

1. THE PRESENT SITUATION OF AGRICULTURAL NSP IN CHINA

Green revolution technologies, including synthetic fertilizers, high-yielding varieties, highly effective pesticides and irrigation water use have greatly contributed to the rapid increase in land productivity and crop production in China. Since 1966, fertilizers have been the most important contributors to yield increase in China, particularly nitrogen. Agriculture has become extremely intensive by using more applications of inorganic fertilizers and chemical pesticides (Li and Zhang, 1999). About 40% of the growth in total grain output in the period from 1986–1990 was accounted for by increased chemical fertilizer application. China has emerged as the largest consumer, the second largest producer and a major importer of chemical fertilizers in the world (Wang et al., 1996). With effective abatement of point source pollution, in recent decades, nonpoint sources of pollution have become a major source of water environmental degradation in China due to the increasing application rate of chemical materials.

Table 1. Consumption of chemical fertilizers in rural areas.

Year	Total sown area[1] (million ha)	Consumption of chemical fertilizers[2] (million kg)	Kg/hectare
1978	150.105	8840	58.892
1980	146.381	12694	86.719
1985	143.626	17758	123.641
1989	146.554	23571	160.835
1990	148.363	25903	174.592
1991	149.586	28051	187.524
1992	149.008	29302	196.647
1993	147.741	31501	213.218
1994	148.241	33179	223.818
1995	149.879	35937	239.773
1996	152.381	38279	251.206
1997	153.969	39807	258.539
1998	155.706	40837	262.270
1999	156.373	41243	263.748
2000	156.300	41464	265.285
2001	155.708	42538	273.191
2002	154.636	43394	280.620
2003	152.415	44116	289.447
2004	153.553	46366	301.954
2005	155.488	47662	306.532

Source: [1]*China Statistic Yearbook* 2007, China Statistic Press, p. 474; [2]*China's Rural Statistical Yearbook*, China Statistics Press, 2007, p.44.

China's home-produced and imported chemical fertilizers are mainly nitrogen fertilizers. With the rapidly increasing application of manufactured nitrogen since the early 1970s, the marginal response ratios to N have dropped because of the unbalanced provision of other crop nutrients, phosphates and potash, which has become a constraint in many areas. For annual crops, the N uptake efficiency is less than 50%, even under good management practices (Ju and Zhang, 2003). "The average annual application rate of N in China was gradually increased from 38 Kg N ha-1 in 1975 to 130 Kg N ha-1 in 1985, and rapidly increased to 236 Kg N ha-1 in 1995 and 262 Kg N ha-1 in 2001"(Liu et al., 2005 p.212). Although marginal productivity of additional fertilizer application has decreased, fertilizer application rate is increasing over time. The fertilizer application rate increased from 58.892 Kg ha-1 in 1978 to 306.532 Kg ha-1 in 2005 (see Table 1). The increased incidence of nitrate contamination of groundwater has been related to the increased use of N fertilizers and irrigation. It is generally believed that intensive farming with high application rates of N leads to more severe groundwater pollution.

No statistical data are available for pesticide consumption. "From 1949 on, the consumption of pesticides in China increased rapidly, 1920 tons in 1952, 537,000 tons in 1980, and 271,000 tons in 1989 after the manufacture of organic chlorinated pesticides ceased at the beginning of the 1980s" (Li and Zhang, 1999, p. 29). Agriculture is a primary and important source of pollution. Agricultural pollution is predominantly nonpoint due to fertilizer runoff, pesticide runoff, and discharges from intensive animal production enterprises. Croplands are increasingly affected by pollution, mostly arising from industrial discharges of contaminated wastewater or through pollution of irrigation water. Less than 20% of wastewater is treated before being discharged into rivers and lakes. So far, about 2.6 million ha of land have been taken out of agricultural production as a result of wastewater pollution and the amount of grain lost is estimated at 5–10 million tonnes (Li et al., 1997).

2. NSP MANAGEMENT—MARKET OR GOVERNMENT REGULATION?

Environmental quality is a pure public good. How best to limit the use of common property resources to ensure their long-term economic viability is a critical issue; over-exploitation of shared resources is referred to by Hardin (1968) as the "tragedy of the commons". Hardin sees two alternatives to common property resource management: one is privatization and the other is state regulation. Hardin's view needs further consideration due to the following aspects.

2.1. Market Failures of Environmental Management

Markets have been presented as an appropriate institutional framework for optimizing the allocation of resources between confliction uses and users. Moreover, it is argued that transferability through a market-based system produces a voluntary reallocation of resources that is politically more acceptable than bureaucratic regulation. Markets are institutions that exist to facilitate exchange. They exist in order to reduce the cost of carrying out exchange transactions. Under the conditions of complete information and perfect competition, markets will achieve first-best allocations.

A situation where the market does not result in an efficient allocation of resources is the core of market failure. There are three basic reasons for market failure. First, individuals may not have sufficient control of a commodity or resource to undertake the necessary exchange. Second, high transaction and information costs can erode the advantages of trade. Finally, the individuals involved in trade may be unable to negotiate and agree upon the terms of mutually advantageous exchange. Crase et al. (2000) outlined five basic factors hindering the development of the permanent water market: poorly defined rights to access and use the resource, variability of supply, infrastructure obstacles, excessive transaction and transfer costs, and hoarding behavior and speculation. All of these factors may lead to market inefficiencies.

2.2. Government failures with environmental management

Ever since Marshall and Pigou, it has been argued that externalities constitute a prima facie case for government intervention in a market economy (Coase, 1988). Public intervention has problems of its own. These include misallocated project investments, overextended government agencies, inadequate service delivery to the poor, neglect of water quality and environmental concerns, and the underpricing of natural resources. The government faces the same problems as the market: incomplete information and high externalities resulting from incomplete information.

Meanwhile, government regulation confronts administrative costs: including the costs of investigation and administration. Sometimes the costs are sufficiently high that the expected gains from governmental intervention are less than the costs involved. "Under public management the dominant incentive to comply is coercion: that is, setting regulations and using sanctions for those who break them. But this type of incentive is only effective if the State detects infractions and imposes penalties. In many cases the state lacks the local information and ability to penalize, e.g., for breaking water delivery structures or for excessive water withdrawals" (Dinar et al., 2001 p.6). Whether government intervention is desirable depends on the costs relative to the expected gains.

Many public and private programs enhance the environment through the direct acquisition of environment amenities. Two issues arise with public payment for environmental services. First, payments based on producers' WTA—rather than potential environmental benefits—will not target the participation of producers who can, by adopting conservation practices, deliver relatively large benefits per dollar of cost (Babcock et al., 1996). Other mechanisms must be used for benefit-cost targeting. Second, information on potential environmental benefits, conservation costs, and producer WTA is needed to target program enrollment and specify payments.

"What one can observe in the world, however, is that neither the state nor the market is uniformly successful in enabling individuals to sustain long-term, productive use of natural resource systems. Further, communities of individuals have relied on institutions resembling neither the state nor the market to govern some resource systems with reasonable degrees of success over long periods of time"(Ostrom, 1991 p.1). Because of market failures and government failures, the operation of a simple market or the control by government may not function efficiently.

3. REQUIREMENTS FOR FARMERS' INCENTIVES TO CONSERVE THE ENVIRONMENT

The problem of nitrate would be less difficult to solve if it could be demonstrated that farmers apply fertilizer in excess of the expected profit-maximizing level. There is evidence that farmers, at least in some areas plagued with nitrate pollution due to over-applied nitrogen fertilizer. Environmental problems became evident during the 1970s and acute during the 1980s, the environmental outcomes were considered to be public goods. Degrading environment needs peasants to turn their environmental consciousness into practice and grassroots environmentalism should play a more prominent role in environmental decision-making. Huge amount of small-scale farmers makes the management of NSP most complex task in the world. How to induce farmers change from over-application to environmental friendly management practices needs to be considered carefully.

3.1. The Necessary Condition for the Optimal Provision of a Collective Good

There is a growing recognition that environmental outcomes are correlated—benefits are jointly produced by the same action, the starting point many resource management problems is one involving a large number of users with imprecisely defined rights, and an aggregate rate of resource use that is environmentally unsustainable. With regard to NSP management, though all farmers have a common interest in obtaining collective benefits, they have no common interest in paying the cost of providing that collective good–contributions to the maintenance of environment. Each would prefer that others pay the entire cost, whilst receiving the collective benefit irrespective.

In practice, the fundamental choices facing farmers are whether to obey the rules concerning responsibility for maintaining environmental quality over time or not. Here, we assume that ex ante rules concerned with the environmental management have been established that have prescribed standards with land use. Each farmer can choose simultaneously to cooperate or not depending on the expected net benefit obtained from the collective good. If a farmer i chooses to cooperate, then he or she will pay some attentions to the maintenance of local environment. In contrast, if the farmer chooses not to cooperate, he or she will make no contribution to the maintenance of shared environment.

For the sake of simplicity, we assume that there are n farmers in a given location. A farmer's conservation propensity function (CPF) is determined by following factors:

CPF=f (Net Benefits/Individual Characteristics; Physical Conditions; Financial Conditions; Policy Variables; etc.).

Assume that farmer i has a direct utility function of net benefits,

$$U_i = U_i(x_i, W), (i = i = 1, \cdots, n), W = W(\sum f_i)$$

Where, x_i denotes the amount of private goods consumed by farmer i, W denotes the situation of environment, f_i is the consumed amount of environmental quality, or maintenance effort contributed by farmer i.

As a rational economic agent, farmer i will choose their own strategy, (x_i, f_i), to maximize her/his utility under the constraint of financial condition:

$$M_i = p_x x_i + p_c f_i$$

Where p_x is the price of a private good, and p_c is the price of a collective good, that is, the opportunity cost of effort spent maintaining environmental quality.

Given that the utility function has the characteristics:

$$\frac{\partial U}{\partial x_i} > 0, \frac{\partial U}{\partial W} > 0, \frac{\partial^2 U}{\partial x_i \partial W} < 0$$

then the Lagrangian function will be:

$$L = U_i(x_i, W) + \mu(M_i - p_x x_i - p_c f_i)$$

The first-order conditions for utility maximization are:

$$\frac{\partial L}{\partial f_i} = 0, \frac{\partial L}{\partial x_i} = 0$$

that is,

$$\frac{\partial U_i}{\partial W}\frac{\partial W}{\partial f_i} - \mu p_c = 0, \frac{\partial U_i}{\partial x_i} - \mu p_x = 0$$

then,

$$\frac{\partial U_i/\partial W}{\partial U_i/\partial x_i}\frac{\partial W}{\partial f_i} = \frac{p_c}{p_x} \text{(i=1,2, ...,n)}$$

Hence, the necessary condition for the optimal provision of a collective good, through the voluntary and independent action of the members within a group is that the marginal utility of additional units of the collective good must provide utility in the same proportion as the additional units of the private goods. Only if this is true will each member find that his or her utility is maximized. These perceptions are based on their individual farming enterprises, not on the broader set of environmental amenities that might be created from participation. Many producers might undertake conservation of their own accord, given the private benefit. However, they may not be aware of the specific activities to achieve both the conservation outcomes and their own financial return.

3.2. Difficulties Faced by a Conservation Program—Estimation of Non-Market Values

A number of studies have suggested that conservation programs using a range of mechanisms, such as grants and taxes, have been inefficient because they have focused on on-site information rather than environmental outcomes (Ribaudo, 1986; Wu and Skelton-Groth, 2002). The estimation of values for preserving environmental assets is more challenging (Rolf and Windle, 2005). Many changes in the provision of environmental assets are not reflected in markets, partly because people hold values for environmental goods without actually using them. These non-use values can be separated into existing values, bequest values and option values.

A conservation program has to recognize a correlation between the production of environmental outcomes and private good production. There is a growing recognition that environmental outcomes are correlated – benefits are jointly produced by the same action, including revegetation may jointly produce carbon, improvements to water quality and wildlife benefits. Key to the success of a conservation program is the gathering of previously 'missing information' linking landholder actions on farm with environmental objectives. If relevant information is gathered and shared between buyers (government) and sellers (landholders) of environmental goods and services, new markets can be created for these products. The productivity of the farm and extent of change to achieve conservation are critical to participation.

It is also critical to offset higher perceived costs to increase participation. Farmers draw up a conservation plan for management practices and infrastructure investment with technical assistance. This is the basis for entry into the program, so the locally available technical assistance is critical. The institutional structure and a history of incentive driven participation in this voluntary program makes the producer's focus on his/her perceived benefits and costs critical to participation. Extension and education programs may be useful in providing this type of information if the government desires more environmental outcomes. It is still facing following issues to induce farmers' conservative activities.

The environmental concerns and the potential environmental benefits, costs, and WTA associated with addressing them may vary widely across regions, producers, soil types, topography, and location. Information requirements are large and costly for environmental conservative programs (Claassen et al., 2005). Agricultural conservation programs could not focus only on on-site physical criteria, such as soil erosion and recharge, rather than the benefit to the environment of a reduction in erosion or recharge. Agricultural conservation programs must account for both physical production relationships between environmental outcomes and the value of those outcomes. The environmental outcomes are considered to be public goods. That is the market would not provide these goods as there is no monetary incentives for farmers to change their land use practices. Also the outcomes have public value and the public needs to make decisions about what they want and how much money to spend.

3.3. Incentive Contracts: Group or Individual Incentives

How best to limit the use of common property resources to ensure their long-term economic viability is a critical issue faced by developed and developing countries in the world. According to Hardin, 'we have several options. We might sell them off as private property. We might keep them as public property, but allocate the right to enter them. The allocation might be on the basis of wealth, by the use of an auction system. It might be on the basis of merit, as defined by some agreed-upon standards. It might be by lottery. Or it might be on a first-come, first-served basis, administered to long queues. These, I think, are all objectionable' (Hardin, 1968). According to Hardin (1968), the term 'private property' refers to the exclusive control of a resource by a single agent, and 'common property' refers to an absence of exclusive rights.

This simple dichtomous classification of property rights was criticized by many writers in the 1970s and 1980s because of the difficulty in grouping property rights exercised by government and collectively by finite groups of people. Challen (1991) believes that "the term 'common property' has been used to refer to almost all situations where a resource is subject to joint exploitation, ranging from well-defined joint ownership and management by a finite set of individuals to open access" (2000, p. 22). Bromley further argues that "Common property is in essence private property for the group and in that sense it is a group decision regarding who shall be excluded" (p. 29).

Hardin's model has been formalized as a prisoner's dilemma game. Rational economic agents choose noncooperative strategy to maximize their own well-being. In the simple prisoner's dilemma game, agents cannot establish binding commitments under the constraint of no communication among the players, but they own complete information about the behavior of each player. Without the capacity to engage in a binding contract, each chooses their dominant strategy, which is not to cooperate. An established binding contract is the necessary condition for agents to adopt collective action (Models, 1975; Lomborg, 1996).

Together with difficulties of credible commitment, a special issue with NSP is efforceability. The defining characteristic of nonpoint source pollution is that it cannot be traced to any one source. Much of the previous work on incentive contracts does not directly address NSP issues. Some contracts have been applied to nonpoint source pollution problems. Heckathorn (1988) argues that "many social groups are subject to collective sanctions, including both collective punishment and collective rewards"(p.535). If output does not meet requirements, one agent is selected for punishment. Or if output does not meet requirements, all but one agent are punished. In a nonpoint source pollution context, that means punishing one farmer for the pollution of his/her neighbors. If that farmer can prove that his/her nitrogen applications could not have caused the level of pollution observed, it is unlikely that this punishment would be legally enforceable. In the current political environment, it seems unlikely that penalties are acceptable. It is critical to develop a voluntary incentive program to induce a reduction in nitrogen fertilization levels that also (a) avoids moral hazard; (b) is politically acceptable to the farm community and legally enforceable.

3.4. The Way out of the Tragedy of Common Pool Resources

Resolving the conflict between individual rationality not to cooperate and collective rationality to cooperate is a major topic in common pool resource management. Olson (1965) specifically sets out to challenge the optimism that individuals with common interests would voluntarily act so as to further those interests. Olson argues that only in a small group can one expects rational and self-interested individuals to contribute voluntarily to the supply of collective goods. On the contrary, in a large group, individuals do not contribute to their provision without selective incentives or private incentives. Olson demonstrates that 'unless the number of individuals in a group is quite small, or unless there is coercion or some other special device to make individuals act in their common interest, rational, self-interested individuals will not act to achieve their common or group interests' (1965, p. 2).

While Olson stresses the influence of the group size and selective incentives on the cooperation of rational, self-interested agents to achieve their common or group interests, Ostrom (1991) focuses on the factors that encourage cooperation within a group. She identifies eight characteristics of robust common property resource institutions. They are: clearly defined boundaries; congruence between appropriation and provision rules and local conditions; collective-choice arrangements; monitoring; graduated sanctions; conflict-resolution mechanisms; minimal recognition of rights to organise; nested enterprises. Hence one reason why some groups fail to cooperate is that "the participants may simply have no capacity to communicate with one another, no way to develop trust, and no sense that they must share a common future"(Ostrom, 1991, p. 21). Rules to ensure cooperation can therefore be a key component to overcoming incentives not to cooperate. The importance of Ostrom's work is that it demonstrates how groups of individuals have relied on internal rules of behaviour, resembling neither state nor market institutions, to govern some resource systems with reasonable degrees of success over long periods of time.

There is a body of literature concerning collective action in late 20th century. In addition to selective incentives, Heckathorn (1993) points out that there is a complex link between group heterogeneity, collective action, and the rules of cooperation. The heterogeneous factors like interest in the public good, resources available to contribute to public goods production, and the cost of those contributions have a complex influence on collective action. "Depending on context, heterogeneity can increase or reduce social cooperation" (p. 347). Hence, "the principal obstacle to collective action is not the free-rider problem but the problem of efficacy" (Macy, 1991, p. 730), rational actors will contribute to a nonexcludable good if their efforts are cost-effective; in other words, it will be a start-up problem, "unless a critical mass of strongly motivated individuals is willing to absorb these costs, collective action never begins" (Heckathorn, 1993, p. 251). It is concluded whether rational actors cooperate depends on a comparison of benefits against costs.

Gould (1993) argues that both fairness of norms and expectations of individual efficacy 'play a central part in the production of collective goods' (p. 184). According to Gould, there are three boundary conditions what affect the production of collective goods. First, efficacy of individual contribution should have potential impacts on the effort of others. Second, the voluntary initial contribution must induce the forthcoming response of other members. Third, 'actors must perceive themselves as members of an identifiable collectivity—even if this collectivity is defined merely as the total number of potential beneficiaries of the collective good' (p. 185).

The inefficiency of resource allocation will be somewhat less serious in groups composed of members of greatly different size or interest in the collective good. In such unequal groups, there is a tendency toward an arbitrary sharing of the burden of providing the collective good. When one can unilaterally exert effects on others, she might tend to bear the costs alone if she prefers this outcome over the risk that the other will not do anything. In the absence of coercion, "rational players should contribute if and only if they believe that their contribution is critical in affecting the outcome of the competition" (Bornstein et al., 1996, p. 490). Such differentiation is typically efficient (Skaperdas, 1991). Heterogeneity of interests, cost, and resource could affect the cooperation of actors. However, collective interests do not necessarily produce collective action.

Macy (1991) suggests that "cooperative propensities are shaped over time by social sanctions and cues", prosocial norms are a consequence rather than a cause of cooperation, which are useful in promoting forgiveness of random deviance (p. 838). Instead of the external rules, "sustainable collective action must be based on internal adjustment mechanisms that could respond effectively to fluctuations in the demand and supply sides of the resource" (Benvenisti, 1996, p. 409). Shared interests seem to consistently outweigh conflict-inducing characteristics. Furthermore, once cooperative regimes are established through treaties, they turn out to be impressively resilient over time (Yoffe and Wolf, 1999). Cooperation into future needs much trust and assurances against defection due to the worries about changing circumstances and shifts in the power relations throughout the life of the contract. The maintenance of mutuality among parties plays a critical role in the development of institutional arrangements.

4. BRIEF CONCLUSIONS AND IMPLICATIONS

While markets are ideal for maintaining incentive to reduce costs and adapt autonomously, the incentives through the market weaken as adaptations require a coordinated response from involved parties. Neither the state nor the market is uniformly successful in enabling individuals to sustain long-term productive use of natural resource systems. Because of market failures and government failures, the operation of a simple market or the control by government may not function efficiently.

In many resource management problems, the starting point is one involving a large number of users with imprecisely defined rights, and an aggregate rate of resource use that is environmentally unsustainable. The necessary condition for the optimal provision of a collective good, through the voluntary and independent action of the members within a group, is that the marginal utility of additional units of the collective good must provide utility in the same proportion as the additional units of the private goods. The environmental outcomes are considered to be public goods. The market would not provide these goods as there are no monetary incentives for farmers to change their land use practices. It is critical to develop a voluntary incentive program to induce farmers' environmental conservative activities.

Resolving the conflict between individual rationality not to cooperate and collective rationality to cooperate is a major topic in organization theory. Limitation of group number, establishment of internal rules, group heterogeneity, fairness of norms, expectations of individual efficacy and maintenance of mutuality can enhance farmers' cooperative behavior of environmental management. Cooperation into the future needs much trust and assurances against defection due to the worries about changing circumstances and shifts in the power relations throughout the life of the contract. The maintenance of mutuality among parties plays a critical role in the development of institutional arrangements.

ACKNOWLEDGMENTS

This research was supported by by 2008 Program for New Century Excellent Talents in University, National Natural Science Foundation of China in 2005 (70573091) and Zhejiang Provincial Natural Science Foundation of China (Z607126) and the key research project funded by a national key research base for humanities and social sciences under the guidance of the Ministry of Education (2007JJD630014).

REFERENCES

Babcock, B.A., Lakshminarayan, P.G., Wu, J.J., and Zillberman, D., 1996. "The economics of a public fund for environmental amenities: a study of CRP Contracts", *American Journal of Agricultural Economics*, Vol.78, pp.961-971.

Benvenisti, E, 1996. "Collective Action in the Utilization of Shared Freshwater: The Challenges of International Water Resources law," *The American Journal of International Law*, Vol.90, No.3, PP.384-415.

Bromley, D., 1991. *Environment and Economy: Property Rights and Public Policy*, Cambridge, Mass., USA.

Claassen, R., Cattaneo, A., Johansson, R., 2005. "Cost-Effective Design of Agri-Environmental Payment Programs: U.S. Experience in Theory and Practice", Paper to be presented at ZEF-CIFOR workshop on payments for environmental services in developed and developing countries Titisee, Germany, June 16-18, 2005, http://www.cifor.cgiar.org/pes/publications/pdf_files/US_paper.pdf.

Coase, R. (1988). *The Firm, the Market, and the Law*, University of Chicago Press, Chicago and London.

Crase, L., Reilly, L., and Dollery, B. (2000). 'Water markets as a Vehicle for Water Reform: the Case of New South Wales', the *Australian Journal of Agricultural and Resource Economics*, 44:2,pp.299-321.

Dinar, A., Rosegrant, M., and Meinzen-Dick, R. (2001). 'Water Allocation Mechanisms—Principles and Examples', World Bank, Agriculture and Natural Resources Department. http://www-esd.worldbank.org/

Dudgeon, D., 2000. "Large-scale Hydrological changes in tropical Asia: prospects for riverine biodiversity", *Bioscience*, Vol.50, No.9, Hyhdrological alternations, pp.793-806.

Gould, R., 1993. "Collective Action and Network Structure," *American Sociological Review*, Vol.58, pp.182-196.

Hardin, G., 1968. The Tragedy of the Commons. *Science*, 162: 1243-1248.

Heckathorn, D., 1988. "Collective Sanction and the Creation of Prisoner's Dilemma Norms", *American Journal of Sociology*, Vol. 94, pp.535-562.

Heckathorn, D., 1993. "Collective Action and Group Heterogeneity: Voluntary Provision versus Selective Incentives", *American Sociological Review*, Vol. 58, pp.329-350.

Ju, X.T., Zhang, F.S., 2003. "Nitrate accumulation and its implication to environment in north China. *Ecol. Environ.* 12, pp.24-28.

Lomborg, B., 1996. "Nucleus and Shield: The Evolution of Social Structure in The Iterated Prisoner's Dilemma," *American Sociological Review*, Vol.61, pp.278-301.

Li, X., Zuo, Ch., and Tschirley, J., 1997. "Sustainable Agriculture and Rural Development in China". http://www.fao.org

Li, Y. Zhang, J.B., 1999. "Agricultural diffuse pollution from fertilisers and pesticides in China", *Wat. Sci.Tech.* Vol. 39, No.3, pp.25-32.

Liu, G.D., Wu, W.L., and Zhang, J., 2005. "Regional differentiation of non-point source pollution of agriculture-derived nitrate nitrogen in groundwater in northern China", *Agriculture, Ecosystems and Environment* 59, pp.211-220.

Macy, M., 1991. "Learning to Cooperate: stochastic and Tacit Collusion in Social Exchange", *American Journal of Sociology*, Vol 97, pp. 808-843.

Models, A., 1975. "Individual Contributions for Collective Goods," *Journal of Conflict Resolution*, Vol.19, pp.310-320.

Olson, M., 1965. *The Logic of Collective Action-Public Goods and the Theory of Groups*, Harvard University Press.

Ostrom, E. (1991). *Governing the Commons: the Evolution of Institutions for Collective Action*, Cambridge University Press, Cambridge.

Rolf, J., and Windle, J., 2005. "Valuing options for reserve water in the Fitzroy Basin", *The Australian Journal of Agricultural and Resource Economics*, Vol.49, pp.91-114.

Ribaudo, M., Agapoff, J.,2003. "Cost to Swine Operations from Meeting Federal Manure Application Standards: The Importance of Willingness to Accept Manure", to be presented at the SERA-IEG 30: Natural Resource Economics Meetings, Held at the University of Kentucky. .

Skaperdas, S., 1991. "Conflict and Attitudes Toward Risk," *The American Economic Review*, Vol.81, No.2, p

Wang, Q.B., Halbrendt, C., and Johnson, S.R., 1996. "Grain production and environmental management in China's fertilizer Economy", *Journal of Environmental Management* 47, pp.283-296.

Wu J.J., Skelton-Groth , K. , 2002. "Targeting conservation efforts in the presence of threshold effects and ecosystem linkages", *Ecological Economics* 42 (2002) 313–331.

Yoffe, Sh., Wolf, A.T., 1999. "Water, Conflict and Co-operation: Geographical Perspectives," *Cambridge Review of International Affairs*, vol.12:2, Univ. of Cambridge, Spring/Summer 1999, pp.197-213.

In: Handbook of Environmental Policy
Editors: Johannes Meijer and Arjan der Berg

ISBN 978-1-60741-635-7
© 2010 Nova Science Publishers, Inc.

Chapter 17

ANALYZING EFFECTIVE ENVIRONMENTAL POLICY-MAKING PROCESS AND EVIDENCE FROM AVIATION SECTOR

Joosung J. Lee[*]

Korea Advanced Institute of Science and Technology
335 Gwahak-ro (373-1 Guseong-dong), Yuseong-gu, Daejeon 305-701,
Republic of Korea

ABSTRACT

This article asks two key questions. What factors should be considered in making environmental policy? And what is the effective process for environmental policy formulation? To answer these questions, this article applies the concept of social demand articulation that has been developed to analyze the drivers and processes of environmental policy-making. Social demand articulation is a systematic approach that stimulates society toward environmental innovation. In particular, knowledge and information flows that raise the technological capability and awareness level of firms and consumers for environmental improvement are analyzed in greater detail than in previous work. Their indicators have been developed and applied to analyze environmental performance improvement cases in the air transportation sector. For effective environmental policy-making, this article emphasizes the steps to establish scientific evidence as well as public awareness regarding an environmental problem. If one of the two steps is missing, mere environmental protection movement could mislead society to achieving the political agenda of a particular interest group. In addition, it is important for environmental policy to contain a clear vision for future society we aim to craft. The philosophy that humans should prosper in harmony with nature must underlie environmental policy-making. With this in mind, institutionalizing knowledge and information flows among firms and societal stakeholders can set a path for the environmentally conscious market in which greener products are valued highly and give competitive advantage to environmentally innovative firms.

[*] jooslee@kaist.ac.kr

1. INTRODUCTION

Today environmental protection is among the central matters for natural conservation, public health and sustainable business. With advanced technologies and changing lifestyles, the consumption of resources and release of wastes and pollutants are fast increasing. This requires policy makers to design environmental policies that properly guide the development of new products and business operations [1]. Overall, the goal of environmental policy is to limit, slow down, reduce or eliminate environmental damages caused by industrial and human activities. Some environmental policies successfully carry out this goal while some others fail to achieve its intended goal. What causes these different outcomes? One reason is the process in which policy is made [2]. Policy-making is "the process by which governments translate their political vision into programs and actions to deliver 'outcomes' – desired changes in the real world [3]." Promoting good practice (i.e. the process of policy-making in particular) in policy making is essential to the delivery of intended policy outcomes for citizens. To do so, policy makers should have available to societal stakeholders the widest and latest information on research and best practice. Thus, good policy is transparent as the processes used to develop policy must be clearly communicated and widely understood. Policy makers should engage those individuals and organizations who will be affected by policy change from the outset [4].

In addition, successful environmental policy must articulate a clear vision for treatment of environmental issues. The policy can communicate the organization's intent to societal stakeholder and encourage their actions. Thus, it should demonstrate to society why certain environmental problems should be dealt with and by solving the problem, what type of future society it aims to craft. This common picture provides consensus and shared value for environmental protection among societal stakeholders. Having a clear vision of what policy, legal and administrative measures may be required becomes a basis for building consensus among the experts in the relevant public and private sectors as well as in non-governmental organizations [5] A clear vision leads to a good policy as it can set the direction from the outset and clear target for the problem to be fixed by policy change. Policy is an instrument of change that must be aligned with government and corporate goals. [6]

Lastly, environmental protection is successful when the root cause of the relevant problem is removed or alleviated. There are many potential solutions for reducing wastes and pollutants. However, simple treatments (i.e. end-of-pipe pollution reduction at a coal-fired power plant) are not a permanent solution to the problem. Changing the fuel types used and power conversion processes should eventually accompany on-going efforts to reduce pollutions at the source [7]. To diagnose and solve the true cause of environmental problems, it is important to accumulate scientific knowledge and technological capability to deal with the problems. In addition, good policy solutions should consider systems (e.g. society, economy, engineered products) impact in the immediate short term and longer terms.

Without a clear goal, right processes and assessment of the root cause (i.e. scientific evidence and knowledge to solve environmental problems), consensus among the stakeholders cannot be established. An environmental policy with one of these components missing may end up with achieving only the agenda of a particular interest group. In this vein, this article elaborates the importance of goal setting, assessment of the root cause, consensus building and weaves them in a holistic framework. The framework can guide policy making processes for a successful outcome as well as the root problem solving.

2. Goals of Environmental Policy

For environmental policy to have effective outcomes, the societal stakeholders (e.g. citizens, government, companies) should clearly understand why it is so important to preserve the environment. However, not many of us seem to have a clear picture for a future society where environmental protection is regarded a core value. Environmental protection is often justified by the logic that humans will exhaust natural resources if we continue the economic development without a significant change in the current industrial activities. Is environmental protection then only an important political or economic agenda to be promoted? Or do we advocate environmental conservation because we want to recover the original nature that existed at the birth of the earth? If the latter is a more important reason, environmental protection can be closely related to a natural desire in human beings or perhaps survival instinct for us not to disappear from the earth. Therefore, one must present a fundamental reason why environmental protection is so important other than just political or economic reasons.

Since many of today's environmental protection movements are justified with economic reasons, environmental protection is a benefit for those who are early entrants in green technologies and products. On the other hand, those nations who do not have the capability to transition their industry to a 'cleaner' one are subject to the environmental power of advanced countries' greener but more expensive resources and technologies. If environmental innovations benefit only a few leading nations, undeveloped countries cannot afford to pay for less polluting but much more expensive technologies and energy sources. Thus we should not pursue environmental protection only for economic reasons. Other important values as humans and considerations for the less developed must be incorporated into environmental policy. To position such factors as one of central values in the future, we should first have the philosophy in environmental policy that addresses not only economic values but also human values. Without such a guiding philosophy, we may make mistakes and rush ourselves into blind environmental protection without knowing where we are going.

A philosophical investigation into the reasons for environmental protection shows that human dignity can be upheld through a harmonious relationship with nature, and self realization is impossible outside of nature. On an ethical side, it can be emphasized that passing down the inherited natural environment to future generations, and enjoying happiness together with other living things, corresponds to the most fundamental moral principle of mankind [8]. In addition, it is emphasized once again that in order to achieve future values such as the happiness of future generations and comfort in advanced society, environmental protection is the central responsibility of mankind. The environment belongs not just to those of us living today but to our ancestors and to our children. The environment is just as important to future generations as to us. Environmental protection means the pursuit of greater environmental value in the future where humanity will prosper harmoniously with the environment.

The role of technology is another important aspect for effective environmental policy-making. Technologies continue to advance and subsequent resource exhaustion and pollution due to industrial activities are increasing. One simple way to stop environmental pollution is to stop all technological and industrial development. This proposal may be justified on the ground that human life seems no better than pre-industrial evolution era (i.e. having such problems as starvation and suicide) even with such fast technological advancement. On the other hand, technologies can certainly be used for environmental protection. More and more energy-saving and environmentally friendly technologies give rise to environmentally innovative products and systems. From this point of view, technological development should not be viewed solely as damaging nature. Then what types of technologies are the ones that contribute to sustainable growth? We should first develop a guideline/metrics of assessment for the future investment of technological development. Technological development projects are then evaluated against the guideline/metrics of assessment, and those projects that meet sustainable growth criteria are invested with priority. Such metrics measure not only economic benefits but also environmental and social benefits. Environmental policy should articulate for corporate top management to use such guideline/metrics to assess the environmental benefits and social welfare from their technological development projects.

3. UNCERTAINTY AND ENVIRONMENTAL POLICY

Global warming is among the central environmental issues. Many advanced and developing countries recognize carbon dioxide is at the bottom of global warming. Also most people have developed their awareness on environment issues, so the phenomenon of global climate change is recognized as the main cause of environmental disruption. As a result, most advanced countries create 'carbon markets' and make green policies to reduce carbon dioxide emissions [9]. Others point to the complex factors including natural variables that affect climate on the other hand. That is, it is not a certain scientific fact that carbon dioxide emission is the main driver of global warming [10]. The purpose of this section is not to determine the truth of climate change. Rather, it emphases the fact that scientific uncertainty should be explicitly considered in environmental policy making.

Thus it is important for policy makers to have a balanced view on actual facts and uncertainty regarding a particular problem. For example, there is uncertainty about how much fossil fuels are left in the reserve. It could be another 30 to 100 years before gasoline and coal resources are exhausted [11]. Considering the fact that alternative energy sources are still expensive and their energy density is low, fossil fuels are still the most preferred energy source for many industries including the air transportation sector [12]. In addition, what if all the carbon reduction effort is less effective for alleviating global warming? What if some other factors are the primary cause of global climate change? In this regard, environmental policy makers must endeavor to assess the primary cause of an environmental problem and cure the root cause that is scientifically proven. For this reason, environmental policy must articulate the importance of reflecting scientific evidence in making policies. The current attention to global warming and reducing carbon dioxide is not wrong as long as the ultimate purpose of environmental policy is creating a better quality environment for humans. Proper strategies which can translate the current attention about global warming issues into broader environmental issues and achieve improved environmental quality while minimizing related uncertainty are essential.

In the real policy-making arena, decisions are often made under uncertainty. From individual decisions whether to bring an umbrella on a cloudy day to government policy decisions reacting to future economic situations, there is limited information about the future. Policy makers are then required to gather as much accurate information and feasible scenarios as possible and make policies that are robust under fast changing environments. To make a robust environmental policy that meets its prime objectives, the large scientific uncertainties in an environmental problem should be communicated to policy makers. As computer models are often used to analyzing and setting a goal for greenhouse gas emissions reduction, the lack of understanding of the uncertainties in the models is an increasing concern of policy makers [13]. For instance, different estimates that arise from uncertainties in different models are fueling some of the global climate change debate [14]. Thus, it is important for policy makers to understand how modeled outcomes may change as a result of applying different policy options, and if the outcomes can be distinguished given the uncertainties of the models used. It is also important to know where models disagree with each other and to identify the modeling assumptions that cause the differences. Therefore, establishing and communicating model fidelity is an important task, which must parallel model development efforts. Identifying the uncertainty associated with model assumptions also provides a rational framework for model development and allows developers to focus efforts on the most important features [15]. This way, the policy scenarios will have a range of possible outcomes and policy makers should then choose one that reduces uncertainty in the intended outcome.

4. Inducing Environmental Innovations

This section explains in detail the process of environmental innovation and then an effective way to induce firms toward environmentally innovative products and services. To do so, this article introduces a Social Demand Articulation (SDA) framework where knowledge and information flows are the key drivers that formulate environmental demand and eventually stimulate corporate intention to adopt environmental protection in their business strategy.

For effective environmental policy-making, the steps to establish scientific evidence as well as public awareness regarding an environmental problem are important. If one of the two steps is missing, mere environmental protection movement could mislead society to achieving the political agenda of a particular interest group. Below is a short discussion on the processes of environmental innovation and SDA framework to induce firms toward environmentally friendly technology/product development [16].

Environmental innovation requires having a technological capacity in order to make changes. The serious lack of knowledge regarding the environmental damage caused by high-solvent paints and alternative technologies is a major obstacle in promoting low-solvent paints in the Netherlands [17]. In the past, firms themselves had to carry out every task from basic research to product development for new environmental technologies. In the future, however, it will be more desirable if environmental knowledge is accumulated and readily available throughout society so that firms can develop an environmental technology at lower cost. This raises the issue of what types of public R&D projects must be performed and how their outputs can be effectively transferred to firms for the purpose of environmental innovation. The current industrial structure, however, does not allow for easy emergence of the requirements for environmental innovation. It appears that firms are most sensitive to market requirements, not environmental considerations. The social demand for environmental improvement is largely external to business.

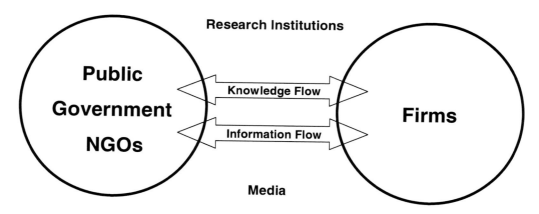

Figure 1. Knowledge and Information Flows Necessary for Social Demand Articulation.

Lee et al [16] first formulated a framework for *social demand articulation*, in which environmental knowledge and information flows connect between firms and societal stakeholders (*e.g.*, the public, government, and NGOs) as in Figure 1. Through this process, the stock of knowledge available to all stakeholders expands, and awareness level of all stakeholders is improved. The White Papers on the Environment in Japan, for example, point out the lack of knowledge and information about environmental issues is often a major obstacle in environmental innovation [18]. Transfers of knowledge and information will induce the adoption and diffusion of new environmental technologies [19]. In doing so, social demands—often called externalities—are internalized into the corporate decision-making process so that external considerations can coevolve with other important functional requirements and market demands from the product development stage. Specific roles and components of knowledge flow and information flow are summarized in Table 1 and explained below.

Table 1. Components of Knowledge and Information Flows.

Knowledge Flow		Information Flow	
Evidential Flow	Scientific evidence and knowledge of environmental problem	*Educational Flow*	Diffusion of information about potential dangers of environmental problem
Innovating Flow	Technological solution for environmental problem	*Evaluational Flow*	Assessment of environmental performance of particular products.

4.1. Knowledge Flow

The knowledge flow contains the science of environmental effects resulting from economic activities. This knowledge flow is important in two aspects. One is that it provides a credible reason for the fact that there is an environmental problem. The other aspect of the knowledge flow is to provide a scientific capability for solving environmental problems. Both are important and discussed below as 'evidential flow' and 'innovating flow.'

4.1.1. Evidential Flow

This form of knowledge flow has a 'proving' role. It confirms the fact that a particular environmental problem is scientifically true. For example, whether global warming is scientifically valid or even worrisome is still the point of debate. Some nations are reluctant to take precautionary actions without a solid scientific evidence for global warming. In Japan, the implementation of environmental measures was postponed during the period 1945-69 because of the insufficient knowledge among enterprises concerning the impact of pollutant emissions from concentrated industrial complexes on human health and the environment [20]. An environmental movement without a solid evidential flow may only help some interest groups achieve their political agenda. Before an environmental policy is made, therefore, regulatory bodies must thoroughly examine the scientific evidence behind environmental issues. News and media must also carefully examine the scientific causes of environmental problems before making them a social issue.

4.1.2. Innovating Flow

This form of knowledge flow has a 'innovating' role. It provides technological knowledge to make environmental innovations. To acquire environmental innovation capability, it is important to conduct environmental R&D. Firms may not have enough R&D resources to deal with environmental issues, so in the early stage, it is useful for government to conduct public R&D to increase the knowledge base. Universities are also in a good position to conduct environmental R&D.

It is necessary that basic scientific knowledge be accumulated in order for firms to perform further research and develop a new product. In this regard, public R&D for environmental science and technology can be utilized as a reservoir of knowledge to be transferred to firms. As previously discussed, firms themselves may not know the new technology or have enough financial and human resources for environmental innovation. Further, such environmental technology development may accompany a high risk. Public R&D with detailed scientific and technological knowledge about a new environmentally friendly product helps reduce uncertainty about the cost and benefits of adopting the product. Public R&D is also advantageous for the development of socially indispensable technologies because it is not driven by corporate profits. Therefore, public R&D may well play an important role with respect to science and technology development for environmental improvement [19]. Research papers, R&D expenditures, and patents are some tangible measures of the knowledge flow.

4.2. Information Flow

The major role of the information flow is to make the general public better informed of the importance of environmental conservation through media including television, radio, newspapers, magazines, bulletins, and films. Events such as environmental week and school education programs are also good carriers of the information flow.

4.2.1. Educational Flow

This form of information flow has a 'promotional' role. It spreads among the public the information about the potential dangers of the environmental problems. However, without scientific evidence, mere information spread can mislead the public. Japanese data show increasing newspaper articles addressing environmental issues raise people's awareness for environmental conservation [18]. The government, civic organizations, and even companies themselves can initiate these information flows. In particular, companies increasingly hold more public relations activities to gain consumer acceptance about their environmental superiority over other firms. The results of a survey conducted for 558 Japanese firms in 1993 and 906 in 1994 show around 15% of the surveyed firms doubled or more the public relations and education funds for environmental protection and company's green image promotion over the period 1993-94. They further believed that such public relations activities ought to be increased to promote eco-labeled products and enlarge individual awareness on the environment [21]. In order to promote these information flows, it is important to establish information supply networks and also support private enterprise activities [22].

4.2.2. Evaluational Flow

The role of this information flow is 'evaluating' the environmental performance of particular products. When information about a certain product's low environmental performance along with a scientific evidence for its environmental hazards spreads out to society, a social demand arises for improvement. For example, the environmental impact analysis statements that were made publicly available through the US National Environmental Protection Act of 1970 caused a number of court cases from citizens [23]. The British Society for Social Responsibility in Science learned about the occupational health hazards of working with such materials as asbestos and polyvinyl chloride (PVC) and acted as a pressure group supplying the information to trade unions and publicizing the dangers [23]. Thus, the cultural atmosphere in which the environment is valued highly puts a credibility pressure on firms as well as on consumers to improve environmentally [17].

This forms a clear signal to the firm that consumers want greener products. Such a signal can lead to R&D efforts of firms and eventually result in innovating knowledge flow. This also conveys information about particular products and their environmental performance. For superior environmental quality products, consumers (especially 'early green adopters') make up a core group of users which can then create a green market. Consumers may also have good information/ideas about new materials or even new designs. From this, they can suggest innovative products for companies to consider.

Green campus initiatives are a good example of utilizing information flows. Recently, Korean Council on Green Campus Initiatives was launched. Sangji University is one of the leading universities in Korea that have pursued eco-friendly green campus with active participation of faculty, staff and students. According to the chairman of green campus committee of Sangji University, two measures were mainly used to increase information flows regarding green campus movement. They first actively used media to promote the green campus tasks and to raise the university-wide awareness. Various forms of media such as campus newspapers, bulletin boards, weekly e-mailing, and web magazines were used persistently to remind the university members about the importance and the current status of the green campus movement. As a result, most faculty, staff and students became very familiar with green campus initiatives and even some of them suggested ideas for it. The other unique measure is that Sangji University introduced in every school (e.g. schools of engineering, management, humanities, etc.) a new curriculum on green, eco-friendly campus. Initially, there were significant objections on having a mandatory green campus-related course in the university curriculum. The university president met with deans and department heads to persuade them about the need of environmental education. By having the environmental curriculum, the students not only learned various ways and benefits of having a green campus, but also they gathered best-practice cases around the world and devised ways to achieve green campus through class assignments and research. Student groups appeared in order to continue their learning and green campus initiatives even outside of class. These student groups also made self-regulations on what should be and should not be done to preserve a green, eco-friendly campus [24]. This case clearly shows that Sangji University effectively created and utilized educational as well as evaluational information flows.

5. INTER-LINKAGE OF KNOWLEDGE AND INFORMATION FLOWS

While all four types of flows are inter-linked and influence one another, it is possible to draw a logical connection between them. Regarding a particular environmental problem, evidential flow is first formed. The environmental impacts are assessed and scientifically proven. With this, educational flow is developed to spread the facts about the environmental problem. This eventually forms a green demand where educated consumers show clear signs for greener products and help bring about environmental innovations. Firms would respond to this by increasing R&D investment in environmental performance improvement of their products.

In this sense, information flow can lead to a demand-pull innovation. Knowledge flow can lead to a science-push role. As known, demand-pull and science-push are intertwined to make environmental innovations. [25] Also note that top management's support is extremely important for successful implementation of green initiatives. To have the active participation of the entire organization, a strong drive by top management is the key to effectively increase the awareness and involvement level of the constituents. The Sangji University case clearly demonstrates this because without the strong support of the university president, the green campus movement would not have had the success.

Then what should be the role of environmental policy to leverage this inter-linkage between knowledge and information flows? One way is to use economic incentives to form a green market directly or indirectly. Firms gain additional profits by making green products. However, such economic incentives are most effective when the knowledge and information flows parallel them. The evidential and innovating knowledge flows as well as educational and evaluational information flows must exist and be raised to a certain level. In other words, economic incentives should become effective when the social demand for environmental protection clearly exists and is accepted by companies.

For successful diffusion of green technology, three step strategies are needed. First step is increasing the brand recognition of green technology in economy and society. Since green technology requires a higher investment cost than conventional technologies, government should provide an incentive to those companies doing R&D of green technology. Also government should advertise the advantages of green technology to the public for increasing participation and consensus on its necessity within society. The second step is changing the standard of choosing national R&D projects. This means government should consider the "greenness" of a new technology when national R&D budget is allocated. The final step is building a concrete roadmap of green technology with particular tasks and objectives. One good example is the Japanese government, which already constructed a concrete roadmap up to 2050 of green technology including 21 core technologies. It has invested the largest amount of money to green technology than any other advanced countries [26]. The roadmap of green technology is the milestone of national green technology diffusion, so building a concrete roadmap is an essential strategy.

6. INDUCING SUSTAINABLE INNOVATIONS FOR AVIATION SECTOR

In this section, the key concepts of the SDA processes are applied to the air transportation sector. The environmental issues of the aviation industry is briefly discussed and the cases of aircraft noise abatement and jet emissions reduction are comparatively discussed from the view point of the SDA framework.

Aviation energy consumption and emissions are expected to increase and constitute a greater proportion of the total anthropogenic climate impact. Air transport growth has outpaced reductions in energy intensity and will continue to do so through the foreseeable future, perhaps by an increased margin. Unless measures are taken to significantly alter the dominant historical rates of change in technology and operations, the impacts of aviation emissions on local air quality and climate will continue to grow [12]. Given the technological, operational and economic constraints in controlling the increase in aircraft energy consumption and emissions impacts, this section examines the question, "How should the aviation sector induce environmental innovations in aircraft technologies and operations that can achieve sustainable industry growth?"

For air transport industry, the word, sustainable would mean increased mobility for people around the world, profitable industry growth, protection of the environment and continuous improvement in safety and security. Regarding protection of the environment, it is important to stabilize, reduce or even eliminate conventional greenhouse gas emissions from aircraft engines. To do so, the aviation sector must consider not only the technological, operational solutions and economic costs but where aviation stands in relation to society. Currently, the public demand to reduce the aircraft emissions impacts on global climate change is not strong. For the cases of aircraft noise or automobile emissions, a clear demonstration of health damages followed by strong public pressure to reduce the environmental nuisances have led to dramatic improvements in both technologies and the way the systems are operated [27]. On the contrary, the public awareness about the effects of aircraft engine emissions on global warming is relatively low today. Also, there are large scientific uncertainties about the potential climate change effects of jet engine emissions discharged at altitude [28].

In this regard, it will be important to continue to advance atmospheric science on the environmental effects of jet engine emissions and raise general public awareness about aviation's impacts on local air quality as well as the global atmosphere. Such efforts along with modeling and assessments of various emissions reduction options will be make an important step toward sustainable air transport. To make this point clear and help formulate a strategy to induce environmental innovations in aircraft technologies and operations, this section derives insight from the case of aircraft noise reduction. Based on the case study, it is possible to understand the drivers of technological, operational change and evolution of effective government regulations to guide policy toward sustainable air transport.

6.1. Drivers of Aircraft Noise Reduction: Strong Social Pressure

Historically, strong public demand supported by the scientific evidence of health damage caused by aircraft noise and subsequent government regulations to limit the operation of noisy aircraft have led to large reductions in community noise around airports. This section first reviews the metrics used to judge the magnitude and scope of aviation noise impacts.

The U.S. Federal Aviation Administration (FAA) employs the day-night noise level (DNL) to determine the compatibility of airport-local land uses with aircraft noise levels. At 55 dB DNL (indoors or outdoors) a community will generally perceive aviation noise as no more important than various other environmental factors with about 3% of the population highly annoyed. At 65 dB DNL, 12% of the population may be highly annoyed and the community will generally consider aviation noise as one of the important adverse aspects of the environment. For comparison, the median outdoor exposure to noise in urban areas is 59 dB DNL with a range of 58 to 72 dB. Corresponding ranges for suburban and wilderness areas are 48 to 57 dB and 20 to 30 dB, respectively. For both commercial and military aviation, most complaints regarding aviation noise come from areas with a DNL less than 65 dB [27].

The historical evolution of noise exposure in these zones and future projections developed by the FAA show over 80% reductions in the population affected by commercial aviation noise. The large reductions have resulted primarily from two factors: low noise aircraft operations enabled by advances in aircraft communication, navigation and surveillance, and air traffic management (CNS/ATM) technology, and the phase-out of high noise aircraft through regulatory action enabled by the availability of improved engine technology (e.g. as increased bypass ratio). The importance of the latter, made possible through international agreement and enacted through the 1990 Airport Noise Control Act (49 USC App. 2151 to 2158), is quite significant. While the total number of aircraft phased-out corresponded to 55% of the fleet in 1990, that portion of the fleet contributed to more than 90% of the total DNL levels at airports. The cost of prematurely retiring these aircraft has been estimated at between $5B and $10B [29]. Over the next 20 years, estimates by the FAA suggest that the number of people affected by commercial aircraft noise in the U.S. will be constant; increases in the number of operations are expected to offset projected improvements in technology within the fleet [27].

What factors drove this drastic reduction in community noise from aircraft? The key drivers were knowledge accumulation and information diffusion about the health damage caused by aircraft noise and subsequent government regulations responding to the public demand seeking relief from jet aircraft noise. It was in the 1960's that people gained scientific knowledge that noise produces a variety of adverse physiological and psychological effects. Common among these are speech interference and sleep disturbance, which may result in reduced productivity for a variety of tasks associated with learning and work. Definitive evidence of other non-auditory health effects as a direct consequence of aviation noise is not available [30], but some studies suggest such connections. In 1965, Rogen and Olin [31] showed that high levels of aircraft noise that commonly exist near major commercial airports increased blood pressure and contributed to hearing loss. Some research also indicated that aircraft noise contributed to heart diseases, immune deficiencies, neurodermatitis, asthma and other stress related diseases [31, 32].

The information about the health damage caused by aircraft noise around the vicinity of airports was quickly spread out via media and environmental groups. The diffusion of such information among the public resulted in civil lawsuits totaling billions of dollars in such cities as New York, Chicago, Los Angeles and Washington, D.C. Newspapers, journals and reports highlighted the potential danger of aircraft noise to human health. In 1975, the U.S. Environmental Protection Agency (EPA) reported, "An estimated 16 million Americans are now subjected to a wide range of aircraft noise. Such noise can interfere with the normal use of homes and yards and poses a particularly serious problem for such institutions as schools and hospitals" [33].

The knowledge accumulation and information diffusion about the health damage of aircraft noise caused U.S. government to initiate the Noise Control Act of 1972 (42 USC 4901 to 4918). Under the NCA, engine-nacelle combinations must be quieted and therefore all civil subsonic turbojet engine powered airplanes must comply with the noise level requirements of FAR 36 [33]. The federal noise regulations in FAR 36 defined aircraft according to four classes and set a phase-out schedule based on the weight and number of engines, and resulting noise level under various operating conditions. Around 20% of the current fleet already achieves a noise target 14dB better than the current (Chapter 3) standards. The majority of aircraft designed in recent years are already quiet enough to attain the impending (Stage 4) standards [34].

The substantial reductions in the noise of individual aircraft are largely due to improvements in aircraft technology. Aircraft noise arises from engines and from the movement of turbulent air over the airframe. To date noise reduction has focused mainly on reducing engine noise, and it is becoming increasingly important to tackle noise from the airframe, which may be more challenging to reduce. To address continued noise concerns the FAA has adopted a 'Balanced Approach'—a combination of operational changes, land-use planning, abatement (e.g., through insulation programs) and technological improvements (e.g., through increasingly stringent noise standards) [35].

In sum, the social demand to reduce aircraft noise was well formulated by scientific evidence and information diffusion of the detrimental effects to human health. It then triggered government regulations, which have driven the technological and operational innovations to reduce the noise impact around the vicinity of airports.

6.2. Drivers of Aircraft Emissions Reduction: Fuel Cost Is Main Reason to Improve Fuel Efficiency

Regarding the historical reductions in aircraft energy use and emissions, the fuel cost has been the main driver for aircraft fuel efficiency improvement. Fuel efficiency gain was strongest during the 1970's when the oil prices were highest [36]. Since the oil prices are reaching the historical high level again, airlines are eager to adopt advanced aircraft types with much improved fuel economy. In order to achieve constant improvement in aircraft fuel efficiency as well as reduction in jet engine emissions adversely impacting global climate and local air quality, there should be a stronger social pressure for the aviation sector. Currently the social demand for low emission aircraft is not strong enough because the general public is not well aware of the effects of aviation emissions on global climate change. At the same time, the effects of aircraft engine emissions on the global atmosphere are not well understood scientifically, either [28].

To address the potential impact of aviation emissions on climate change, international dialogues have taken place. In the 1944 Chicago Convention, the International Civil Aviation Organization (ICAO) was first created as the UN specialized agency with authority to develop standards and recommended practices regarding all aspects of aviation, including certification standards for emissions and noise. Since 1977, ICAO has promulgated international emissions and noise standards for aircraft and aircraft emissions through its Committee on Aviation Environmental Protection (CAEP) [28].

The ICAO CAEP is primarily responsible for monitoring the aviation sector's emissions and noise reduction efforts and seeking further options to mitigate the impacts of aviation on community noise, local air quality and the global atmosphere. Over the years, CAEP has set aircraft engine certification standards and phase-outs of noisy aircraft. In its recent meeting (CAEP/6), increasing the stringency of NOx emissions standards was one of the issues under consideration [37]. The ICAO CAEP has also developed International Standard and Recommended Practices for the control of fuel venting and of emissions of carbon monoxide, hydrocarbons, nitrogen oxides and smoke from aircraft engines over a prescribed landing/takeoff (LTO) cycle below 3,000 feet. While there is no regulation or standard for aircraft emissions during cruise, these LTO standards also contribute to limiting aircraft emissions during cruise [28].

In a broader perspective of climate change, the UN Framework Convention on Climate Change seeks to stabilize atmospheric greenhouse gases from all sources and sectors, but it does not specifically refer to aviation. The Kyoto Protocol to the Convention, adopted in December 1997, is the first international initiative to include two provisions that are particularly relevant to aviation. First, the Kyoto Protocol requires industrialized countries to reduce their total national emissions by an average of 5% for the average of the period 2008 to 2012 compared to 1990 the level. Second, the Kyoto Protocol's Article 2 contains the provision that industrialized countries pursue policies and measures for limitation or reduction of greenhouse gases from aviation bunker fuels [12].

As a result of the continued effort of international organizations including ICAO, aviation-related energy and environmental issues have now brought greater awareness among the air transport industry stakeholders. This is changing the way current aircraft are operated and new aircraft are developed. European airports and airlines are particularly concerned with the increasing environmental impacts of jet engine emissions. For example, Zurich Airport has imposed an emissions surcharge to its landing fee based on engine certification information. An aircraft engine is classified within one of five groups subject to an emission charge in 0 to 40% to the landing fee. This Zurich emission charge intends to provide an incentive to airlines to fly their lowest NOx emitting aircraft into Zurich and accelerate the use of the best available technology [28]. Virgin Airways has earmarked all of the profits to financing research and development of alternative fuels. In early 2008, it would conduct a demonstration flight of its Boeing 747 jets using biofuel — the first airborne test of a renewable fuel by a commercial jet [38].

Airlines in North America, too, are more conscious of environmental impact of jet engine emissions than ever. According to the Air Transport Association of America, many of its member carriers have already adopted such operational practices as continuous descent approaches, which have the potential to significantly reduce noise, fuel burn and emissions on every landing [39]. Some non-profit organizations are also working to circulate information on the environmental impacts of jet engine emissions and deliver public opinions to policy makers. For example, Sustainable Travel International is an advocacy group for eco-friendly travel, is actively involved with educating air carriers and travelers on environmental issues and suggesting ways the aviation sector can offset its emissions discharges.

In sum, this section discussed that the main reason for aircraft fuel economy improvement has been to lower fuel cost, and that aircraft manufactures and airlines are increasingly more conscious of global climate change due to jet engine emissions. However, the scientific knowledge and public awareness about the impacts of aviation emissions on the global atmosphere are still at a low level. This is the key difference from the case of aircraft noise reduction where the scientific evidence and strong public demand have induced a large decline in the level of community noise from aircraft. To examine this difference quantitatively, the next section presents a short data analysis on the social factors that drive the environmental innovations of aircraft systems.

6.3. Quantitative Comparison of Social Drivers of Aviation Noise and Emissions Reduction

This section offers some empirical evidence that knowledge accumulation and information diffusion were stronger for the reduction of aviation noise than emissions. The level of knowledge accumulation was measured by the number of research papers and that of information diffusion by the number of newspaper articles that contain 'aviation (or airport) noise' and 'aviation (or aircraft) emission' as keywords. 'Health' and 'environment' were also included in the keyword search of the research papers and newspaper articles. The EBSCO research database containing selected full text for 25 U.S. and international newspapers was used to search the keywords for the period 1999-2007. The ISI Web of Knowledge database was used to browse through published research papers including articles, reviews, letters, and editorial materials for the period 1996-2007.

Figure 2 shows the annual data for the number of research papers that correspond to the keyword search. While the research papers on both aviation noise and emissions are generally increasing in number, it is clear that those research papers that deal with aviation noise are about 50 percent more than those that examine aviation emissions. This indicates that knowledge accumulation has been significantly greater for aviation noise-related areas than for aviation emissions-related ones.

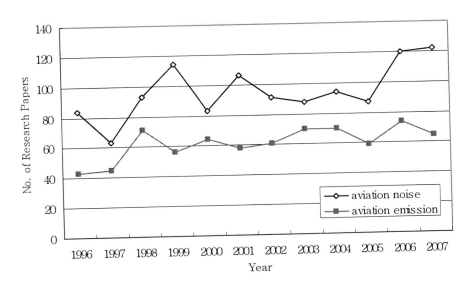

Figure 2. Trends in Knowledge Accumulation for Aviation Noise vs. Emissions.

In terms of information diffusion, Figure 3 displays the annual trend in the number of newspaper articles matching the keyword search. While the newspaper coverage on both aviation noise and emissions is growing, the newspaper articles that feature aviation noise issues are much greater in number than those that cover aviation emissions issues.

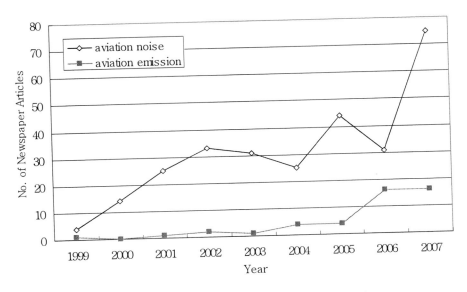

Figure 3. Trends in Information Diffusion Regarding Aviation Noise vs. Emissions.

While an in-depth empirical analysis requires examining a broader range of data, current results show knowledge accumulation and information diffusion have been stronger for aviation noise-related fields. This trend corresponds to the fact that a greater level of aviation noise reduction was possible in the past as the knowledge and information available to societal stakeholders have a significant impact on inducing the environmental performance improvement of industry.

To expedite environmentally conscious innovations for low emission aircraft, increased amount of knowledge and information should flow among aviation firms and the rest of societal constituents, such as citizens, government, and non-government organizations. The knowledge accumulation is important in two aspects. One is that it provides a credible reason for the fact that there is an environmental problem. The other aspect of knowledge is to provide a scientific capability for solving environmental problems. It is necessary that basic scientific knowledge be accumulated in order for firms to perform further research and develop a new product (Lee et al., 2006). The major role of information diffusion is to make the general public better informed of the importance of environmental conservation through media including television, radio, newspapers, magazines, bulletins, and films. Events such as environmental week and school education programs are also good carriers for information diffusion [16].

For knowledge accumulation, the impacts of jet engine emissions on global climate change must be scientifically better understood. Research and development should continue to create technological and operational solutions to the environmental concerns caused by jet engine emissions. To raise public awareness level (i.e., information diffusion), much more active dissemination of and education on the environmental impacts of aircraft engine emissions are needed. All of this will construct an environmentally conscious market for the aviation sector and eventually give aviation firms a corporate social responsibility that environmental performance improvement be adopted as part of their business strategy.

7. SUMMARY AND CONCLUSIONS

Environmental policy is an expression of human value to preserve the natural environment. Since there are more than economic, political reasons for environmental protection, environmental policy must covey a clear picture why the environment should be restored and protected and what kind of future society it aims to create. Environmental policy without proper knowledge or information flows regarding a particular environmental issue cannot effectively achieve its intended outcome. For effective environmental policy-making, this article discussed the social demand articulation process by which knowledge and information flows mobilize the requirements needed for a successful environmental innovation. Institutionalizing knowledge and information flows among firms and societal stakeholders can form an environmentally conscious market in which greener products are valued highly and give competitive advantage to environmentally innovative firms.

ACKNOWLEDGMENTS

The author would like to thank Tae Hoon Jeong and Bo Sung Kim for their contribution on discussing the goals of environmental policy and Miyoung Nam and Bonjin Koo for their contribution on discussing the strategies to diffuse green technology.

REFERENCES

[1]   Rondinelli, D.A. & Berry, M.A. (2000). Environmental Citizenship in Multinational Corporations: Social Responsibility and Sustainable Development, European Management Journal, 18(1), 70-84.

[2]   Sutton, R. (1999). The Policy Process: An Overview, Overseas Development Institute, Chameleon Press, London.

[3]   Modernising Government. Presented to Parliament by the Prime Minister and the Minister for the Cabinet Office, United Kigdom, March 1999.

[4]   Better Policy Making, Government of South Australia, http://www.nationalschool. gov.uk/policyhub/better_policy_making/

[5]   Resetar, S.A., Camm, F.A. & Drezner, J.A. (1998). Environmental Management in Design: Lessons from Volvo and Hewlett-Packard for the Department of Defense, Office of the Secretary of Defense, U.S. National Defense Research Institute.

[6]   An Internal Guide To Policy Making, Government of South Australia

[7]   Welford, R. & Starkey, R. (eds), (1996). Business and the Environment, Taylor & Francis, Bristonl, PA.

[8]   Jeong, T.H. & Kim, B.S. (2008). Study on the Rationale of Environmental Protection, Engineering and Technology Management Class Paper, December.

[9]   Crampton, T. (2007). "More in Europe worry about climate than in U.S., poll shows". International Herald Tribune. http://www.iht.com/articles/2007/01/04/ news/poll.php. January 4.

[10]   Global warming controversy, http://en.wikipedia.org/wiki/Global_warming_ controversy

[11]   Shafieea, S. & Topal, E. (2009). When Will Fossil Fuel Reserves Be Diminished?, Energy Policy, 37(1), 181-189.

[12]   Lee, J.J., Lukachko, S.P., Waitz, I.A., & Schafer, A. (2006). Historical and Future Trends in Aircraft Performance, Cost and Emissions, Annual Review of Energy and the Environment, Volume 26.

[13]   Cipra, B. (2000). Revealing Uncertainties in Computer Models. Science 287 (February), 960–961.

[14]   Reilly, J., Stone, P.H., Forest, C.E., Webster, M.D., Jacoby, H.D., & Prinn, R.G. (2001). Climate Change: Uncertainty and Climate Change Assessments. Science 293 (July), 430–433.

[15]   Lee, J.J., Waitz, I.A., Kim, B.Y., Fleming, G.G., Holsclaw, C.A., & Maurice, L. (2007). System for Assessing Aviation's Global Emissions (SAGE), part 2: Uncertainty Assessment. Journal of Transportation Research, Part D, 12(6), 381-395.

[16] Lee, J.J., Gemba, K., & Kodama, F. (2006). Analyzing the Innovation Process for Environmental Performance Improvement, Technological Forecasting and Social Change, 73(3), 290-301.

[17] Kemp, R. (1997). Environmental Policy and Technical Change: A Comparison of the Technological Impact of Policy Instruments, Edward Elgar, Cheltenham, United Kingdom.

[18] The White Paper on the Environment (in Japanese), Ministry of the Environment, Japan, 1997.

[19] Jaffe, A.B., Newell, R.G. & Stavins, R.N. (2000). Technological Change and the Environment, National Bureau of Economic Research, Working Paper 7970.

[20] Business and Environmental Governance, Institute for Global Environmental Strategies, 1999, Japan.

[21] Earth–Human Environment Forum: Environmentally Friendly Activities of Firms (in Japanese), Report for 1994, Japan, 1995.

[22] Government of Japan, Quality of the Environment in Japan, Ministry of the Environment, Japan, 1989.

[23] Coombs, R., Saviotti, P. & Walsh, V. (1987). Economics and Technological Change, Macmillan Education, Hampshire, the United Kingdom.

[24] Korean Council on Green Campus Initiatives, Kick-off Meeting, November 25, 2008.

[25] Chung, S.Y. (2006). Technology and Management (in Korean), KyungMoon Publishers, pp. 53-55.

[26] World Economy Update, Korea Institute for International Economic Policy (KIEP), August 14, 2008.

[27] Waitz, I.A., Lukachko, S.P. & Lee, J.J. (2005). "Military Aviation and the Environment: Historical Trends and Comparison to Civil Aviation." Journal of Aircraft, 42(2), p. 329-339.

[28] Lee, J.J. (2009). "Can we accelerate the improvement of energy efficiency in aircraft systems?" Energy Conversion and Management, DOI: 10.1016/j.enconman.2009.09.011

[29] Morrison, S.A., Winston, C. & Watson, T. (1999). "Fundamental Flaws of Social Regulation: The Case of Airplane Noise." The Journal of Law and Economics, 42, p. 723-743.

[30] Federal Interagency Committee on Aircraft Noise (FICAN), Federal Agency Review of Selected Airport Noise Analysis Issues, 1992.

[31] Rosen, S. & Olin, P. (1965). "Hearing loss and coronary heart disease." Archives of Otolaryngology, 82, p. 236.

[32] World Health Organization (WHO). Guidelines for Community Noise, Edited by B. Berglund, T. Lindvall, and D.H. Schuela. Cluster of Sustainable Development and Healthy Environment. Department of the Protection of the Human Environment, Occupational and Environmental Health, 1999.

[33] Environmental Protection Agency, "EPA Proposes Quieting of Jet Airplanes." Press Release, January 31, 1975.

[34] Bearden, D.M. (2006). Noise Abatement and Control: An Overview of Federal Standards and Regulations. Environmental Policy Resources, Science, and Industry Division. CRS Report for Congress, RS20531.

[35] Parliamentary Office of Science and Technology, Aircraft Noise. U.K. Parliament. Number 197, June, 2003.

[36] Greene, D.L. (1992). Energy-efficiency improvement potential of commercial aircraft. Annu. Rev. Energy Environ. 17, p. 537–573.

[37] Lee, J.J., Waitz, I.A., Kim, B.Y., Fleming, G.G., Holsclaw, C.A. & Maurice, L. (2007). System for Assessing Aviation's Global Emissions (SAGE), part 2: Uncertainty Assessment. Journal of Transportation Research, Part D, 12(6), p. 381-395.

[38] Clark, N. (2008). "Virgin Atlantic Plans a Biofuel Flight." The New York Times, January 15.

[39] May, J. ATA Letter to Senator Lieberman and Senator Warner Raising Concerns to S. 2191. The Air Transport Association of America, November 2007.

In: Handbook of Environmental Policy
Editors: Johannes Meijer and Arjan der Berg

ISBN 978-1-60741-635-7
© 2010 Nova Science Publishers, Inc.

Chapter 18

REVIEW OF ENVIRONMENTAL GOVERNANCE IN KENYA: ANALYSIS OF ENVIRONMENTAL POLICY AND INSTITUTIONAL FRAMEWORKS

*Caleb Mireri** and Sammy Letema*
Department of Environmental Planning & Management,
Kenyatta University, Nairobi, Kenya

ABSTRACT

The main aim of this paper is to show that Kenya has made progress in institutionalising environmental governance, particularly following the Rio Conference on Environment and Development. Prior to the Rio Conference, environmental management was scattered in the line ministries with no clear focus on sustainable development. The paper shows that the country has elaborate legislative framework with instruments that can significantly contribute to sustainable environmental management. However, the implementation of the legislation faces a number of challenges including lack of policy on environmental management and weak capacity. It is evident that serious concerted efforts must be directed at capacity building to make the legislative intentions of a good and healthy environment for all in Kenya a reality.

1. INTRODUCTION

Rio Conference on Environment and Development in 1992 gave environmental management major impetus. Member states of the United Nations committed themselves to the principles of sustainable development as per Agenda 21. Sustainable environmental management is an integral part of Agenda 21. Following the conference, member states of the United Nations of which Kenya is one, committed themselves to initiate processes to institutionalise good environmental governance for sustainable development. As part of its commitment to the UN Conference, Kenya implemented National Environment Action Plan

* E-mail calebmireri@yahoo.com.

(NEAP) in 1994 to provide a basis for up-scaling environmental management in Kenya. NEAP process culminated into Environmental Management and Co-ordination Act (EMCA) of 1999, which came into force in 2002.

Prior to the promulgation of EMCA in 1999, environmental issues had no *Locus Standi*. Environmental management issues were dealt with by several sectors in different ministries and good environmental governance was not highly prioritised. In cases of violations to the environment affecting the public generally, then the person who had an interest and right to sue was the Attorney General and not a private citizen. EMCA has created instruments for good environmental governance with potential to create a clean and healthy environment if only it is effectively and efficiently implemented.

Environmental Management and Co-ordination Act (EMCA, 1999) states that environment includes the physical factors of the surroundings of human beings including land, water, atmosphere, climate, sound, odour, taste, the biological factors of animals and plants and the social factor of aesthetics and includes both the natural and the built environment. The Act defines environmental management includes the protection, conservation and sustainable use of the various elements or components of the environment. The National Environment Management Authority (NEMA), however, have officers up to district levels and thus in the divisions, locations and sub-locations are absent. Moreover, NEMA officers lack transportation facilities, which is central to their effective in enforcement, compliance and violation of environmental rights or damage to the environment.   This paper, therefore, examines environmental policies and institutional framework in Kenya since independence. Despite good progress made in environmental governance in Kenya, especially passage of Environmental Management & Co-ordination Act, the implementation of the Act is faced with serious challenges of weak capacity and lack of policy framework.

2. HISTORICAL PERSPECTIVE OF ENVIRONMENTAL GOVERNANCE IN KENYA

Since independence in 1963, Kenya has formulated a whole range of policy and legal requirements geared towards environmental management. The environmental issues prior to 2002 were however handled by over 20 line ministries and departments dealing with environmental matters. These included among others Ministry of Environment and Natural Resources (MENR), the National Environment Secretariat (NES), the Forest Department, the Kenya Wildlife Service (KWS), and the Permanent Presidential Commission on Soil Conservation and Afforestation. Currently there were more than 70 different laws that either directly or indirectly apply to environmental management including Forest Act, Wildlife Management and Conservation Act, the Factories Act, Water Act, Agriculture Act, and Chiefs Act. Moreover, the policies and regal requirements are not to be found in one body of policy document or statute or law, but in various national development plans, sessional papers, presidential decree/statements and Acts of Parliament.

Table 1. Major Environmental Events and Laws in Post Independent Kenya.

Events	Year
Sessional Paper No. 10 on African Socialism and its Application to Planning in Kenya	1965
Presidential Decree of 390Km^2 area in Amboseli area for wildlife conservation, which culminated in declaration of Amboseli as National Park in 1975	1971
National Development Plan (1974-1978)	1974
National Environment Secretariat	1974
Wildlife Conservation and Management	1976
Sessional Paper No. 3 on National Policy for Wildlife Conservation and Management	1976
Ministry of Environment and Natural Resources	1979
Forest Act of 1942, Amended in 1982	1982
Presidential Decree declaring aloe species protected	1986
Wildlife Conservation and Management (Amendment)	1989
Green Towns Project aimed at localising Agenda 21 (1992-1996)	1992
National Environmental Action Plan	1994
Physical Planning Act	1996
National Development Plan (1997-2001	1997
Environment and Development Policy	1999
Environmental Management and Coordination Act	1999
Ministerial Position Paper on Environmental Considerations in Land-Use Planning and Management	2000
Waste Management Regulations	2006
Water Quality Regulations	2006

The principal environmental agency prior to commencement of National Environment Management Authority (NEMA) in 2002 was NES. NES, however, lacked legislative mandate, autonomy and technical competence to lead an effective national environmental policy formulation. Furthermore, NES was largely ineffective in implementing many of its mandate including promoting and enactment of laws and regulations, enhancing enforcement of legislation, and encouraging scientific research (Juma et al, 1996 in Mugabe et al, 1997). Attempts were made to enhance the coordination role of NES via establishment of Inter-Ministerial Committee on Environment (IMCE), which was coordinated and chaired by MENR. However, IMCE operated in an *ad hoc* basis and lacked legislative mandate. Furthermore, the MENR, which housed and chaired the committee's activities had limited capacity and did not have much influence over other ministries, which were guided by their sectoral policies and interests.

Sessional papers and national development plans were major sources of policy since their was no environmental policy in place. The Sessional Paper No. 10 of 1965 on African Socialism and its Application to Planning in Kenya stated that "heritage of future generations depends on the adoption and implementation of policies defined to conserve natural resources and create physical environment in which progress can be enjoyed. The thoughtless destruction of forests, vegetation, wildlife, and productive land threatens our future and must be brought under control". The 1974-78 National Development Plan articulated the need to manage the environment for ecological, socio-cultural and economic reasons. It recognised the lack of appropriate institutional arrangements and policies as the main factor limiting environmental management: "Not only are the various arms of the government in disagreement or in confusion but there are no clear policies providing for environmental management and well established and coherent institutional system to implement the policies" (Juma et al, 1996). The 1997-2001 National Development Plan recognises the underlying causes of environmental degradation. It observed that: envireonmental management tools, including laws relating to the management of internationally shared resources, cross-border issues, environmental economics and accounting , and environmental impact assessments, have not been adequately developed for effective environmental management. In the plan the government makes a commitment to achieve successful environmental management through: implementation and enforcement of environmental laws; provision of economic incentives and penalties to encourage sustainable use of natural resources and ecological functions; increasing resource allocation for environmental management; making adjustments in taxation to promote sustainable use of natural resources; and instituting pollution charges (RoK, 1996).

Serious efforts to improve environmental governance in Kenya followed the Rio Conference on Environment and Development in 1992. Kenya as a signatory of Agenda 21 on sustainable development initiated the process of better environmental management beginning with National Environment Action Plan (NEAP) in 1994. The key components of 1994-1999 National Environment Action Plan (NEAP) are:

- Formulation of a national environment policy correlating the environment with economic development;
- Involving local communities and local authorities in formulating renewable resource management policy;
- The need for environmental impact assessment; monitoring of all development projects, including agriculture, irrigation, land allocation and tourism; and
- Review of tax laws to incorporate conservation measures.

The numerous central government agencies with overlapping, poorly coordinated, and limited capacities; and lack of strong local government environmental institutions hamper the development, implementation and enforcement of environmental policies and laws (Mugabe et al, 1997). The inherent weaknesses in sectoral policies led to the formulation of National Environmental Management and Coordination Act (EMCA) of 1999 in order to harmonise all the sectoral laws. However, EMCA did not lapse the existing sectoral laws. Moreover, National Environment Management Authority is located in the MENR, and thus may face the same problems as NES, which when it was housed in the MENR, it lost its powers to coordinate other ministries as well.

3. STATUS OF ENVIRONMENTAL POLICY

There is no comprehensive and coherent environmental protection policy in Kenya. The legislation exist without the policy. At the moment the government is in the process of formulating environmental policy. Whereas the policy should precede the legislation, in the case of Kenya the legislation is being implemented in the absence of policy framework. Nevertheless, some local authorities have prepared by-laws on environment management in their areas of jurisdictions. Moreover, the private sector enterprises i.e. industries, hotels, real estates developers as well as corporate bodies have prepared environmental management plans to meet ISO certification as well as comply with NEMA regulations by undertaking environmental impact assessment and audit for new and existing projects respectively.

Existing policies are fragmented and sectoral in nature, thus do not provide an enabling environment for other players i.e. line ministries, lead agencies and civil society to orient their activities towards common national environmental policy framework. Whereas we have a comprehensive environmental legislation promulgated in 1999, however, other sectoral statues that touch on environment were not amended and revised accordingly. Each sector or agency, therefore, sought to expand its authority leading to power struggles which result in the disruption of environmental management process. As such, environmental issues are uncoordinated and viewed as sector specific resulting in duplication of efforts, thus their effectiveness is marred. As such, Kenya has been unable to provide the necessary inter-sectoral policy for integration and co-ordination of environmental management.

4. HIGHLIGHT OF KEY LEGISLATIVE PROVISIONS OF ENVIRONMENTAL MANAGEMENT AND CO-ORDINATION ACT (EMCA)

General Principles of Environmental Management

Kenya's environmental legislation also known as Environmental Management and Co-ordination Act (EMCA) was passed into law in 1999 and came into force in 2002. EMCA provides for the creation of National Environment Management Authority to co-ordinate the various environmental management activities being undertaken by the lead agencies and promote the integration of environmental considerations into development policies, plans, programmes and projects with a view to ensuring the proper management and rational utilization of environmental resources on a sustainable basis for the improvement of the quality of human life in Kenya.

The general principle of EMCA as provided for in Section 3 of the legislation states that every person in Kenya is entitled to a clean and healthy environment and had the duty to safeguard and enhance the environment. The entitlement to a clean and healthy environment under subsection (1) includes the access by any person in Kenya to the various public elements or segments of the environment for recreational, educational, health, spiritual and cultural purposes. Other principles governing environmental management in Kenya are:

(a) public participation in the development of policies, plans and processes for the management of the environment;

(b) the cultural and social principle traditionally applied by any community in Kenya for the management of the environment or natural resources in so far as the same are relevant and are not repugnant to justice and morality or inconsistent with any written law;

(c) international co-operation in the management of environmental resources shared by two or more states;

(d) inter-generational and intra-generational equity;

(e) the polluter-pays principle; and

(f) the precautionary principle.

Administration of Environmental Legislation

EMCA (1999) Section 9 (1) creates National Environmental Management Authority. The object and purpose for which the Authority is established is to exercise general supervision and co-ordination over all matters relating to the environment and to be the principal instrument of Government in the implementation of all policies relating to the environment.

The Act, sections 24 and 25 respectively creates National Environment Trust Fund and Restoration Fund to facilitate environmental Management. The object of the Trust Fund shall be to facilitate research intended to further the requirements of the environmental management, capacity building, environmental awards, environmental publications, scholarships and grants. The object of the Restoration Fund shall be as supplementary insurance for the mitigation of environmental degradation where the perpetrator is not identifiable or where exceptional circumstances require the Authority to intervene towards the control or mitigation of environmental degradation.

The Act (Section 28) creates instruments to cushion the environment from environmental degradation. The Environmental Authority shall create a register of those activities and industrial plants and undertakings which have or are most likely to have significant adverse effects on the environment when operated in a manner that is not in conformity with good environmental practices. The Minister responsible for finance may, on the recommendations of the National Environment Council, prescribe that persons engaged in activities or operating industrial plants and other undertakings identified under subsection (28.1) pay such deposit bonds as may constitute appropriate security for good environmental practice. The Environmental Authority may, after giving the operator an opportunity to be heard, confiscate a deposit bond where the operator is responsible for environmental practice that is in breach if the provisions of this Act, and the Authority may in addition cancel any licence issued to the operator under this Act if the Authority is satisfied that the operator has become an habitual offender.

Section of 30 of the Act creates Provincial and District Environment Committee, which shall – (a) be responsible for the proper management of the environment within the province or district in respect of which they are appointed. (b) perform such additional functions as are prescribed by this Act or as may, from time to time, be assigned by the Minister by notice in the Gazette. The creation of environmental governance structures only up to the district level makes it difficult to reach the grassroots.

The formation of Complaints Committee under Section 32 of the Act is one of the important instruments for public participation in environmental management. The functions of Complaints Committee includes to investigate (i) any allegations or complaints against any person or against the NEMA in relation to the condition of the environment in Kenya; and (ii) on its own motion, any suspected case of environmental degradation, and to make a report of its findings together with its recommendation thereon to the National Environment Council.

Environmental Planning, Protection and Conservation

EMCA part IV creates instruments for environmental planning, specifically National Environment Action Plan Committee, preparation of National and District Action Plans after every five years to guide environmental management. EMCA part V has specific provisions to protect and conserve the environment, especially river, land, and wetlands, protection of traditional interests, hill tops, hill sides, mountain areas and forests, and conservation of biodiversity.

The Act (Section 57.(1)) provides for the use of fiscal instruments for the protection and conservation of the environment. The government can levy tax and other fiscal incentives, disincentives or fees to induce or promote the proper management of the environment and natural resources or the prevention or abatement of environmental degradation.

Environmental Impact Assessment and Monitoring

Part VI of the Act makes provisions for the preparation of an environmental impact assessment (EIA) for projects, which are likely to have negative impacts on the environment. Section 58 (1) states that notwithstanding any approval, permit or license granted under this Act or any other law in force in Kenya, any person, being a proponent of a project, shall, before submit a project report to the Environmental Authority before undertaking or causing financing, initiating or implementing by another person as specified in the Second Schedule (list of projects requiring EIA) to this Act. Environmental monitoring is an integral component of environmental management. Section 69. (1) states that, the Environmental Authority, in consultation with the relevant lead agencies, monitors:- (a) all environmental phenomena with a view to making an assessment of any possible changes in the environment and their possible impacts; or (b) the operation of any industry, project or activity with a view to determining its immediate and long-term effects on the environment.

Environmental Restoration, Conservation and Easement Orders

Part IX of the Act provides for environmental restoration orders, environmental conservation orders and environmental easements empowers the Environmental Authority to issue environmental restoration order to: restore the environment; prevent any action that can cause harm to the environment; award compensation for environmental damage; and levy a charge for environmental restoration. An environmental restoration order may contain such terms and conditions and impose such obligations on the persons on whom it is served as will, in the opinion of the Environmental Authority, enable the order to achieve all or any of the purposes specified in the order.

An environmental restoration order shall be issued to – (a) require the person on whom it is served to restore the environment as near as it may be to the state in which it was before the taking of the action which is the subject of the order; prevent the person on whom it is served from taking any action which would or is reasonably likely to cause harm to the environment; (b) award compensation to be paid by the person on whom it is served to other persons whose environment or livelihood has been harmed by the action which is the subject of the order; (c) levy a charge on the person on whom it is served which in the opinion of the Authority represents a reasonable estimate of the costs of any action taken by an authorised person or organisation to restore the environment to the state in which it was before the taking of the action which is the subject of the order. An environmental restoration order shall specify clearly and in a manner which may be easily understood:- (a) the activity to which it relates; (b) the person or persons to whom it is addressed; (c) the time at which it comes into effect; and (d) the action which must be taken to remedy the harm to the environment.

The object of an environmental easement is to further the principles of environmental management by facilitating the conservation and enhancement of the environment through the imposition of one or more obligations in respect of the use of the burdened land. An environmental easement may be imposed on and shall thereafter attach to the burdened land in perpetuity or for a term of years or for an equivalent interest under customary law.

5. CRITIQUE OF THE ENVIRONMENTAL LEGISLATION

EMCA has useful instruments for environmental management. Important instruments provided are: EIA, environmental restoration orders, environmental easement orders and environmental conservation. Effective implementation of the legislation can significantly contribute towards a clean and healthy environment. However, the legislation suffers from slow rate of implementation. The environmental legislation was passed into law in 1999 and took effect in 2002, but up to now it is yet to be fully operationalised. This is blamed in part on inadequate financial allocation for the establishment of structures. As a result, the environmental offices lack adequate number of personnel with the right mix of skills and knowledge. It is not uncommon to find district environment offices run by one environment officer lacking transport and other basic services to effectively oversee environmental management in an often expansive area. Despite good legislative provisions, its full benefits cannot be realised in the face of weak enforcement.

EMCA is a co-ordination framework relying on the goodwill of lead agencies for implementation. Environmental Authority co-ordinates environmental management, while implementation of implementation of environmental management is undertaken by lead agencies. NEMA does not have a mechanism to compel lead agencies to implement environmental management decisions. In many cases, lead agencies have their own legislative mandates to implement as such environmental management is seen as an incidental activity. It becomes even more difficult in cases where decisions of environmental authority conflicts with that of lead agencies. In such cases, it is unlikely to expect much needed co-operation of the lead agencies.

NEMA generates revenue as provided for in the Act, but it cannot directly spend such monies on environmental management. This occurs even in circumstances where state financial allocations for environmental management falls far below projected expenditure. The Environmental Authority generates revenue from sources such as Environmental Impact Assessment application and waste management. At the moment, NEMA is required to submit its revenue to the central government, and it can only spend allocations from the exchequer. One way of improving environmental governance is to allow the environmental authority to spend most of its revenue. This will even motivate to it to harness resources for better environmental management.

There is evidence of conflicting mandate of the environmental authority. The law states that NEMA is a co-ordination agency, but it is also involved in actual implementation of projects in some cases. This has been a source of conflict with some of the lead agencies. For example, for example EMCA empowers NEMA to management waste water while at the same time Water Act empowers Water Resources Management Authority to manage not only water resources but also waste water. This can potentially cause conflict among state department, thus hamper effective implementation of the policies and legislation.

EMCA creates environmental management structures only up to the district level making it difficult to co-ordinate grass root organizations. EMCA provides for District Environmental Committees as the lowest environmental management unit while Kenya's administrative structure evolves to Divisional, Locational, Sub-locational and Village levels. It implies that lawfully, the government cannot create environmental management structures at these lower levels making it difficult to effectively reach grass root communities.

The environmental legislation provides for community participation in environmental management. However, there are no mechanisms to ensure that community participation is effectuated. This is mainly because regulations, guidelines and standards for the each of the legislative provisions have been developed. For example, the Act states that those likely to be affected by a proposed project should be fully consulted during an EIA process, but there is no mechanism to ensure that this is realised.

6. CONCLUSION

Following the Rio Conference on Environment and Development, efforts have been made by the Kenya government to institutionalise environmental management. As a result, the Kenya government passed into law Environmental Management & Co-ordination Act in 1999. The environmental legislation has useful instruments, which when fully implemented can contribute to sustainable environmental management. Efforts have been to implement the Act and integrate environmental concerns in the development process. The implementation of environmental legislation is ongoing and it is unknown when it will be fully implemented. The Act is not yet fully implemented. Effective implementation of the Act is hampered by weak capacity and inadequate financial allocation by the state.

In order to scale up environmental management in Kenya, there is need to focus on capacity building of environmental authority, lead agencies and other key stakeholders. Capacity building should focus on recruiting and retaining adequate supply of personnel with the right mix of qualification; supply of requisite equipment and other facilities and popularising good environmental governance among the lead agencies. These will require long-term commitment of political leadership.

REFERENCES

Institute for Law and Environmental Governance (2003). Community Guide to Environmental Management in Kenya. Nairobi, Institute for Law and Environmental Governance.

Juma et al (1996) In Land We Trust: Environment Private Property and Constitutional Changes.

Nairobi, African Centre for Technology Studies

Mugabe, j. and Clark, N. (1998). Managing Biodiversity: National Systems of Conservation and Innovation in Africa. Nairobi, African Centre for Technology Studies.

Mugabe, J. Seymour, F. and Clark, N. (1997). Environmental Adjustments n Kenya: EmergingOpportunities and Challenges. Nairobi, African Centre for Technology Studies.

Ministry of Lands and Settlement (2000). Ministerial Position Paper on Environmental Considerations in Land-Use Planning and Management. Nairobi, Environmental Management Unit, Ministry of Lands and Settlement.

Republic of Kenya (1965) Sessional Paper No. 10 on African Socialism and its Application to Planning in Kenya. Nairobi, Government Printer.

Republic of Kenya (1974). National Development Plan 1974-78. Nairobi, Government Printer.

Republic of Kenya (1994). National Environmental Action Plan. Nairobi, Government Printer.

Republic of Kenya (1997). National Development Plan 1997-2001. Nairobi, Government Printer.

Republic of Kenya (1999). Environmental Management and Co-ordination Act. Nairobi, Government Printer.

In: Handbook of Environmental Policy
Editors: Johannes Meijer and Arjan der Berg

ISBN 978-1-60741-635-7
© 2010 Nova Science Publishers, Inc.

Chapter 19

STAKEHOLDER ASSESSMENT IN ENVIRONMENTAL POLICY ANALYSIS

Zhenghong Tang[1], Feng Xu[2] and Christopher Hussey[1]*

[1]Community and Regional Planning Program, College of Architecture, University of
Nebraska – Lincoln, Lincoln, NE, USA
[2]Bureau of Comprehensive Development, Ministry of Water Resources, P.R.China

ABSTRACT

Stakeholder assessment is an important approach in environmental policy analysis. It helps to identify the critical environmental issues in controversy in a given situation, the affected interests, and the appropriate forms of handling the conflicts. Although many studies have highlighted the major prevailing environmental theory of coordination and communication, few have linked this theory with practical environmental policy analysis. The purpose of this paper is to provide a conceptual framework of stakeholder assessment to reach the goal of environmental dispute resolution and assess the major stakeholders' roles in potential environmental conflicts, particularly in resource-dependent local jurisdictions. A case study was conducted to further examine the theoretical model of stakeholder assessment in environmental conflicts and the policy analysis process. The stakeholder assessment process involves potentially interested stakeholders in order to: assess the causes of the conflict; identify the entities and stakeholders who would be substantively affected by the conflict's outcome; assess those stakeholders' interests and identify a preliminary set of relevant issues; evaluate the feasibility of using a consensus-building or other collaborative process to address these issues; educate interested persons on consensus and collaborative processes; and design the structure of a negotiating committee or other collaborative process to address the conflicts. The results highlighted that stakeholder assessment has been proven valuable as a first step in consensus-building processes and in finding constructive approaches to resolving environmental conflicts.

Keywords: stakeholder assessment, environmental conflicts, environmental policy, resource-dependent, local jurisdictions.

---

* Email: ztang2@unl.edu, Phone: (402) 472-9281.

1. BACKGROUND

Stakeholders are defined as *"those effecting change as well as those affected by it"* (Randolph, 2004). Stakeholder involvement has been identified as the heart of collaborative environmental planning (Beierle and Konisky, 2001; Brody, 2003a; Brody, 2003b; Brody et al., 2003c; Brody et al., 2003d; Burby and May, 1998; Daniels and Walker, 1996; Daniels and Walker 2001; Innes, 1996; Selin and Carr, 2000; Wondolleck and Yaffe, 2000). Stakeholder involvement helps develop a "shared value" that incorporates collaborative efforts intended for the stakeholders to come up with an agreement. It also helps environmental conflict resolutions. Potential environmental conflicts can be resolved through negotiation and mediation. Furthermore, it also helps develop creative solutions that may not have emerged from traditional environmental management. Stakeholder assessment can lead to more effective management of environmental systems because of the characteristics of cross-boundary management. It also builds up participation-increased ownership-commitment to implement plan or policies, which results in less need for costly enforcement.

Stakeholder involvement brings several important components that influence plan quality. Involvement of large landowners ensures resulting plans cover the complete ecosystem, improving the quality of the plan. Further, stakeholder involvement improves the resources such as volunteer time, as well as improves the base of knowledge that can be drawn upon which will improve the depth and quality of the plan. (Brody, 2003b).

A study performed by Samuel Brody found that simply bringing stakeholders to the table does not improve plan quality. Multiple competing interests voicing their opinions results in the plan catering to the "lowest common denominator" in terms of plan quality. However, the presence of certain individual key stakeholders such as resource-based industry groups and large landowners has a strong positive influence. Brody concludes these conflicting findings: "Despite the broad theoretical support for representation as a basis for sound planning, the empirical evidence suggests that having all of the stakeholders and community members present during the decision-making process does not necessarily guarantee the adoption of a strong plan."(Brody, 2003b).

What are the benefits of stakeholder involvement? Stakeholder involvement and citizen participation helps people to recognize that they are active members of the community with the ability to affect changes in their community and in their environment. Instead of complaining to city administrators to find a solution, citizens feel that they can help to form the solution, making them active members in the community and contributing to the public good. Stakeholder involvement also provides an educational background to citizens about the issues the community is dealing with, and also to help them understand the differing interests that must be taken into account in developing a consensus plan. (Day, 1997).

There are often problems and barriers to involving stakeholders in the planning process. Much has been much written on the subject of citizen participation in local planning efforts. However empirical studies as well as the theoretical literature on the subject differ widely on their findings due to differences in methodologies and techniques, as well as definitional problems. (Day, 1997).

There are problems that arise with involving stakeholders in planning decisions. Involving citizens can be unwieldy due to the large number of potential stakeholders whose participation heighten conflict rather than build consensus. Further, many stakeholders may lack the expertise needed to constructively add to the process, and many experts may see their contributions as amateurish. Citizen participation may also be selective depending on who decides to participate, and may not accurately represent the aggregate of citizen interests or preferences. (Day, 1997).

What is the best way to ensure stakeholder involvement in the planning process? A case study by Samuel Brody, et al. (Brody, et al. 2003c) concludes that mandates requiring planners to have citizen participation included in the planning processes "do indeed affect local government attention to citizen involvement." They further report that it is possible for planners to overcome citizen apathy by engaging stakeholders in participation programs. (Brody, 2003c).

2. ENVIRONMENTAL CONFLICTS AND ENVIRONMENTAL LAWS FOR COAL MINING

Coal mining has significant impacts on the environment. Coal mining is one of the most extensively regulated industries in the United States. It is a complicated, time-consuming process for miners and coal producers to obtain local, state and federal permits to mine. The preparation of the procedures can take around ten years. More than three dozen federal environmental laws and regulations cover all aspects of mining today.

Mining and mineral extraction is a transitory land use that takes many forms. It includes exploration, site preparation, mining, milling, waste management, decommissioning or reclamation, and even mine abandonment. Coal has been mined for more than 1,000 years, and large-scale mining was practiced as early as the 18th century. Two principal systems of coal mining are used: surface or strip mining and underground or deep mining. Strip mining is possible only when the coal seam is near the surface of the ground. In large surface mines, power shovels and draglines are used to remove the earth and rock from above the seam. Smaller shovels then load the coal directly into trucks. The chief advantage of strip mining over underground mining is time and labor savings. In underground (deep) mining, the coal seam is reached through vertical shafts or level tunnels. The coal deposit is usually marked out in rooms. The coal is cut and blasted away, with pillars of coal left to support the roof.

Among the chief problems in underground mines are ventilation and roof support. Ventilation is important because of the presence in coal mines of dangerous gases such as methane and carbon dioxide. Large fans and blowers must be used to maintain the circulation of pure air. In order to prevent the spread of coal dust, which can be highly explosive, mine interiors are frequently sprayed with limestone dust, a process known as rock-dusting. To provide support for the roofs of tunnels and workspaces, steel roof bolts that bind together the overlying rock layers are inserted into the mine ceiling.

Surface mining eradicates surface vegetation, can permanently change topography, alters soil and subsurface geological structure, and interrupts surface and subsurface hydrologic regimes. Two main results of such impact is erosion and acid mine drainage. Some secondary mining impacts are urban development to support mining and construction of off-road networks for exploration activities. Also, off-site impacts such as stream pollution can occur. Water quality impacts can be controlled during active mining, but many coal reserves cannot be mined using current technology without "residual acid seepage" requiring treatment in order to protect large river systems.

It is important to remember that the scale of surface mining impacts depends on mining technology employed, extent of disturbance, chemical and physical composition of the mineral, and method of reclamation.

Erosion or sediment carried by rainfall is one of the greatest potential environmental hazards of surface mining. The principal erosion processes is divided into three steps. First, detachment or dislodging of soil particles from the soil mass by erosive forces such as raindrop impact, water flow and, wind occurs. Then the soil particles are transported from its original location. Finally, sedimentation or deposition of transported sediment finishes the process. The major factors affecting erosion are hydrology, soil erodibility, topography, soil cover and land management. Soil cover has the greatest impact because it can act as interceptor of rainfall lowering raindrop energy and preventing raindrop detachment. Also, soil cover slows runoff velocity reducing available energy for soil detachment and transport. Finally, land management has an impact on soil cover which makes it an important factor affecting erosion. In order to prevent such effects, under the Surface Mining Law all surface water flowing off disturbed areas of the mine must be routed through sediment ponds. The ponds are designed to slow the flow of water.

Acid mine drainage is drainage flowing from surface mining, deep mining or coal refuse piles that is typically highly acidic with elevated levels of dissolved metals. It occurs when the mineral pyrite (FeS_2) is exposed to air and water, resulting in the formation of sulfuric acid and iron hydroxide.

Pyrite is commonly present in coal seams and in the rock layers overlying coal seams. Acid mine drainage formation occurs during surface mining when the overlying rocks are broken and removed to get at the coal. The products of acid mine drainage formation can destroy water resources by lowering the pH and coating stream bottoms with iron hydroxide, forming the familiar orange colored "yellow boy" common in areas with abandoned mine drainage. Nevertheless, acid mine drainage is also a microbially mediated process. The acid load produced from a mine site depends on the following factors: Chemistry of Pyrite Weathering, Microbiological Controls, Depositional Environment, Acid Base Balance and Reaction Rates, Lithologic Controls, Mineralogical Controls, and Mine site Hydrology.

Techniques for mitigation of acid mine drainage include two methods: controlled placement of overburden materials and water management. Controlled placement is a preventative measure where pyritic or alkaline material is placed during mining to minimize or neutralize the formation of acid mine drainage. Oxygen and water are necessary to initiate acid formation for pyrite oxidation. The exclusion of either reactant should prevent acid production6. Water management strategies are another option for reducing acid generation. Water management can include the following: routing of surface drainage away from pyritic material or through alkaline material, prompt removal of pit water, polluted pit water can be isolated from non-contaminated sources (no commingling) to reduce the quantity of water requiring treatment and constructed under drain systems can be used to route water away from contact with acid forming material.

3. MAJOR ENVIRONMENTAL LAWS ON COAL MINING

The most important law about coal mining and reclamation is the Surface Mining Control and Reclamation Act (SMCRA) of 1977. SMCRA is the first comprehensive national surface mining law that mandates strict regulation of surface mining. Besides the SMCRA, other major federal laws with significant impact on coal mining and reclamation are the following:

- The National Environmental Policy Act (NEPA) of 1969: It requires interdisciplinary approach to environmental decision-making.
- The Clean Air Act (CAA) of 1970: It sets air quality standards in mining and reclamation activities.
- The Clean Water Act (CWA) of 1977: It directs standards to be set for surface water quality and for controlling discharges to surface water.

Furthermore, the following federal laws and regulations are all related to govern mining and reclamation activities. In addition, each state usually has its own laws and regulations that mining companies must follow as well.

- American Indian Religious Freedom Act of 1978
- Antiquities Act of 1906
- Archeological Nd Historical Preservation Act of 1974
- Archeological Salvage Act
- Bald Eagle Protection Act of 1969
- Comprehensive Environmental Response, Compensation and Liability Act
- Endangered Species Act of 1963
- Federal Land Policy and Management Act
- Federal Mine Safety and Health Act
- Fish and Wildlife Coordination Act of 1934
- Forest and Rangeland Resources Planning Act of 1974
- Historic Preservation Act of 1966
- Laws to Regulate Sale, Transport and Storage of Explosives
- Migratory Bird Treaty Act of 1918
- Mining Law of 1872
- Mining and Minerals Policy Act of 1970
- Multiple Use - Sustained Yield Act of 1960
- National Forests Management Act of 1976
- National Historic Preservation Act
- National Trails System Act
- Noise Control Act of 1976
- Resource Conservation and Recovery Act
- Rivers and Harbors Act
- Safe Drinking Water Act of 1974
- Soil and Water Resources Conservation Act of 1977
- Toxic Substance Control Act

- Solid Waste Disposal Act
- Wild and Scenic Rivers Act
- Wilderness Act of 1964

4. Environmental Policy Analysis Process: Stakeholder Representation and Participation

Because the environmental policy making process, even at the local level, involves collaboration across multiple jurisdictions, agencies and owners, representation and participation of key stakeholders is widely recognized as the most important element of successful environmental management. Additionally, public participation is often an important aspect of the planning process. Many states mandated public participation for some environmental projects and local land use planning.

Stakeholder assessment should identify the key stakeholders, establish authority, structure the process, build up trust, share authority, assign roles, and engage in collaborative learning. Stakeholder assessment should have some prerequisites: 1) good information, 2) adequate time to participate, 3) commitment of participants, 4) willingness to share, 5) responsibility for decisions.

Typically, stakeholder groups in the environmental policy decision-making process include 1) community and neighborhood groups, 2) public service providers, 3) educational institutions, 4) industry and business, 5) community service organizations, 6) non-governmental organizations, 7) religious communities, 8) governments, 9) media, and 10) other public agencies.

5. A Case study of environmental assessment

This study uses a case study as an example to examine the theory of stakeholder involvement and further identify the role of environmental assessment in environmental policy analysis. The environmental conflicts occurred in West Virginia. Their causes and effects are many and their histories are much more complex than expected. For over a century, coal mining in these hills has fed the people and anchored the towns while gutting the mountains around them. The mining industries have been important economy resources, which provided tens of thousands of other jobs in this region. However, according to the study by the Environmental Protection Agency (EPA) and other federal regulator, Appalachia's future is being threatened by mountaintop mining. At issue is the future of so-called mountaintop removal, a profitable and drastic method of extracting high-grade coal by decapitating peaks. Its use has reshaped American mining over the last two decades as dramatically as it has altered the landscape of West Virginia. At least 560 miles of streams have been lost under valley fills since 1985 in West Virginia, Kentucky, Virginia and Tennessee. Research data has predicted that mining would eliminate another 230,000 acres of forest in the coming years. Environmentalists and miners are debating the future of mountaintop removal in this region. That prospect has revealed a divide throughout the region between those environmentalists and their supporters who say the mountaintops have been overburdened by coal mining, and others who wonder how they would survive without mining. Indeed, the link between coal resources and the outbreak of environmental conflicts has been recognized for decades. Environmental conflicts by nature do not disappear in West Virginia. It should be understood that a conflict is usually the playing out of human needs and fears in society. West Virginia's potential for sustainable development remains unfulfilled, and is now increasingly threatened by environmental devastation. Environmental conflicts occurred. For coal industries, workers and some governments, the life without mining was a nightmare to them. However, it was a dream for the environmentalists and some local people. This study uses this case as a theoretical model to illustrate a stakeholder assessment in environmental policy decision-making.

There are many debates between environmentalists and miners for the future of mountaintop removal. One school says that the weary landscape can bear no more mining, and the other school wonders how they would survive without it. The two sides involved in the case disagree on whether the ruling also applies to mining in this area.

6. STAKEHOLDERS INVOLVED IN THE CONSENSUS-BASED PROCESS

The objective of the stakeholder analysis is to identify means and build capacities to manage inevitable conflicts among stakeholders at various scales, including mechanisms to compensate local people for foregone opportunities. This environment conflict include the following stakeholders: (1) coal industries, (2) mining workers and other relative labors, (3) local governments, (4) environmentalists and environment organizations, eg. Coal River Mountain Watch, (5) federal agencies, (6) residents, and (7) legislature: federal court, local court. Each stakeholder has it's own interests in this environment conflict. Each group in this conflict needs to work with various stakeholders to devise methods they can use to monitor and understand the impacts of ongoing change and to develop workable responses under dynamic conditions.

Stakeholders have a variety of motivations for involvement in the discussions. The interests, positions and roles of stakeholder analysis are explained below:

Coal mining industries: The coal mining companies are the key stakeholders in the conflicts. For over a century, coal mining has been the important economic support in mining regions in West Virginia. If the decision will be made to bar the coal industries from mining coal from the mountaintops and filling in Appalachian rivers and valleys with rock and dirt from the mountain top mines, their economic interests will be damaged greatly. The coal companies oppose this decision firmly and they are against an emerging coalition of environmental and citizens' groups. They defend mountaintop mining as ecologically responsible. Coal mining companies who were wary of environmental regulation are explicitly for self-protection in the conflicts.

Workers: Imaging life without mining was a nightmare to the coal mining workers and other related workers. Many of the coal companies will be closed and a lot of workers will be unemployed if the legislatures pass the bill to ban mountaintop coal mining. Although no worker likes to see the mountains destroyed, but they considered more with their own jobs. Whatever decision will be made, the laborers are always in a relatively disadvantaged position in the conflicts.

Local governments: In the position of the local governments, on the one hand, they do not support the environmental devastation by mining; on the other hand, they are focused on local economies. If mining dies, the town dies. Finding an alternative solution for the coal mining industries is vital for local governments' interests. Local governments will support the environmental protection plans if there are some practical alternative plans for local development without mining. The local governments will use their political power to solve the related disputes in the conflicts.

Environmentalists and environment organizations: They think the mountaintops are overburden, so they strongly oppose the mountaintop coal mining. The critics contend the mountaintop mining method is spoiling wildlife habitats and pouring millions of tons of "valley fill" onto quiet hollows once home to mountain communities (Evan Osnos from the Newspaper). They provide the scientific data for mountaintop protection. They suggest the legislature end the right to mine coal in the areas endangering citizen health and safety. For example, Sen. Joseph Lieberman, (D, Conn), who led a subcommittee of the Environment and Public Works Committee, presided over a hearing and introduced legislation to clarify that dumping waste in waterways, is not allowed. Environmentalists pay more attention to the future of the environment, rather than how the future unemployment due to mining closes.

Federal agencies: The major federal agencies involved with mining reclamation are as follows: *A) Environmental Protection Agency (EPA)*: EPA is concerned with the NEPA compliance process for any mining reclamation activity. *B). Office of Surface Mining (Department of the Interior)*: The Surface Mining Control and Reclamation Act (SMCRA) passed by the United States Congress established national coal mining and reclamation standards and created the Office of Surface Mining Reclamation and Enforcement (OSM) within the U.S. Department of Interior. OSM is responsible for making sure that the requirements of SMCRA are met. *C). U.S. Fish and Wildlife Service (Department of the Interior):* The Act establishes a program for the regulation of surface mining activities and the reclamation of coal-mined lands, under the administration of the Office of Surface Mining, Reclamation and Enforcement, in the Department of the Interior. Mine operators are required to minimize disturbances and adverse impacts on fish, wildlife and related environmental values and achieve enhancement of such resources where practicable. Restoration of land and water resources is ranked as a priority in reclamation planning. *D). U.S. Army Corps of Engineering:* U.S. Army Corps of Engineers requires the application of nationwide permit No.21 for discharges of dredged or fill material into waters of the US associated with surface coal mining and reclamation operations provided the coal mining activities are authorized by the DOI, Office of Surface Mining (OSM), or by states with approved programs under Title V of the Surface District Engineer in accordance with the "Notification" General Condition. To be authorized by this NWP, the District Engineer must determine that the activity complies with the terms and conditions of the NWP and that the adverse environmental effects are minimal both individually and cumulatively and must notify the project sponsor of this determination in writing. For discharges in special aquatic sites, including wetlands, and stream riffle and pool complexes, the notification must also include a delineation of affected special aquatic sites, including wetlands.

Residents: Residents from West Virginia to southern Pennsylvania have sued coal companies in recent years, blaming mountaintop removal for fouling the drinking water and aggravating flooding. But the issue also separates neighbors throughout this valley where prosperity is legendarily scarce. Here, 30 miles south of the capital city of Charleston, the debate over land use reflects deep tensions as the Mountain State strains to recast its priceless terrain as a destination for tourists as well as a home to miners. The residents always work together with the environmentalists, but they also consider local economic development relative to the mining industries.

Legislature: The legislatures include federal court and local court. A federal court ruling may sharply reduce coal mining in Appalachia as well as in West Virginia. The District Court wants to bar the coal industry from filling in Appalachian rivers and valleys with rock and dirt from the mountain top mines. The District Court also rebuked the White House for having regulators rewrite mining rules to ease the way for mountaintop removal. The legislatures think the federal government is only concerned with the mining industry and its employees, which was "contrary to the spirit and the letter of the Clean Water Act". Before a strategy other than lawsuits and hard nose political wrangling will be chosen, the legislature is consider a consensus-based decision-making process. That means a chance for each stakeholder to reach an agreement for the mining conflicts.

7. THE ANALYSIS OF THREE MAIN DIFFERENT ROLES

Here the three main different roles are analyzed: the role of negotiation power, the role of technical information, and the role of politics.

The role of negotiation power: There are many sources of negotiation power in the mining conflicts. The legitimate (federal and local court) power has authority, reputation, and performance in the negotiations of the coal mining conflicts. The role of the powers will be to develop a good working relationship, promote mutual understanding, and invent an elegant option among the different stakeholders with various interests in the conflicts between mountaintop mining and environmental protection (Fisher, 1991).

The role of technical information: The role of providing technical information is always played by environmentalists or environmental organizations. The technical knowledge of the environmentalists may be the most important contribution to the policies and practices of mountaintop protection and to improve coordination among the agencies. The process of developing technical information provided mutual learning opportunities among scientists, companies, residents, laypersons, legislatures and governments (Susskind, 1999a). However, many environmentalists are reluctant to participate because of their view that consensus building allows too much compromise; the other interests weaken the goal of environmental protections (Susskind, 1999a; Susskind, 1999c).

The role of politics: The role of politics involves governments and legislatures. Politics can powerfully affect the negotiation of the conflicts. They insist on including a set of management options. Politics can establish an interactive and cooperative procedure that creates additional pressures on other parties to remain in the process until they reach a settlement (Susskind, 1999b). For example, although the EPA's voting power gave it the upper hand in early negotiations, the agency will not prevail throughout. The EPA strongly favored a high level of public participation, including more hearings and a public advisory committee; some governments will be against further public participation and will want their studies to remain confidential until the findings have been reviewed by the parties involved (Susskind, 1999b; Susskind, 1999d). The legislatures could use lawsuits and political wrangling to solve the conflicts.

8. Census Building Process

Impetus for consensus building: Consensus building involves a number of collaborative decision making techniques in which a facilitator or mediator is used to assist diverse or competing interest groups to reach agreement on the mountaintop mining conflicts (Godschalk, 1994). Working through these coal-mining conflicts related to the environment, natural resources, and public or private lands involves using a range of methods of alternative dispute resolution. Unlike traditional litigation, in which a judge or jury may impose a judgment or make a final determination, alternative methods of assisted negotiation — such as facilitation, mediation, and conflict assessment — allow all stakeholders in the mining conflicts to reach a mutually satisfactory agreement on their own terms. Unless workable interventions can be identified and disseminated, these conflicts over coal mining and environmental protections are likely to intensify in the future throughout much of the topics (Susskind, 1999e). Stakeholders' skepticism regarding the potential success of a consensus building effort is not a reason to recommend against proceeding. Most people have not been involved in consensus building efforts before and have difficulty judging whether or not such an effort is likely to be successful. The process of consensus building provides stakeholders in coal mining with an opportunity to work out their differences and arrive at joint solutions. Doing so can save time and avoid many of the costs of traditional legal proceedings. In addition, stakeholders in the coal mining conflicts who work toward a shared, positive outcome can often achieve better results than they would have received in court.

The obstacles for reaching agreement: Due to the controversy's size and complexity, there were other obstacles to reaching a successful agreement. The first main obstacle will be polarization (Bacow, 1984). Any stakeholder who is unwilling to compromise or remains hostile to the idea of mediation will cause mediation to fail. For example, the mining companies maintained that their mountaintop mining plans were the only technically satisfactory alternative. However, the environmentalists and the EPA were opposed to the mining plans on all mountaintop mining sites; the mining companies will think no benefit in pursing mediation. This polarization will cause obstacles for the mediation. The polarization of the mining conflicts grew over time, and the personal animosity between environmentalists, companies, governments and residents was intense. The environmentalists received fierce public attack from the mining companies, workers, local politicians, and other mining supports. They claim that the EPA staff was incompetent, irresponsible, and politically motivated.

The second main obstacle will be the scope of the dispute (Bacow, 1984). The conflicts among environmentalists, residents, companies, and workers will be exacerbated by a major strategic error. If a mediator will be introduced in the conflicts without careful consideration about taking a public stand for or against the mountaintop mining, some parties will promptly rejected this suggestion for a number of reasons. Furthermore, instead of privately selling the idea of the mediation to each party individually, which made it appear as if agreeing to mediation was agreeing to her environmental position or industries position, other parties will oppose the ideas. In addition, it made many of the contesting parties believe that they are primarily seeking good press coverage and had little real interest in resolving the conflict. A mediation effort involving various special interest groups would simply tend to remove the controversy from the forum where it belongs, and substitute talk for proper agency action.

That would contribute nothing to the progress of the case now before the Court, and would probably delay resolution of the entire matter with consequent mounting expense (Bacow, 1984). Though conflict is unavoidable and full of opportunities, resolving conflicts can be a very stressful and frustrating experience.

9. POSSIBLE NEGOTIATION PROCESS

Committing to and following the negotiation process, including a set of ground rules and some practical participation techniques, can help ensure that the conflict resolution process is as positive, fair, and equitable as possible.

9.1. Ground Rules to Facilitate Agreement

After the stakeholder analysis had been completed and presented to the coalition, the first task is to help the group reach agreement on ground rules that ensure the negotiation process will be constructive. Ground rules should address the following (Susskind, 1999f): 1) Representation: including selection, role of members, role of alternates, role of advisers, and role of other members of the public. 2) Primary responsibilities of members and alternates. 3) Decision making: The purpose of the process is to share information, discuss concerns and viewpoints, and build consensus. 4) Communication: Participation in discussions will be restricted to the members seated at the table, unless the facilitator sets aside time on the agenda for others to speak. In order to facilitate an open and collaborative discuss, members should communicate concerns, interests, and ideas openly and to make the reasons for their disagreements clear. 5) Role of facilitators: The facilitation team should formulate the agenda for all meetings and facilitate these proceedings. 6) Working groups: Working groups may be established to undertake more in-depth discussion or carry out discrete tasks. 7) Media: All meetings will be open to the public and the media. Press conference will not be held in conjunction with these meetings. However, the facilitators may, periodically produce press releases, for approval by the members, to keep the media informed of the deliberations (Susskind, 1999f). In addition, ground rules should address other issues about which stakeholders might be concerned. Usually, at the early stage of the negotiations, most people in the room will be more interested in debating the real issues about the mining rights than debating how to debate the issues. Over time, however, the group will begin to take more interest in ground rules.

9.2. Participation Techniques to Ensure Effective Negotiation

To facilitate the negotiation process, some practical participation techniques are needed to ensure effective communication. The main `alternative' or `collaborative' techniques for resolving disputes are direct negotiation, conciliation, facilitation and various combinations of techniques such as negotiated rule making.

9.3. Direct Negotiation

Negotiation is a process in which the parties to the dispute meet to reach a mutually acceptable resolution. Since each party represents its own interests, we could try to let the two parties talk to each other for some special issues. The coal mining companies could discuss with the EPA about how to deal with the waste materials. Also the residents could talk with the local government about their responses to the air pollution caused by mining coal at the mountaintop.

9.4. Multi-Party Collaboration

The multiple parties who share the same interests could work together to produce some alternative solutions or reach some common agreements on some specific problems. Environmentalists have historically spearheaded research on factors that promote and hinder cross-sectoral collaboration in a number of different arenas ranging from waste materials, groundwater resources, and air pollutions caused by the mountaintop mining. The local governments, residents and the mining companies could work together to determining how to arrange the workers from the possible closing of some special mining sites.

9.5. Conciliation

Any public or private entity that wishes to initiate conciliation procedures must submit an application to the court, which will send a copy of this application to the other parties. A good example may be that the coal mining companies provide conciliation for the local residents. Once both parties have accepted intervention by the court, a commission comprising an odd number of conciliators will be appointed in consultation with the coal mining companies, local residents and the members of the court. The commission will have to clarify the points of controversy between the mining companies and local residents and strive to achieve an agreement between them, under conditions acceptable to both sides. If, at any time during the procedure, the commission decides that there is no possibility of achieving an agreement between the parties, it will declare the procedure closed and draw up a document, making note of the fact that the controversy has been submitted to conciliation without an agreement having been reached.

9.6. Facilitation

In the processes of negotiation for the mountaintop mining removal, how communication channels affect coalitional bargaining should be examined, which could facilitate the communication. Facilitation is a collaborative process in which a neutral party seeks to assist the stakeholders to discuss constructively a number of complex, potentially controversial issues, such as how to reduce the mountaintop mining, and how to deal with the potential unemployed workers in mining. The facilitator typically works with participants before and during these discussions to assure that appropriate persons are at the table to help the parties set ground rules and agendas, enforce both, assist parties to communicate effectively, and help the participants keep on track in working toward their goals.

9.7. Negotiated Rule-Making

Negotiated rulemaking is a multi-party consensus process in which a balanced negotiating committee seeks to reach agreement on the mountaintop mining and proposed environmental protection. The negotiating committee is comprised of representatives of those interests that will be affected by, or have an interest in, the ruling, including the rulemaking agency itself. Affected interests that are represented in the negotiations are expected to abide by any resulting agreement and implement its terms. This agreement-seeking process usually occurs only after a thorough mining conflict assessment for the mountaintop mining has been conducted, and is generally undertaken with the assistance of a skilled, neutral mediator or facilitator. Furthermore, social, economic and political mechanisms also are needed to address the inevitable conflicts among the interests of these stakeholders, who range from coal industries and mining workers, to local residents and government policymakers, to environmental advocacy groups, federal or local legislatures.

10. CONCLUSIONS

Stakeholder assessment will prove to be valuable as a first step in the consensus-building processes and in finding constructive approaches to resolving the conflicts between mountaintop mining and environmental protection in West Virginia.

REFERENCES

Bacow, L.and M.Wheeler. 1984. Environmental Dispute Resolution. New York: Plenum Press, *Mediating Large Disputes*, ch. 9, pp 199-244

Brody, S. D. 2003a. Examining the role of resource-based industries in ecosystem approaches to management: An evaluation of comprehensive plans in Florida. *Society & Natural Resources* 16:625-641.

Brody, S. D. 2003b. Measuring the effects of stakeholder participation on the quality of local plans based on the principles of collaborative ecosystem management. *Journal of Planning Education and Research* 22:407-419.

Brody, S. D., D. R. Godschalk, and R. J. Burby. 2003c. Mandating citizen participation in plan making - Six strategic planning choices. *Journal of the American Planning Association* 69:245-264.

Brody, S. D., W. Highfield, and V. Carrasco. 2004d. Measuring the collaborative planning capabilities of local jurisdictions to manage ecological systems in southern Florida. *Landscape and Urban Planning* 69:33-50.

Burby, R., May, P., 1998. Intergovernmental environmental planning: addressing the commitment conundrum. *J. Environ. Plan. Manage*. 41 (1), 95–110.

Daniels, S., Walker, G., 1996. Collaborative learning: improving public deliberation in ecosystem-based management. *Environ. Impact Assess*. Rev. 16, 71–102.

Daniels, S. E., and G. Walker. 2001. *Working through environmental conflict: The collaborative learning approach*. Westport, CT: Praeger.

Day, Diane. 1997. Citizen Participation in the Planning Process: An Essentially Contested Concept? *Journal of Planning Literature* 11, 3: 421-434.

Fisher, Roger and William Ury. 1991. *Getting to Yes*. New York: Penguin Books, pp.95-149.

Godschalk, David, David Parham, Douglas Porter, William Potapchuk, Steven Schukraft. 1994. Pulling Together: A Planning and Development Consensus Building Manual. Washington, D.C.: Urban Land Institute, *Participation and communications*, pp. 50-59.

Innes, J. 1996. Planning through consensus building: A new view of the comprehensive planning ideal. *J. Am. Plan. Assoc*. 62:460.

Lewiski, Saunders, and Minton, 2001. *Essentials of Negotiation*, pp.1-150

Moor, 1996, *The Mediation Process: Practical Strategies for resolving conflict, How Mediation Works*, Ch.2, pp. 41-77.

Selin, S., Carr, D., 2000. Modeling stakeholder perception of collaborative initiative effectiveness. *Soc. Nat. Resour*. 13, 735–745.

Susskind, Lawrence, Sarah Mckearnan, and Jennifer Thomas-Larmer, (eds.). 1999a. *Consensus Building Handbook: A comprehensive Guide to Reaching Agreement. California Sage Publications: San Francisco Estuary Project*, pp801-825

Susskind, Lawrence, Sarah Mckearnan, and Jennifer Thomas-Larmer, (eds.). 1999b. Consensus *Building Handbook: A comprehensive Guide to Reaching Agreement. California Sage Publications: Joint-Fact-finding and the use of technical experts*, 375-400

Susskind, Lawrence, Sarah Mckearnan, and Jennifer Thomas-Larmer, (eds.). 1999c. Consensus Building *Handbook: A comprehensive Guide to Reaching Agreement. California Sage Publications: Activating a policy Network: The case of Mainport Schipohl*, pp.685-708

Susskind, Lawrence, Sarah Mckearnan, and Jennifer Thomas-Larmer, (eds.). 1999d. Consensus Building *Handbook: A comprehensive Guide to Reaching Agreement. California Sage Publications: Northern Oxford Country Coalition*, pp.711-746

Susskind and Mckearnan, 1999e, Journal of Architectural and planning Research, 16:2 (Summer) *The Evolution of Public Policy Dispute Resolution*, pp.96-115

Susskind, Lawrence, Sarah Mckearnan, and Jennifer Thomas-Larmer, (eds.). 1999f. Consensus Building Handbook: A comprehensive Guide to Reaching Agreement. California Sage Publications: ch.2. *Conflict assessment*, pp99-137

Beierle, T. C., and D. Konisky. 2001. What are we gaining from stakeholder involvement? Observation from environmental planning in the Great Lakes. *J. Environ*. Plan. C Gov. Policy 19:515–527.

Wondolleck, J., and S. Yaffee. 2000. *Making collaboration work: Lessons from innovation in natural resource management*. Washington, DC: Island Press.

INDEX

B

C

F

G

I

M

N

S

T